D0213257

Applied Wetlands Science and Technology

Edited by
Donald M. Kent

LEWIS PUBLISHERS
Boca Raton   Ann Arbor   London   Tokyo

Library of Congress Cataloging-in-Publication Data

Applied wetlands science and technology / edited by Donald M. Kent.
   p.   cm.
Includes bibliographical references and index.
ISBN 0-87371-749-X
   1. Wetland conservation. 2. Ecosystem management. 3. Wetlands
Water quality management. I. Kent, Donald M.
QH75.A44 1994
333.91'8—dc20   94-11056
                                                                  CIP

No official support or endorsement by the Environmental Protection Agency or any other agency of the Federal Government is intended or should be inferred.

© 1994 by CRC Press, Inc.
Lewis Publishers is an imprint of CRC Press

No claim to original U.S. Government works
International Standard Book Number 0-87371-749-X
Library of Congress Card Number 94-11056
Printed in the United States of America 1 2 3 4 5 6 7 8 9 0
Printed on acid-free paper

Compared to other ecosystems, wetlands have received an unprecedented amount of attention. Much of this attention has occurred as a result of, and subsequent to, passage of Section 404 of the Clean Water Act in 1977 which recognized the importance of wetlands to the societal good. That is not to suggest that wetlands values were institutionally unrecognized prior to this time. Beginning in the 1930s, wetlands were recognized as valuable for the production and protection of wildlife, especially waterfowl and furbearers. In the 1960s wetlands were recognized as important for attenuating flood waters. Now, wetlands are recognized for providing these and other functions, including nutrient and contaminant retention and transformation, groundwater recharge and production export.

Ironically, and coincident with this recognition of the value of wetlands, is an awareness that wetlands are disappearing at an alarming rate. Between 1780 and 1980, an estimated 117 million acres of wetlands were lost in the contiguous United States. Seventy five percent of northern central United States wetlands, largely prairie potholes, were lost between 1850 and 1977. Bottomland hardwood forests were cleared at a rate of 165,000 acres per year between 1940 and 1980. Gulf coast wetlands disappeared at a rate of 25,000 acres per year. Of those wetlands which remain, many are degraded from channelization, damming, and agricultural and urban runoff. As well, remaining wetlands are fragmented or isolated.

This awareness of wetlands loss and degradation, as well as the promulgation of laws and regulations for protecting wetlands, and regulating their use, has spawned the development and growth of the wetlands professions. This book is for working wetlands professionals and nonprofessionals alike. It is intended for managers, regulators and consultants responsible for effective decision making, as well as those interested in how wetlands function, how wetlands can be protected, and how wetlands can be managed. In some ways, it is a 'how to' book in that it is a guide for working with wetlands. However, it recognizes that each and every situation is unique, and therefore requires a specific solution. As such, it seeks to provide the guidelines for effective decision making.

The book was written by practicing wetland professionals including consultants, academicians and regulators. In this manner, the most relevant, up to date information on applied wetlands science and technology is made available to those who need it. Sixteen chapters are provided. The first chapter consists of an introduction which defines wetlands and provides a federal regulatory context. Chapter 2 through 9 discuss fundamental issues of applied wetlands science including offsite and onsite identification of wetlands, functions and values, ecological assessments, avoidance and minimization of impacts; enhancement, restoration and creation of freshwater and coastal wetlands, and monitoring. Chapters 10 through 12 discuss the use of wetlands for renovating wastewater, stormwater and acid mine drainage. An additional chapter on wetland treatment of agricultural runoff was omitted, and may be included in subsequent editions as the knowledge base expands. The final four chapters reflect selected management issues. Chapter 13 discusses the relevance of modern conservation principles to the design of wetlands wildlife preserves, and Chapter 14 discusses the management of wetlands specifically for wildlife. The management of coastal marshes, a wetland type that has been severely impacted and continues to be threatened perhaps more than any other wetland type, is discussed in Chapter 15. Finally, the book concludes with a chapter on wetlands education. The chapter is included to emphasize that effective management and regulation of wetlands depends ultimately on educating the general populace on the value of wetlands functions.

Donald M. Kent, Ph.D.

ACKNOWLEDGMENTS

A book of this type is particularly dependent upon the generosity of others. The research which provides the basis for the chapter on acid mine drainage was supported by the U. S. Department of the Interior Bureau of Mines through the National Mine Land Reclamation Center at West Virginia University. Dr. Skousen, Dr. Sexstone, Dr. Garbutt and Dr. Sencindiver would like to especially thank Paul Ziemkiewicz, Director of the Center, and the chapter constitutes contribution #2362 of the West Virginia Agricultural and Forestry Experiment Station.

Ms. Drake and Mr. Del Vicario, authors of the education chapter, wish to thank all of the educators who responded to their inquiries, and who generously provided copies of their educational materials. A special thanks goes to Dr. Lundie Spence and Mr. Brian Lynn for their time and review. We thank all the educators who responded to our inquiries and provided copies of their educational materials. We thank Dr. Lundie Spence and Mr. Brian Lynn for their time and review.

I am personally grateful to a number of individuals who have either directly or indirectly contributed to this book. Mr. Robert A. (Skip) DeWall of Lewis Publishers recognized the need and potential for a book of this nature, and provided me with the freedom to design the content, seek the contributors and conduct the technical editing. I am grateful for your support and confidence, as well as your publishing expertise. To all those contributing authors who recognized the importance of a book of this nature, and who generously made their experience available to others, thanks for your efforts and the opportunity to get to know you a little better. Thanks also to Walt Disney Imagineering, especially to Ben Schwegler, for encouraging this effort. Others too numerous to mention, both past and present, have stimulated and influenced my thinking, and our discussions are reflected in this book.

Finally, any project of this magnitude, particularly one conducted at night and on weekends, requires the patience and tolerance of loved ones. To my family Debra, Brianna and Andrew, thanks for allowing me the time to pursue this project. Your support and confidence meant the world to me.

Donald M. Kent, Ph.D.

ABOUT THE AUTHORS

Mr. David R. Bingham is a Vice President and Director of the Environmental Quality Division of Metcalf & Eddy, Inc., specializing in water resources projects. He has worked on projects involving combined sewer overflow and stormwater, point and nonpoint source pollutant evaluations, mathematical modeling and engineering design for coastal areas, estuaries, rivers, lakes and groundwater resources. Mr. Bingham's experience includes management for the Massachusetts Water Resources Authority's combined sewer overflow master plan and facilities plan for the Boston Harbor cleanup, and preparation of the U. S. Environmental Protection Agency's Needs Survey of combined sewer overflow. He is preparing an urban runoff control handbook for the U. S. Environmental Protection Agency.

Dr. Paul T. Bowen is a Vice President and Senior Staff Consultant in Metcalf & Eddy, Inc.'s Corporate Technology Group. Following degrees in Chemistry/Natural Science and Environmental Systems Engineering, Dr. Bowen was an assistant professor in the School of Civil Engineering and Environmental Science at the University of Oklahoma. He currently serves as chairman of the Water Pollution Management Technical Committee of the Environmental Engineering Division of the American Society of Civil Engineers, and also serves the society on the National Environmental Systems Policy Committee.

Dr. Robert Buchsbaum is Massachusetts Audubon Society's coastal ecologist and is responsible for applied research on coastal habitats and providing technical analysis on coastal issues. He has published technical papers on a variety of topics including herbivory by Canada geese on salt marsh plants, nitrogen dynamics in decomposing marsh plants, and on eelgrass wasting disease, and nontechnical papers for the lay public. Dr. Buchsbaum's current research includes studies of human impacts on wildlife and water quality in salt marshes, and investigations into the role of the environment in the ability of eelgrass to resist disease.

Mr. Robert A. Corbitt is an Associate in the Atlanta office of Metcalf & Eddy, Inc. Having worked initially for the Georgia Water Quality Regulatory Authority, Mr. Corbitt has been a consulting environmental engineer for almost 26 years. He has served as chair of the Professional Coordination Committee, and later the Executive Committee of the Environmental Engineering Division of the American Society of Civil Engineers. Mr. Corbitt is a diplomat of the American Academy of Environmental Engineers, and the author of numerous technical writings covering a wide range of environmental issues.

Mr. Mario Del Vicario is the Chief of the Marine and Wetlands Protection Branch of the U. S. Environmental Protection Agency, Region II, and is responsible for wetlands programs in New York, New Jersey, Puerto Rico and the Virgin Islands. Prior to his present position, Mr. Del Vicario in the Environmental Analysis Branch of the U. S. Army Corps of Engineers, where he ran the environmental wetland program for the New York District. Throughout his career, Mr. Del Vicario has been active in the Society of Wetland Scientists, having served as the president and vice president of the North Atlantic Chapter.

Ms. Kathleen Drake is in the Marine and Wetlands Protection Branch of the U. S. Environmental Protection Agency, Region II. Ms. Drake's responsibilities include organizing the agency's outreach efforts for the region. Prior to her present position, Ms. Drake participated in large scale wildlife inventory and riparian habitat replacement projects in the Colorado River Valley. When not working, Ms. Drake devotes her time to her children, pets and garden, and visits swamps, marshes, tidal flats, aquariums and museums.

Dr. Keith Garbutt is Associate Professor of Biology at West Virginia University. His research interests relate to plant ecology, plant ecophysiology, and plant evolution and development. A primary area of his expertise has been in planting patterns in wetlands, wetland species interactions and competition, and plant response to water quality.

Dr. James F. Hobson is Director of the Environmental Toxicology Division for Technology Sciences Group, Inc. in Washington, D. C. Dr. Hobson has significant experience in environmental toxicology, including testing programs in support of existing and new chemical product registrations under United States federal, state, Canadian and European regulations. He is a diplomat of the American Board of Toxicology and has frequently spoken and chaired industrial and academic conferences on environmental toxicology issues.

Dr. Kenneth D. Jenkins is Director of the Molecular Ecology Institute at California State University and a principle in the consulting firm of Jenkins, Sanders & Associates in Long Beach, California. He has been directly involved in numerous ecological risk assessments, especially for hazardous waste sites. Dr. Jenkins is widely published in the field of ecological assessments.

Mr. David J. Kent is an environmental toxicologist with Technology Sciences Group, Inc. in Washington, D. C., and was previously manager of the aquatic toxicology laboratory and project manager on field assessments for the International Technology Corporation. He has extensive experience in performing and managing ecological assessments for a wide variety of regulatory programs including ecological risk assessments for the pesticide industry, and RCRA and Superfund sites. Mr. Kent has numerous scientific presentations and publications to his credit in the areas of aquatic toxicology and ecological risk assessment.

Dr. Donald M. Kent is a member of the Walt Disney Imagineering Research and Development group, specializing in environmental science and technology issues. Prior to joining Disney, he was a consultant with Metcalf & Eddy, Inc. and Chief Biologist Water Quality Laboratory of the New England Division of the U. S.

Corps of Engineers. Dr. Kent's broad experience includes studies of water quality and fisheries in limnological systems, functional ecological assessments, alternative impact analyses, and the design, implementation and monitoring of wetland mitigation projects. He is presently investigating and developing various applied ecological and biotechnological techniques for integrating land use and natural resource protection.

Mr. Roy R. (Robin) Lewis is the founder and president of Lewis Environmental Services, Inc., an environmental consulting firm located in Tampa, Florida. Mr. Lewis's expertise includes the ecology, restoration and creation of fresh and saltwater marshes, mangrove forests and eelgrass meadows. He has served on the editorial board for *Wetlands*, has been a convenor and editor for the Conference on Creation and Restoration of Wetlands, and is a member of the Tampa Bay Regional Planning Council's Agency on Bay Management and the National Research Council's Committee on the Role of Technology in Marine Habitat Protection. Mr. Lewis's clients have twice received the Florida Audubon Society Corporate Award.

Mr. Kevin McManus is presently Manager of Technical Services for the Massachusetts Water Resources Authority's Toxic Reduction and Control Program. Mr. McManus has 12 years of experience in the environmental field, working in both the public and private sectors. He has specialized in environmental impact assessment, wetlands permitting and mitigation, facility siting, NEPA compliance, pollution prevention, and oil spill response technologies during this period.

Dr. Robert J. Reimold is a Vice President and the National Director of Environmental Quality for Metcalf & Eddy, Inc., a ninety year old environmental engineering firm owned by parent corporation Air and Water Technologies, Inc. Having trained under ecologists Eugene P. Odum and Franklin C. Daiber, Dr. Reimold has thirty years experience in directing applied wetlands ecology projects in a diversity of east, south and west coast, as well as inland, locations. He has authored numerous technical and scientific publications and trade journal articles related to wetlands.

Dr. John C. Sencindiver is Professor of Soil Science at West Virginia University. His major areas of interest and experience are land reclamation, and soil genesis and classification. Dr. Sencindiver has published extensively in the area of land reclamation and minesoil development. He helped to initiate some of the early wetland research at West Virginia University.

Dr. Alan J. Sexstone is Associate Professor of Environmental Microbiology at West Virginia University. His academic training is in chemistry, and his interests include microbial degradation of hydrocarbons, nitrogen cycling and microbial transformations, and soil microbial ecology. Recent projects have focused on anaerobic and aerobic wetland processes mediated by microorganisms.

Dr. Jeffrey G. Skousen is Associate Professor and Soil Science and Extension Reclamation Specialist at West Virginia University. He has experience in Range Science, Range Ecology and Agronomy. Dr. Skousen's interests involve land reclamation, revegetation, acid mine drainage and other soil and water quality problems. He publishes regularly in journals, symposium proceedings and coal magazines on reclamation and environmental subjects.

Mr. Carl Tammi is a Project Wetlands Scientist and Project Specialist with ENSR Consulting and Engineering in their Ecology and Eco-Risk Group in Acton, Massachusetts. He is responsible for directing wetlands delineation, mitigation design and monitoring, functional analysis, regulatory negotiation, and wetlands permitting. Mr. Tammi has conducted numerous wetland delineations throughout the United States, and is Certified as a Wetlands Delineator by the Baltimore District of the U.S. Army Corps of Engineers.

Mr. John Zentner is a principal with Zentner & Zentner, a professional consulting firm with offices in California which specializes in planning and restoration throughout the western United States. Mr. Zentner specializes in wetland science, land planning, permit processing and restoration of natural resources. Among other accomplishments, Mr. Zentner is President of the Western Chapter of the Society of Wetlands Scientists, and has participated in various federal, state and local wetlands working groups.

Contents

FUNDAMENTALS

WETLANDS FOR WATER QUALITY RENOVATION

SELECTED TOPICS IN WETLANDS MANAGEMENT

CHAPTER 1

INTRODUCTION

Donald M. Kent

DEFINING WETLANDS

Wetlands have been defined directly or implicitly in a variety of ways in recent years. Several factors, including personal perspective, position in the landscape, diversity and function contribute to the tractable nature of the definition.

Each individual or group brings to the definition their own perspective, based upon cumulative experience and personal needs. For example, the lay person when asked to define wetlands may envision a deep water marsh teeming with ducks, or alternatively, a dark swamp. To an engineer a wetland may be a place which will require a specialized construction design to accommodate poorly drained soils. The scientist likely has a functional perspective, defining a wetland as a place where anaerobic processes occur, and plants are specially adapted for living in saturated or inundated conditions. Finally, those charged with regulating wetland use are likely to have a structural perspective, defining wetlands by characteristic soil, hydrology and plants so as to facilitate permit decision-making.

Defining wetlands is further complicated by their position in the landscape. Wetlands are transitional habitats in the sense that they are neither terrestrial nor aquatic, but exhibit characteristics of both. Their boundaries are part of a continuum of physical and functional characters, and may expand or contract over time depending upon factors such as average annual precipitation, evapotranspiration and modifications to the watershed. The transitional nature of wetland characteristics, and the shifting of wetland boundaries, renders precise identification of wetland boundaries difficult, if not impossible.

The diversity of wetland types also contributes to the tractable nature of the definition. Five wetland systems are recognized, including marine, estuarine, riverine, lacustrine and palustrine (Table 1, Cowardin et al. 1979). These encompass familiar habitats such as salt marsh, freshwater marsh and hardwood swamp, as well as less familiar seasonal wetlands (Figures 1-4). Systems are further subdivided into 10 subsystems which reflect tidal characteristics of

Table 1. Classification hierarchy of wetlands and deepwater habitats (Cowardin et al. 1979).

System	Subsystem	Class
Marine	Subtidal	Rock Bottom
		Unconsolidated Bottom
		Aquatic Bed
		Reef
	Intertidal	Aquatic Bed
		Reef
		Rocky Shore
		Unconsolidated Shore
Estuarine	Subtidal	Rock Bottom
		Unconsolidated Bottom
		Aquatic Bed
		Reef
	Intertidal	Aquatic Bed
		Reef
		Streamed
		Rocky Shore
		Unconsolidated Shore
		Emergent Wetland
		Scrub-Shrub Wetland
		Forested Wetland
Riverine	Tidal	Rock Bottom
		Unconsolidated Bottom
		Aquatic Bed
		Rocky Shore
		Unconsolidated Shore
		Emergent Wetland
	Lower Perennial	Rock Bottom
		Unconsolidated Bottom
		Aquatic Bed
		Rocky Shore
		Unconsolidated Shore
		Emergent Wetland
	Upper Perennial	Rock Bottom
		Unconsolidated Bottom
		Aquatic Bed
		Rocky Shore
		Unconsolidated Shore
	Intermittent	Streambed
Lacustrine	Limnetic	Rock Bottom
		Unconsolidated Bottom
		Aquatic Bed
	Littoral	Rock Bottom
		Unconsolidated Bottom
		Aquatic Bed
		Rocky Shore
		Unconsolidated Shore
		Emergent Wetland
Palustrine		Rock Bottom
		Unconsolidated Bottom
		Aquatic Bed
		Unconsolidated Shore
		Moss-Lichen Wetland
		Emergent Wetland
		Scrub-Shrub Wetland
		Forested Wetland

Figure 1. Salt marsh is an emergent, intertidal estuarine wetland system characterized by persistent plant species such as cordgrass *(Spartina alterniflora)*.

Figure 2. Freshwater marsh is a type of palustrine wetland. Dominant vegetation may be emergent, as is the cattail *(Typha latifolia)* in the photograph, or comprised of aquatic bed species such as water lily.

Figure 3. Another type of palustrine wetland is hardwood swamp, such as this deciduous forested wetland in New Jersey. The trees are adapted to seasonal saturation or inundation.

Figure 4. Seasonal palustrine wetlands such as this vernal pool in California will only be inundated or saturated for a short time in early spring.

marine, estuarine and riverine systems; gradient, water velocity and permanence of water for riverine systems; and water depth for lacustrine systems. Subsystems are divided into 55 classes which reflect substrate composition and vegetation type. Finally, wetlands can further be described by water regime, water chemistry, soil and special modifiers.

Wetlands also defy a unifying functional definition. Each wetland is unique with respect to its size, shape, hydrology, soils, vegetation and its position in the landscape. As such, wetlands exhibit a wide range of functional attributes, including provision of aquatic and wildlife habitat, retention of sediments and toxicants, flood attenuation, nutrient metabolism, groundwater recharge and production export. Individual wetlands may exhibit some of these attributes, all of these attributes, or in rare instances, none of these attributes. Moreover, individual wetlands of similar attributes are likely to provide functions to differing degrees.

Despite the difficulty in singularly defining wetlands, several formal definitions have been proposed. The earliest definition was intended for managers and scientists, particularly those concerned with waterfowl and wildlife (Shaw and Fredine 1956). Largely a structural definition, it uses language understandable to the lay person.

> The term wetlands ... refers to lowlands covered with shallow and sometimes temporary or intermittent waters. They are referred to by such names as marshes, swamps, bogs, wet meadows, potholes, sloughs, and river-overflow lands. Shallow lakes and ponds, usually with emergent vegetation as a conspicuous feature, are included in the definition, but the permanent waters of streams, reservoirs, and deep lakes are not included. Neither are water areas that are so temporary as to have little or no effect on the development of moist-soil vegetation.

The definition established two parameters essential for a habitat to be considered a wetland: the presence of surface water and the development of moist-soil vegetation. Twenty three years later at a workshop of the Canadian National Wetlands Working Group, a definition evolved which recognized a third parameter, hydric soils, and which noted the functional attributes of wetlands (Tarnocai 1979). Furthermore, it expanded the previous definition of wetlands to include not only those habitats with surface water, but also those having saturated soils.

> Wetland is defined as land having the water table at, near, or above the land surface or which is saturated for a long enough period to promote wetland or aquatic processes as indicated by hydric soils, hydrophilic vegetation, and various kinds of biological activity which are adapted to the wet environment.

That same year, the U.S. Fish and Wildlife Service adopted a definition which also recognized wetland hydrology, hydric soils and hydrophytic vegetation as defining parameters (Cowardin et al. 1979). Intended for wetlands scientists, the definition is distinguished from the Canadian definition in that a wetland need not exhibit characteristics of all three parameters.

> Wetlands are lands transitional between terrestrial and aquatic systems where the water table is usually at or near the surface or the land is covered by shallow water. For purposes of this classification wetlands must have one or more of the following three attributes: (1) at least periodically, the land supports predominantly hydrophytes; (2) the substrate is predominantly undrained hydric soils; and (3) the substrate is nonsoil and is saturated with water or covered by shallow water at some time during the growing season each year.

The three parameter approach, which was developed for scientists and managers, is reflected in Section 404 of the Clean Water Act, forming the basis for regulatory decision-making.

> The term "wetlands" means those areas that are inundated or saturated by surface or ground water at a frequency and duration sufficient to support, and that under normal circumstances do support, a prevalence of vegetation typically adapted for life in saturated soil conditions. Wetlands generally include swamps, marshes, bogs, and similar areas.

FEDERAL WETLANDS RELATED LEGISLATION

Numerous federal statutes have been enacted which impact activities in and around wetlands (Table 2). These statutes encompass regulation, acquisition and restoration of wetlands, incentives and disincentives to use of wetlands, and other programs. Far from being a cohesive collection of synergistic laws, the statutes have been initiated and enacted piecemeal over the years by various federal agencies.

Federal authority to regulate wetlands derives principally from Section 404 of the Federal Water Pollution Control Act of 1977 (33 U.S.C. 1344). Section 404 requires landowners and developers to obtain permits prior to dredge and fill activities in navigable waters. The definition of navigable waters has been extended to include adjacent wetlands. However, the Water Pollution Control Act exempts normal agriculture, silviculture and ranching activities, provided these activities do not convert areas of U.S. waters to uses to which they were not previously subject, do not impair the flow or circulation of such waters, or reduce their reach.

Due in part to the aforementioned exemptions, Section 404 regulates only about 20 percent of the activities that impact wetlands. The Food Security Act of 1985 (P. L. 99-198 Statute 1354) compensates in part for this lack of regulation through two provisions: Swampbuster and the Conservation Reserve Program. Swampbuster denies federal farm program benefits to producers who plant an

Table 2. Significant federal wetlands related legislation.

Legislation	Date	Effect on Wetlands
Section 10, Rivers and Harbors Act	1899	Requires permits from U.S. Army Corps of Engineers for dredge and fill activities in navigable waterways and wetlands
Migratory Bird Hunting and Conservation Stamp Act	1934	Proceeds of duck stamps used to acquire habitat
Federal Aid to Wildlife Restoration Act	1937	Assistance to states and territories for restoring, enhancing and managing wildlife
Fish and Wildlife Act	1956	Established U.S. Fish and Wildlife Service
U.S. Fish and Wildlife Coordination Act	1958	Requires all federal projects and federally permitted projects to consider wildlife conservation
Land and Water Conservation Fund Act	1965	Purchase of natural areas at federal and state levels
National Wildlife Refuge System Administration Act	1966	Established National Wildlife Refuge System
National Flood Insurance Act	1968	Requires communities to develop floodplain management programs
National Environmental Policy Act	1969	Requires Environmental Impact Statements for federal actions
Water Bank Act	1970	Purchase easements on wetlands
Endangered Species Act	1973	Prohibits federal agencies from undertaking or funding projects which threaten rare or endangered species
Resource Conservation and Recovery Act	1976	Controls disposal of hazardous waste, reducing threat of contamination to wetlands
Section 402, Federal Water Pollution Control Act	1977	Authorized national system for regulating sources of water pollution

Table 2 (cont.). Significant federal wetlands related legislation.

Section 404, Federal Water Pollution Control Act	1977	Regulates dredge and fill activities
Coastal Barrier Resources Act	1982	Prohibits federal expenditure or assistance for development on coastal barriers
Food Security Act	1985	
Swampbuster		Discourages farming on wetlands
Conservation Reserve Program		Removes erodible cropland from use
Emergency Wetlands Resources Act	1986	Promotes conservation through intensified cooperation and acquisition efforts
Agricultural Credit Act	1987	Preserves land reverting to Department of Agriculture's Farmers Home Administration
Everglades National Park Protection and Expansion Act	1989	Increased water flow to park and acquired more land
North American Wetlands Conservation Act	1989	Increased protection and restoration of wetlands under the North American Waterfowl Plan
Coastal Wetlands Planning, Protection and Restoration Act	1990	Restoration of coastal wetlands and funds North American Waterfowl Management projects
Coastal Zone Management and Improvement Act	1990	Sets guidelines and provides funding for state coastal zone management programs
Food, Agriculture, Conservation and Trade Act	1990	Established wetlands reserve program which purchases easements on wetlands
Water Resources Development Act	1990	Requires federal agency development of action plan to achieve no-net loss

agricultural commodity on wetlands that were converted after December 23, 1985. Although Swampbuster is the only legislative provision which directly affects eligibility for other federal benefits, the policy allowed producers to plant a commodity when prices were high enough to make federal farm program benefits unnecessary, and to plant converted wetlands with a noncommodity crop in years when federal program benefits might be needed. To overcome this deficiency, the Food, Agriculture, Conservation and Trade Act of 1990 was enacted to make noncomplying producers ineligible for federal benefits for that year and all subsequent years.

By contrast, the Conservation Reserve Program authorizes the federal government to enter into contracts with agricultural producers to remove highly erodible cropland from production for 10-15 years in return for annual rental payments. Producers are required to implement a conservation plan that usually includes planting cover, such as grass or trees, to hold soil in place and to reduce erosion. The program was expanded in 1989 to make cropped wetlands eligible for enrollment.

Another program which removes land from use is authorized by the Water Bank Act of 1970 (16 U.S.C. 1301). The Water Bank Program provides funds to purchase 10 year easements on wetlands and adjacent areas for the purpose of preserving, restoring and improving the wetlands. Private landowners enter into agreements with the federal government in which they promise not to drain, fill, level, burn or otherwise destroy wetlands, and to maintain ground cover essential for the resting, breeding or feeding of migratory waterfowl. Implementation of the program is concentrated in the prairie pothole region of the United States.

Two other programs having a significant affect on the preservation of wetlands are the Migratory Bird Hunting and Conservation Stamp Act of 1934 (16 U.S.C. 718) and the Coastal Barrier Resources Act of 1982 (16 U.S.C. 3501). The Stamp Act uses proceeds from duck stamps to preserve wetlands and adjacent waterways important to waterfowl through purchase or perpetual easement. The Coastal Barrier Resources Act attempts to minimize the loss of human life, wasteful expenditure of federal revenues, and damage to fish, wildlife and other natural resources by prohibiting federal expenditures and financial assistance for development of coastal barriers.

ROLE OF FEDERAL AGENCIES IN WETLANDS PROTECTION AND REGULATION

Army Corps of Engineers

The Corps is responsible for issuing Section 404 permits authorizing dredge or fill activities in Waters of the United States and adjacent wetlands. Approximately 15,000 project-specific permit applications, and 40,000 minor activities associated with regional and nationwide general permits, are evaluated each year (United States General Accounting Office 1991). Wetland determinations and delineations are also the responsibility of the Corps, including verification of the accuracy of delineations performed by consultants for permit applicants. Public interest reviews are conducted to determine the efficacy of permits. Consideration is given to economics, aesthetics, historic value, fish and wildlife value, value for attenuating floods, navigation, recreation, water supply and quality and other needs and welfare of the public. Compliance inspections are conducted following permit issuance to ensure permit conditions are met. The Corps has the authority to seek civil or administrative remedies for violation of permit conditions, or for other unauthorized discharges into wetlands.

Environmental Protection Agency

The Environmental Protection Agency has statutory enforcement authority to deal with unpermitted dredge and fill activities. In addition, the Agency determines the scope of navigable waters and interprets the scope of exemptions under the Section 404 Program. In consultation with the Corps, the Agency developed the guidelines for selection of sites for disposal of dredged or fill materials. The Agency has veto authority under subsection 404(c) if disposal of dredged or fill material will have an unacceptable adverse effect on municipal water supplies, shellfish beds and fishery areas, wildlife, or recreational areas.

Fish and Wildlife Service

The Fish and Wildlife Service is an advisor to the Corps with regard to the Section 404 Program, making recommendations for approval or disapproval of permits, and recommending conditions for permits to be approved. In addition, the Fish and Wildlife Service is active in a number of programs designed to protect, restore and enhance wetlands. The Fish and Wildlife service assists the Soil Conservation Service in mapping agricultural wetlands and in selecting and managing wetlands protected under the Farmers Home Administration Conservation Program and the Wetlands Reserve Program, enters into agreements to implement restoration projects on highly erodible cropland under the Conservation Reserve Program, and assists the Department of Agriculture and individual farmers in designing wetlands conservation plans necessary to qualify for Farm Bill incentives. Other activities of the Fish and Wildlife Service include management of the National Wildlife Refuge System and research and development of National Wetlands Inventory maps.

Agricultural Stabilization and Conservation Service and Soil Conservation Service

The Agricultural Stabilization and Conservation Service administers and enforces the Swampbuster provision. This includes providing wetlands information to producers and third parties, monitoring compliance with regulations, responding to public complaints and producers' appeals of decisions and dealing with violations.   Operations are carried out in conjunction with state and county committees.

The Soil Conservation Service is responsible for identifying wetlands subject to the Swampbuster provision and for granting certain exemptions. Soil Conservation staff conduct wetland delineations, notify producers of the presence of wetlands and process appeals of Soil Conservation Service delineations.

National Marine Fisheries Service

The National Marine Fisheries Service is active in coastal wetland issues, and makes recommendations to the Corps regarding Section 404 permits under authority of the Fish and Wildlife Coordination Act. Field staff also work closely with state fish and wildlife agencies and water quality agencies.

REFERENCES

Cowardin, L.M., V. Carter, F.C. Golet and E.T. LaRoe. 1979. Classification of Wetlands and Deepwater Habitats of the United States. U.S. Department of the Interior, Fish and Wildlife Service Biological Services Program FWS/OBS-79/31. 103 pp.

Shaw, S.P. and C.G. Fredine. 1956. Wetlands of the United States, Their Extent, and Their Value for Waterfowl and Other Wildlife. U.S. Department of Interior, Fish and Wildlife Service, Circular 39, Washington, D.C. 67 pp.

Tarnocai, C. 1979. Canadian Wetland Registry. Pp. 9-38 In Rubec, D.D.A. and F.C. Pollett (eds.) Proceedings of a Workshop on Canadian Wetlands Environment.   Canada Land Directorate, Ecological Land Classification Series, No. 12.

United States General Accounting Office. 1991. Wetlands Overview: Federal and State Policies, Legislation and Programs. GAO/RCED-92-79FS. Washington, D.C. 44 + pp.

CHAPTER 2

OFFSITE IDENTIFICATION OF WETLANDS

Carl E. Tammi

THE NEED FOR OFFSITE IDENTIFICATION

Identifying the location and determining the areal extent of jurisdictional wetlands on a national level has become an extremely controversial subject as federal wetlands regulatory agencies consider incorporating new provisions into a unified method for delineating federal jurisdictional wetlands (U.S. Environmental Protection Agency 1991). Defining wetlands limits and boundaries is primarily driven by comprehensive federal, state and local land-use laws and regulations. Section 404 of the Clean Water Act is the principal tool that the U.S. Army Corps of Engineers and the U.S. Environmental Protection Agency use to regulate the *discharge of dredged or fill material into waters of the United States, including wetlands* (33 CFR 320-330). At the federal level wetlands are further defined from a regulatory viewpoint as:

> *Those areas that are inundated or saturated by surface or groundwater at a frequency and duration sufficient to support, and that under normal circumstances do support a prevalence of vegetation typically adapted for life in saturated soil conditions. Wetlands generally include swamps, marshes, bogs, and similar areas* (33 CFR 328.3).

In identifying and delineating federal jurisdictional wetlands, three essential technical criteria or parameters are applied. The presence of wetlands hydrology (surficial or groundwater), a prevalence of wetlands vegetation (hydrophytes) which have specialized morphological and physiological adaptations to tolerate saturated or inundated conditions, and wetland soils (hydric soils) which in their undrained condition exhibit characteristics of somewhat poorly drained, poorly drained, or very poorly drained soils.

Other major federal legislation that drives wetland identification includes Section 401 Water Quality Certification (delegated to the individual states), Section

10 of the Rivers and Harbors Act of 1899 and the National Environmental Policy Act (NEPA).

At the state level, most northeastern, east coast and Pacific coast states have promulgated and adopted wetland protection legislation for inland, and where applicable, coastal wetlands. Identification and delineation techniques vary slightly from state to state, although most have adopted the principles of the federal methodology.

Given the regulatory framework behind wetlands protection, it is incumbent upon project proponents to determine, locate and identify wetlands resources on a subject parcel. Furthermore, it is important to adequately and accurately determine the location and approximate areal extent, as well as the predominant wetland cover type, early in project planning stages. This action can streamline the permitting process during more advanced stages of project design through avoidance and minimization of wetland impacts. The net result of conducting macroscale offsite wetland determinations from a project planning viewpoint is:

- A positive or negative wetlands determination for a subject parcel
- Approximate location of wetland and deepwater areas
- Approximate areal extent and distribution of wetlands and deepwater areas
- Determination of predominant wetland cover type (Cowardin et al. 1979)
- Determination of the need for continued analysis and approximate level of effort associated with such an analysis.

In addition, information relative to the potential presence of hydric soils, surficial hydrology and site disturbance can be determined from offsite wetlands determinations. Historical and current land use as it pertains to wetlands resources can also be ascertained.

By making initial determinations and preliminary conclusions regarding the aforementioned parameters, a project proponent can make informed decisions regarding natural resources (wetlands) on a subject parcel, save valuable time and expense relative to unanticipated project design and permitting delays, and determine if detailed onsite investigations are necessary, along with the anticipated level of effort. The level of effort to conduct offsite investigations can vary greatly, and can be tailored to suit individual site or project requirements.

OBJECTIVE OF OFFSITE IDENTIFICATION

In beginning discussions on the objectives of offsite wetlands identification, a good starting point is conceptualizing a working definition of offsite wetlands identification. For the purposes of this chapter, offsite identification of wetlands is defined as *assembling and interpreting readily available natural resource mapping and reports and other documents, both published and unpublished, from existing sources, for the sole purpose of identifying, locating and describing wetland resources on a given site or parcel of land.*

By applying existing resource document information, the researcher can make initial determinations relative to the perceived presence or absence of one, two or sometimes three of the parameters necessary for an area to be considered a jurisdictional wetland. In instances where onsite inspection is not necessary, or is beyond the scope of the investigation (e.g. National Environmental Policy Act wide range alternatives analyses, or limited environmental assessments), offsite wetlands determinations may be the only source of information for environmental planning decisions.

The overall accuracy of offsite wetlands determinations is a function of the quality of the information (sources) used, and the ability of an individual(s) to interpret the data.

The keys to conducting an effective and technically valid analysis include:

- Define the project scope and goals prior to conduct of the analysis
- Ensure that a wide range of sources are investigated and used
- Emphasize comparison and corroboration between different sources for the same site
- Obtain recent data, but also data that covers many different years to assist in understanding the site history
- Understand individual resource documents symbology and interpretation keys

The primary objective of conducting offsite wetlands determinations is identifying and determining whether wetlands exist on a parcel, followed by the approximate distribution and areal extent. In determining and quantifying these parameters, the key is correspondence between different sources. That is, not only locating wetlands on a subject parcel from a single source, but corroborating the identification through multiple sources.

Another important objective of conducting offsite determinations is documenting the dominant wetland cover type on parcels that have been preliminarily determined to have wetlands within their boundaries. Depending on the source, an interpreter can determine whether the wetlands onsite are either forested, scrub-shrub, emergent, aquatic bed or open water. Detailed interpretation requires a greater level of effort and expertise, but can result in greater detail, such as evergreen forest vs. deciduous forest, or persistent emergent vs. nonpersistent emergent, or artificially created vs. naturally occurring. Classification schemes can be tailored to an individual state's system, or the widely accepted federal system developed by the U.S. Fish and Wildlife Service (Cowardin et al. 1979).

Site soil characterizations and surficial hydrological features can also be recognized and described from offsite resources. Published sources exist which reveal site soils mapping to varying levels of detail and accuracy. Determining the hydrological regime, or simply the hydrology of a wetland, is a significant feature in determining the areal extent of a wetland, both in the field and from mapped sources. Offsite interpretation can reveal a wetland's hydrological source, as well as its drainage features.

IDENTIFICATION OF SOURCES

The initial step in conducting offsite wetlands interpretation studies is determining and obtaining readily available sources of information which will provide relevant details on wetland resources for a given site. Sources are generally diverse with varying levels of accuracy. The following sources are readily available and will provide a baseline of information from which to work.

1. United States Geological Survey (USGS) *Topographic Maps*, Standard Edition and Provisional Edition (7.5 minute or 15 minute quadrangles, scales 1:24000 or 1:25000, Continental United States, 1:20000 Puerto Rico, 1:63360 Alaska), United States Department of the Interior Geological Survey National Mapping Division.
2. United States Department of the Interior/Fish and Wildlife Service (USFWS) *National Wetland Inventory Maps* (scale 1:24000, Continental U.S., 1:63360 Alaska), interpreted and adapted from High Altitude Aerial Photography and superimposed on U.S. Geological Survey Topographic Maps.
3. United States Department of Agriculture/Soil Conservation Service *County Soil Surveys*, in cooperation with individual state agriculture experiment stations. Used in conjunction with the *Hydric Soils of the United States*, 1991, National Technical Committee for Hydric Soils, United States Department of Agriculture/Soil Conservation Service.
4. Aerial Photography (Stereo-paired, Black & White, Color, Color Infrared; Positive Transparency/Aero Negative; Various scales and dates), Federal, State and Commercial Suppliers.
5. United States Geological Survey *Surficial Geologic Map Quadrangles* (7.5 minute quadrangles, scale 1:24000), U.S. Department of the Interior Geological Survey.
6. Individual State Wetland Maps (limited coverage and level of accuracy).

OFFSITE WETLANDS IDENTIFICATION AND INTERPRETATION

This section describes in detail analysis and interpretation of the sources described above for offsite wetlands identification. Although the level of detail and accuracy varies with each source, a first-time evaluator should be able to extract sufficient information to reasonably determine if wetlands are present onsite, and the approximate historical or current distribution of wetlands.

United States Geological Survey Topographic Maps

The U.S. Department of the Interior, Geological Survey National Mapping Division has generated 7.5 minute and 15 minute Topographic Maps through the National Mapping Program. Two separate editions are available, the Standard

Edition Maps and the Provisional Edition Maps, both produced at 1:24000 (conventional units) or 1:25000 (metric units) in the Continental United States. Standard Edition Quadrangles represent a finished product, with the earth's topographic relief depicted by contours. Provisional Edition Quadrangles represent an updated draft format, including hand lettering and limited descriptive labeling of some physical features. Both editions are produced by stereoplotting high altitude aerial photographs. Stereoplotted information is field verified. Some quadrangles are mapped by a combination of orthophotographic images and map symbols, with orthophotographs derived from aerial photographs by removing image displacements due to camera tilt and terrain relief variations (USGS 1991).

The use of USGS Topographic Maps for offsite wetlands identification is often the first step to evaluate a site's physical features. In addition to topographic, hypsographic, infrastructure and other physical features, the USGS Topographic Maps provide detailed information relative to vegetational cover types, surface features, coastal features, hydrographic features such as rivers, lakes and canals, and submerged areas and bogs. Figure 1 is a section of an USGS Quadrangle and depicts some of these features. Figure 2 is a section of an USGS Index and provides the interpretation keys for select features that relate to wetlands location and identification. For example, it can be seen in Figure 1 that the site includes wooded marsh or swamp in the western and southern parts of the site, perennial ponds or lakes in the central part of the site, and perennial streams associated with cranberry bogs in the northeastern part of the site. Although most of the wetlands and open water interpretation keys are self-explanatory, the individual submerged areas and bogs keys require a little elaboration to distinguish among different wetland cover types.

Marsh or Swamp is a wetland resource characterized by saturated soil conditions in the root zone (as opposed to inundation), with emergent, herbaceous or aquatic bed vegetation as the dominant cover class. An example would be a rush (*Juncus spp., Scirpus spp.*) and sedge (*Carex spp.*) dominated wet meadow. *Submerged Marsh or Swamp* is an inundated root zone condition with emergent, herbaceous or aquatic bed vegetative dominants. A typical example is a broad-leaved cattail (*Typha latifolia*) marsh. *Wooded Marsh or Swamp* is a wetland characterized by saturated soil conditions with shrub, sapling or mature forest as the dominant cover class. A saturated red maple (*Acer rubrum*) swamp is an example. *Submerged Wooded Marsh or Swamp* indicates root zone inundation (ponding) as the dominant water regime with shrub, sapling or mature forest as the dominant cover class. A bottomland hardwood forest dominated by cypress (*Taxodium spp.*) trees is an example. *Land Subject to Inundation* can be floodplain and flood-prone areas that may support wetland hydrology and wetlands vegetation (hydrophytes). *Rice Fields* and *Cranberry Bogs* are examples of anthropogenically influenced wetland areas.

Some of the advantages in using the USGS Topographic Maps include the relative accuracy of the topographic contours in undisturbed areas, photointerpretation documentation is groundtruthed at regular frequencies, and individual

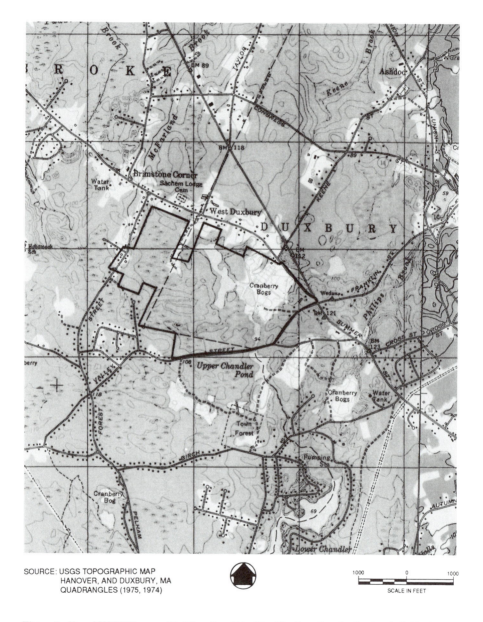

SOURCE: USGS TOPOGRAPHIC MAP
HANOVER, AND DUXBURY, MA
QUADRANGLES (1975, 1974)

1000   0   1000

SCALE IN FEET

Figure 1. Use of USGS Topographic Maps for offsite identification of wetlands of an individual site.

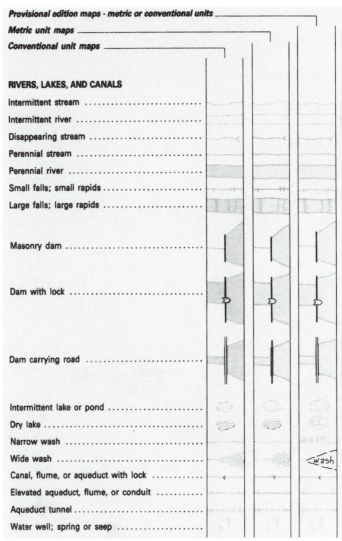

Figure 2. USGS topographic map key symbology for offsite wetlands identification.

quadrangles are periodically photorevised, which assists in chronological evaluation of a site's history. The limitations in using USGS Topographic Maps include interpretation problems associated with the small scale (1 inch equals 610 m) of the maps, and smaller wetlands often are frequently unmapped. In some parts of the country quadrangles may be too outdated to be of use.

United States Fish and Wildlife Service National Wetland Inventory Maps

The USFWS initiated the National Wetland Inventory (NWI) program and mapping in 1975 in an effort to assess, measure and characterize the extent of

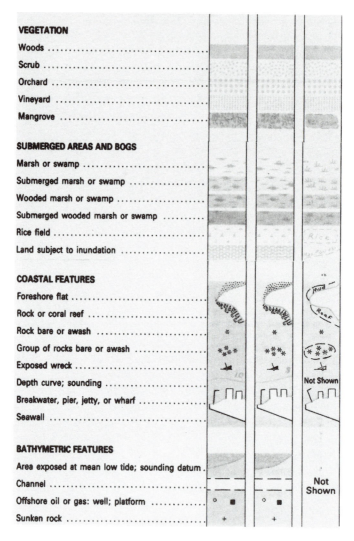

Figure 2. (continued) USGS topographic map key symbology for offsite wetlands identification.

wetlands and open water areas throughout the United States. The NWI Maps are produced from photointerpretation of high altitude stereo aerial photographs. High altitude aerial photographs were selected over satellite imagery at the time due to the problems satellite imagery had in capturing optimum water conditions for wetland detection, detecting smaller wetlands and identifying forested wetlands (Tiner and Wilen 1983). The NWI Maps are developed from 1:60000 color-infrared aerial photographs. Photointerpretation of the aerials provides a 3-dimensional image, thus allowing the interpreter to identify trees from shrubs, while considering shade and slope. Different wetland and open water types are differentiated based on their characteristic photographic signatures.

NWI Maps are developed according to a comprehensive evaluation process (Tiner and Wilen 1983). The preliminary field investigations and photointerpretation of high altitude aerial photographs is the initial step, with review of existing wetlands information and quality control of the interpreted photographs completing the first phase. Draft map production is initiated, with a subsequent interagency review of draft maps and final map production. Two series of NWI Maps are available, the 1:100,000/1:250,000 scale and large scale 1:24000. The USGS Topographic Map Quadrangle is used as a basemap with wetland and deepwater areas depicted as overlays.

A new wetland classification was developed by USFWS to correspond with the NWI Maps. *Classification of Wetlands and Deepwater Habitats of the United States* (Cowardin et al. 1979) describes individual wetland ecological attributes and arranges them in a hierarchical system that facilitates resource management and inventory. The three key components in the ecological hierarchy are hydrophytes, hydric soils and hydrology.

The use of NWI Maps in offsite wetlands identification typically provides the greatest level of detail and accuracy relative to the identification of wetland and open water cover types with the least amount of interpretation effort and expertise. The USFWS Classification System is a comprehensive and progressive inventory that groups wetlands into one of five major systems, marine, estuarine, lacustrine, riverine and palustrine, which all combine hydrologic, geomorphologic, chemical or biological factors (Cowardin et al. 1979). The hierarchy progresses through Subsystems, Classes and Subclasses which further refine and describe specific wetlands structural (vegetation, hydrology, dominant life form, etc.) components. Figure 3 depicts a representative section from an NWI Map which corresponds with Figure 1, and which indicates several different wetland classes within the palustrine system. Figure 3A explains the legend and symbology for the palustrine, riverine and estuarine systems found on NWI Maps. In comparing Figure 3 with Figure 1, the interpreter is able to see that the wooded swamp or marsh of the USGS Map has been further defined as palustrine forested broad-leaved deciduous wetland. An interpreter can become familiar with this system with a little practice, resulting in quick characterizations of site conditions relative to wetland types.

The accuracy of NWI Maps varies between systems and classes, with the highest degree of accuracy occurring for large marine, lacustrine and estuarine systems. Less accurate are smaller mapped units for palustrine wetlands, specifically palustrine forested wetlands, which can be misstated due to photointerpretation difficulties encountered as a result of "leaf-in" periods, where the interpreter cannot accurately describe the forest floor (MacConnell et al. 1989). NWI Maps provide the greatest diversity of all offsite references, with the possible exception of aerial photographs, however the latter requires a greater degree of photointerpretation expertise. Through use of the USFWS Classification System, an interpreter can characterize a wetland's system, the dominant vegetative structural life form (i.e., forested, emergent, aquatic bed, etc.) which can be further refined (evergreen vs. deciduous), and its hydrological regime (intermittent vs. perennial) and substrate (rock bottom or unconsolidated bottom). The taxonomy

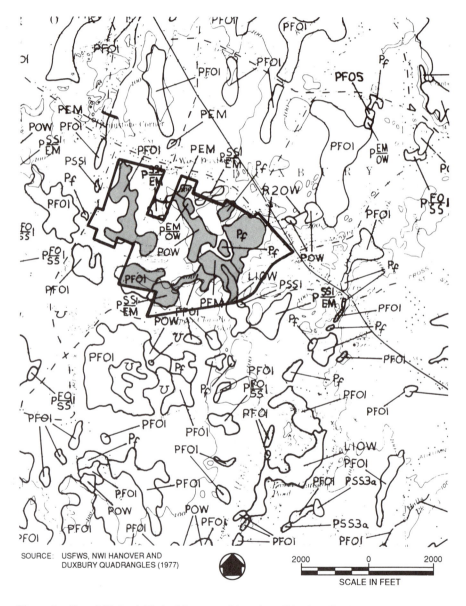

SOURCE: USFWS, NWI HANOVER AND
          DUXBURY QUADRANGLES (1977)

2000   0   2000

SCALE IN FEET

Figure 3. Use of National Wetland Inventory Maps for offsite identification of wetlands on an individual site.

also has provisions for documenting anthropogenic influence on created or farmed wetlands (palustrine farmed cranberry bogs and palustrine open water artificially excavated).

The limitations of NWI Maps for offsite wetlands identification include the small scale (1 inch equals 2000 feet), errors associated with photointerpretation of select cover types (principally deciduous forest), limited field verification, and the fact that the maps produced have not been photorevised since initial production. NWI Maps are beneficial as a qualitative reference, and are the only federally produced and readily available documents for the sole purpose of identifying, inventorying and characterizing wetlands.

United States Department of Agriculture/Soil Conservation Service Soil Surveys and the Hydric Soils of the United States List

The USDA/SCS produces County Soil Surveys in cooperation with the individual state's agricultural experiment station. The programs which have been undertaken have mapped individual soil series based on comprehensive field investigations conducted by SCS and State Soil Scientists. To produce the maps, soil scientists observe the steepness, length and shape of slopes, the size and velocity of streams, the kinds of native plants and rocks, and investigated many soil profiles (USDA/SCS 1978). Soil profiles are examined down to the parent material and are compared to soil profiles examined in other counties for the purpose of comparing and contrasting known soil series. A unified soil taxonomy, the USDA Soils Classification, is used across the nation to characterize and classify soil types. Soil Series and Soil Phase are the most common terms used in describing individual soil types. A soil series is a grouping of soils which have similar profiles and major horizons, and are named after towns in which the series was first discovered (USDA/SCS 1978). A soil phase is a division within the soil series. It further refines the series based on the texture in the surface layer, slope or stoniness (USDA/SCS 1978). Soil mapping units and boundaries are depicted as overlays on high altitude aerial photography, and are originally drafted by the field soil scientists. These boundaries are further refined following laboratory analysis of soils properties. The finished product indicates soil boundary delineations, soil series descriptions and bio-physicochemical properties. Additional sections of the soil survey provide information about recommended use and management of the soils, soil properties and soil formation.

Use of Soil Surveys for offsite wetlands identification is limited to the identification and distribution of hydric soils on an individual site. Hydric soils, one of the three essential characteristics of a federally jurisdictional wetland, have unique physical properties that set them apart from nonhydric soils. A hydric soil is defined as:

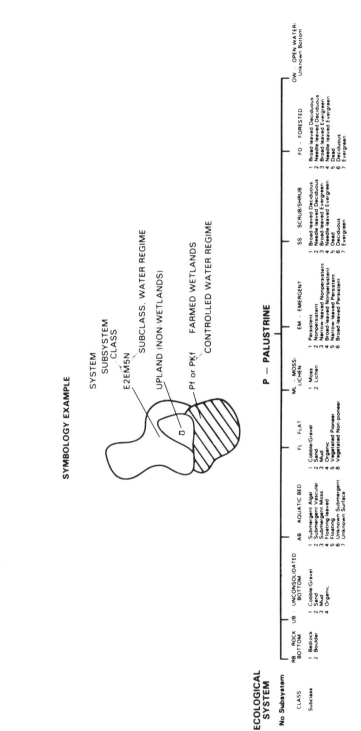

Figure 3a.   National Wetland Inventory Legend for palustrine, riverine and estuarine systems.

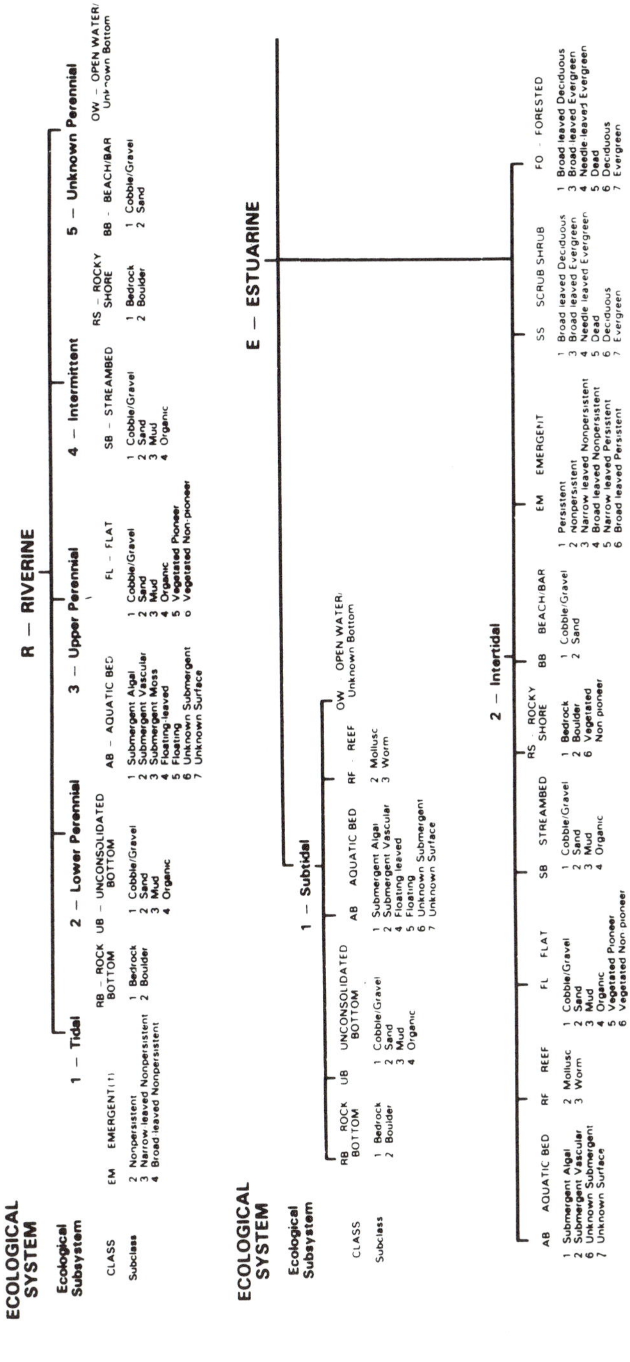

Figure 3a. (continued) National Wetland Inventory Legend for palustrine, riverine and estuarine systems.

SOURCE: U.S. DEPT OF INTERIOR, USFWS, 1980

A soil that is saturated, flooded, or ponded long enough during the growing season to develop anaerobic conditions in the upper part (USDA/SCS 1991).

In addition, the development of hydric soils is ultimately driven by the presence of wetland hydrology, and under sufficiently wet conditions (root zone saturation and inundation), hydric soils support the growth of hydrophytic vegetation. The USDA/SCS has developed a list of hydric soils entitled *Hydric Soils of the United States*, which was originally developed in 1987 and revised in 1991. The list was developed through applying criteria (drainage class, organic vs. mineral etc.) developed by the National Technical Committee for Hydric Soils (USDA/SCS 1991). Included in this list are most of the somewhat poorly drained soil series, and all of the poorly drained and very poorly drained soils.

Through the use of individual soil surveys, an interpreter can determine all mapped soil series onsite and then cross-reference the list of soils with *Hydric Soils of the United States* to verify the presence of mapped hydric soils onsite. Figure 4 is a Soil Survey Map of the previously referenced site. Examination of the site, and cross-reference with the hydric soils list indicate that Sanded Muck (SB), Peat (Pe), Freshwater Marsh (Fr), Scarboro Sandy loam (ScA), Muck (Mv), and Au Gres and Wareham loamy sands (AuA) are hydric soils, and that their occurrence on the site corresponds with the wooded swamp or marsh, perennial lakes or ponds and cranberry bogs of the USGS Map, as well as the palustrine forested broad-leaved deciduous, palustrine farmed wetlands and palustrine open water of the NWI Map.

Although soil mapping involves an intensive field effort, the accuracy of the soil maps is quite variable, and areas mapped as hydric soils (a hydric soil series) can contain inclusions of nonhydric soils. Conversely, areas of nonhydric soils may contain hydric inclusions. The soil mapping information is best used as a macroscale assessment tool and should not be used for definitive boundaries of hydric soils. An advantage in using soil surveys and the list of hydric soils is that the interpreter does not need to spend time learning the USDA Soils Classification and taxonomy system to be able to locate areas of mapped hydric soils on a site. Although, it is recommended that the interpreter understand the basic principles underlying the criteria for listing a soil as a hydric soil.

Comparison and Corroboration

USGS Topographic Maps, USFWS NWI Maps and USDA/SCS Soil Surveys are typically the most accessible sources and require the least amount of technical knowledge to interpret. The effectiveness and level of accuracy in conducting offsite wetlands interpretation studies using these sources are a functions of the time needed for obtaining each source, and interpreting the information. Emphasis should be placed on comparing and contrasting individual sources. That is, comparing and corroborating the results from USGS to NWI to USDA/SCS, being sure to consider the year in which each source was initially produced. For example, the USGS indicates wooded marsh or swamp throughout the western and southern sections of the reference site (Figure 1), the NWI indicates palustrine forested broad-leaved deciduous wetland in the western and southern sections of the site (Figure 3) and the USDA/SCS indicates hydric soils in the western and southern sections of the site (Figure 4).   Therefore, the interpreter can be reasonably confident that wetlands, and most likely forested wetlands, exist at the site.   In an effort to further refine the information already obtained from these sources, additional sources can be consulted and evaluated.

Other Sources

Aerial Photographs

Aerial photography has been used since the 1860's for remote sensing land-use patterns and activities through the use of hot-air balloons (McKnight 1987). These actions spawned the development of photogrammetry, the science of obtaining reliable and defensible measurements from photographs and mapping from aerial photographs (Ritchie et al. 1988).   Historically, the interpretation of aerial photography has fallen under the term remote sensing, with the net result being that aerial photographs were the only tool used in remote sensing. Contemporary views have altered the term to include a wide range of tools and analytical devices.   One recent definition stated *remote sensing* is the *measurement of reflected, emitted or backscattered electromagnetic radiation from Earth's surface using instruments stationed at a distance from the site of interest* (Roughgarden et al. 1991).   Nonetheless, aerial photography is used today as a powerful source for remote sensing land-use, including wetlands identification, characterization, and perturbation as a result of anthropogenic activities.

Stereo-paired vertical contact prints provide the most useful and scientifically defensible information, as a three-dimensional image of the earth's surface is presented to the interpreter via a stereoscope.   This three-dimensional image is obtained through photographs taken in stereo pairs with end overlap.   The photographs are interpreted with a stereoscope which allows the interpreter to closely examine site conditions by adding depth of field, and provides the ability to distinguish wetland cover types.   Different cover types have characteristic signatures which can be quantified based on observable color, tone or hue, shadow, texture and depth of field. Color signatures can be further refined using the Inter-Society Color Council and National Bureau of Standards (ISCC-NBS) method of color description, using Centroid Color Charts (Smith and Anson 1968).

Permanently and seasonally inundated forested, scrub shrub, emergent and open water areas are relatively easy to distinguish from upland habitats, as surface water has a different reflectance pattern than dry areas. The difficulty comes when trying to identify seasonally saturated wetlands, particularly seasonally saturated forested wetlands.

Aerial photographs are available in many formats, scales, and geometry (Figure 5).   The most common photograph types used for wetland identification are

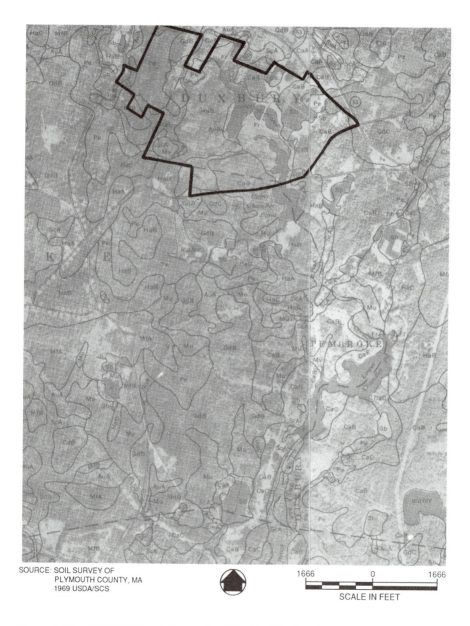

SOURCE: SOIL SURVEY OF
        PLYMOUTH COUNTY, MA
        1969 USDA/SCS

1666   0   1666

SCALE IN FEET

Figure 4. Use of USDA/SCS Soil Surveys for offsite identification of an individual site.

SOURCE: MARKHURD AERIAL, 1986
MINNEAPOLIS, MN

Figure 5. Examples of vertical black and white aerial photographs.

vertically oriented black & white, color and color infrared using aerographic film. Aerial photograph geometry (i.e. basis of their angular relationship to the earth's surface) can be divided into two major categories: oblique, which includes low oblique and high oblique, and vertical (Figure 6) (Hudson and Lusch 1990). Oblique aerial photographs have an advantage to the interpreter in that ground features can be interpreted from a familiar point of view (McKnight 1987). However, due to measurement inadequacies and scale development, vertically oriented photographs are used for determining quantitative information and they provide more useful and defensible information for wetlands identification. In addition, due to the orientation of the camera for oblique photos, the stereoscopic effect is reduced, thereby negating the three-dimensional effect.

Film types are also an important consideration in identifying wetlands. Although Panchromatic black & white photographs are the least expensive and most common, color infrared photographs have a demonstrated and significant advantage in that discrimination of living and dead vegetative matter is possible, along with enhanced abilities to discriminate open water areas, and discriminating natural features from man-made features.

Larger scale, low-altitude aerial photography is recommended over smaller scale, high-altitude aerial photography for clarity, ease of interpretation, distinguishing ground features and general resolution. The most useful and accurate interpretations are made through chronological analysis of a site.

Aerial photographic coverage and availability significantly limit its use for offsite wetlands identification. Federal and state agencies have inventories with spotty coverage and usually are not available for purchase. The National Archives (in Utah) maintains an active database and inventory of aerial photographs (Black & White, Color Infrared, Stereo-paired) covering the entire continental United States for select years, and are available for purchase. These photographs are generally high-altitude, small scale aerial photographs. Private commercial suppliers often maintain inventories, and usually specialize within a region (i.e., New England, Northwest, Southeast). Coverage is largely unpredictable varying from complete chronological coverage over several years to no coverage at all. Purchase costs can be quite high, with some firms charging access fees for database reviews.

As aerial photography provides the basemap and framework for most other offsite sources, it is apparent how important it can be when used in its unrevised form. Stereo aerial photographic interpretation has evolved into a scientific and technical discipline of its own, and in some instances, requires considerable expertise to extract valid information. Nonstereo-paired aerial photographs can be used to supplement the other offsite information in a qualitative manner. There is much variability among nonstereo aerial photographs, especially relative to geometry, scales, film types and coverage. When conducting preliminary qualitative offsite wetlands reviews, interpretation of aerial photographs may not be necessary. However, if the goal is to quantify wetland site conditions over time, aerial photographs may prove to be an indispensable tool for evaluation.

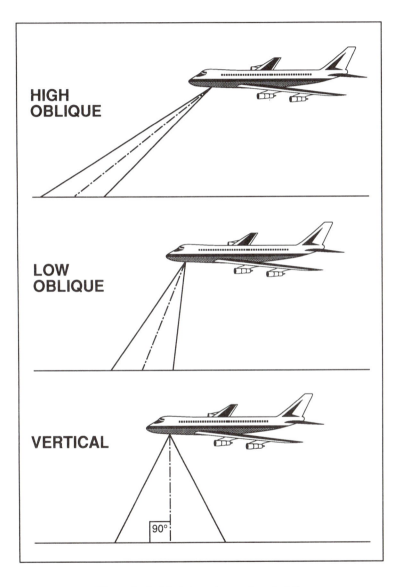

Figure 6. Aerial photography angular orientation.

United States Geological Survey Surficial Geologic Maps and Individual State Wetland Maps

The U.S. Geological Survey produces Surficial Geologic Maps primarily for use as mapping indicators of geologic zonation above bedrock. These maps provide some detail relative to soil and subsurface composition and are helpful in locating and identifying Swamp Deposits, Alluvium, surface water bodies and other wetland features. Figure 7 indicates the location of surface waterbodies, Swamp deposits (Qs) and cranberry bogs on the reference site. The Swamp deposit designations are indicative of organic matter, clay, silt and sand accumulating in swamps (USGS 1967).

Several individual states have produced wetland maps for use in macroscale planning and at least in one instance (MacConnell et al. 1989), for jurisdictional purposes. Most of the maps are developed based on interpretation of stereo-paired aerial photography, similar to the process used by the National Wetland Inventory. States that have produced wetland maps include: Maine, Vermont (based on NWI), Massachusetts (Wetlands Restriction Program), New York and New Jersey. Coverages, interpretation keys (classification systems) and accuracy are variable from state to state, and in some states (e.g. Maine, Massachusetts), statewide mapping is currently an ongoing process (Foote-Smith 1992).

SUMMARY

This chapter has presented some of the sources to be investigated for use in offsite wetlands identification along with their availability and individual use. Emphasis is placed on the use of multiple sources for a given site, and the review of historical information. The ability to document wetland site conditions without detailed onsite investigations has a demonstrated need from a natural resources planning perspective, as well as from a jurisdictional perspective. Documentation of anthropogenic influence on wetlands is another demonstrated need for using offsite materials for wetlands identification. By using the sources and methods discussed in this chapter, a reviewer can make a positive or negative determination regarding the presence or absence of wetlands, estimate the areal extent of wetlands, and in some cases determine major cover types. Offsite wetlands identification is not a substitute for onsite wetland delineations when the goal is a definitive demarcation or delineation of wetlands for site development and project planning purposes.

REFERENCES

33 CFR 320 - 330. 1986 Code of Federal Regulations. U.S. Army Corps of Engineers.

SOURCE: USGS SURFICIAL GEOLOGIC MAP
HANOVER, MA QUADRANGLE (1967)

2000   0   2000

SCALE IN FEET

Figure 7. Use of USGS Surficial Geologic Maps for offsite identification of wetlands on an individual site.

Cowardin, L.M., V. Carter, F.C. Golet and E.T. LaRoe. 1979. Classification of Wetlands and Deepwater Habitats of the United States. U.S. Fish & Wildlife Service - FWS/OBS-79/31. 103 pp.

Foote-Smith, C. 1992. Personal Communication, April 22, 1992, University of Massachusetts at Lowell.

Hudson, W.D. and D.P. Lusch. 1990. Airphoto/Satellite Imagery: An Introduction. Remote Sensing and GIS Applications to Nonpoint Source Planning Proceedings. Spons. USEPA and NE Illinois Planning Commission. Chicago, Illinois.

MacConnell, W., D. Goodwin, K. Jones, J. Stone, D.B. Foulis, G. Springston and D. Swartout. 1989. Wetland Restriction Program: Mapping Standards for Wetlands Restriction Maps in Massachusetts. Technical Memorandum published by the Massachusetts Department of Environmental Protection.

McKnight, T. 1987. Physical Geography: A Landscape Appreciation. Prentice Hall, Englewood, NJ. Second Edition. 539 pp.

Ritchie, W., M. Wood and R. Wright. 1988. Surveying and Mapping for Field Scientists. Aerial Surveying Techniques, Ch. 3. Longman Scientific & Technical.

Roughgarden, J., S.W. Running and P.A. Matson. 1991. What Does Remote Sensing Do For Ecology? Ecology Vol 72, No. 6. pp. 1918-1922.

Smith, Jr., J.T. and A. Anson (eds.). 1968. Manual of Color Aerial Photography, First Edition. American Society of Photogrammetry, Virginia.

Tiner, R.W. and B.O. Wilen. 1983. The U.S. Fish and Wildlife Service's National Wetland Inventory Project. Technical Memorandum. USFWS, Washington, D.C. and Newton Corner, Massachusetts.

U.S. Department of Agriculture/Soil Conservation Service - National Technical Committee for Hydric Soils, 1991. Hydric Soils of the United States.

U.S. Department of Agriculture/Soil Conservation Service. 1978. Soil Survey of Kennebec County, Maine. USDA/SCS Cooperative Publication with the Maine Agricultural Experiment Station.

U.S. Environmental Protection Agency. 1991. Proposed Revisions to the Federal Manual for Identifying Wetlands. Office of Wetlands, Oceans and Watersheds. August 14, 1991.

U.S. Geological Survey. 1967. Surficial Geologic Map. Hanover, MA Quadrangle.

U.S. Geological Survey. 1975. Topographic Map. Duxbury, MA Quadrangle.

U.S. Geological Survey. 1974. Topographic Map. Hanover, MA Quadrangle.

U.S. Geological Survey. 1991. Topographic Map Index.

U.S. Fish and Wildlife Service. 1977. National Wetland Inventory Maps. Duxbury and Hanover, MA Quadrangles. USFWS/Department of the Interior.

CHAPTER 3

ONSITE IDENTIFICATION AND DELINEATION OF WETLANDS

Carl E. Tammi

The need to identify jurisdictional wetlands and delineate wetland and upland boundaries is principally driven by Section 404 of the Clean Water Act, as well as state and municipal wetlands protection statutes. These wetlands protection statutes, and the associated regulatory policies, dictate that wetland boundaries be established prior to project planning, design and construction. Therefore, project proponents and regulators are required to characterize and quantify the differences between wetlands and uplands so that the boundary can be identified with some certainty and repeatability. This is accomplished through the use of field indicators including morphological attributes, visual observations and recorded data.

Through consensus, three characteristics or parameters have been selected to distinguish wetlands from uplands. First, wetlands are characterized by the presence of water, typically from a surface or groundwater source. Water levels in wetlands are typically dynamic, with the frequency of saturation and inundation varying among wetland types, and varying temporally within wetland types. Secondly, wetlands are characterized by the presence of unique soils which are diagnostic of wetland conditions. These soils display properties which indicate anaerobic conditions in the root zone resulting from prolonged saturation or inundation. Finally, wetlands are characterized by the presence of wetlands vegetation which possess morphological adaptations that enable them to tolerate frequent root zone saturation or inundation, and anaerobic conditions.

There are many factors which influence the presence of these parameters in communities. Topographical relief is a significant physical factor which often dictates the source for wetlands hydrology. For example, topographical depressions often correspond closely with water table elevation in glaciated wetlands of the northeastern United States. Headwaters of streams and rivers are often the result of sheet runoff from watersheds which emanate from mountainous regions. Palustrine and riverine wetlands are often associated with these surface water bodies. Other factors which influence the formation of the three wetland parameters include stratigraphy, surficial and bedrock geology, precipitation and watershed characteristics.

In identifying and delineating wetlands, it is important to establish in advance the overall goal and scope of the wetlands investigation. The investigator should determine if it is necessary to conduct a comprehensive onsite delineation of the entire wetland and upland boundary, or simply confirm the presence or absence of wetlands. A wetlands determination is the process by which the evaluator makes a positive or negative assumption that wetlands are extant on a site. This assumption is based on identifying whether or not wetlands characteristics are present anywhere within the site's boundaries.

Wetlands delineation is the process by which the investigator identifies and locates wetlands, then quantitatively assesses the areal extent of wetlands on the site through consideration of hydrological field indicators, soil profiles and vegetation sampling and inventory. Wetland delineation techniques and methodologies vary from place to place in response to local and state jurisdictional requirements. Nevertheless, almost without exception, local and state mandated procedures are predicated upon parameters defined by the federal agencies. These parameters, and their associated technical criteria, are expressed in the *Corps of Engineers Wetland Delineation Manual* (Environmental Laboratory 1987), the *Wetland Identification and Delineation Manuals, Volumes I and II* (Sipple 1988), and the *Federal Manual for Identifying and Delineating Jurisdictional Wetlands* (Federal Interagency Committee for Wetland Delineation 1989). As federal legislation has evolved, resultant modifications to the jurisdictional definitions and limits have occurred, and likely will continue to occur, with ensuing performance standard modifications (U.S. Army Corps of Engineers 1991, U.S. Army Corps of Engineers 1992). Other sources (Tiner 1991, Sipple 1992) discuss the importance of the dynamic nature (seasonality, degree of wetness, etc.) of wetlands as it relates to the individual parameters and the ability to effectively recognize wetland boundaries. Tiner (1993) proposed an innovative approach to delineating wetlands based on identifying primary indicators of hydrophytes and hydric soils in undrained wetlands.

This chapter discusses the three parameters associated with wetlands, with an emphasis placed on conducting field assessments which evaluate and characterize each parameter. Techniques are presented which assist the investigator in assessing each parameter, including recognizing field indicators of wetlands hydrology, hydric soils and hydrophytes. In conducting these analyses, it is important that the methodology be practical, reproducible, efficient and cost-effective.

WETLANDS HYDROLOGY

Wetlands hydrology is the single greatest impetus driving wetlands formation (Mitsch and Gosselink 1986, Federal Interagency Committee for Wetland Delineation 1989, Tiner and Veneman 1989, Tiner 1993). Wetlands hydrology is characterized by permanent, temporary, periodic, seasonal or tidally-influenced inundation or soil saturation within the root zone. This water may derive from surface water (streams, rivers, ponds, lakes, ocean, etc.), groundwater, overbank flooding, precipitation, sheet flow or tidal flooding.

Wetlands hydroperiod is a term used to characterize the hydrological conditions of wetlands, and is a function of flood duration and flood frequency. All wetlands are dynamic systems from a hydrological viewpoint. The hydrology of perennial wetlands varies irregularly on an annual basis. Ephemeral wetlands such as vernal pools typically have seasonally varying hydroperiods. And tidally influenced wetlands experience daily periodic hydrologic fluctuations. Keeping in mind this inherent variation, a wetland's net hydroperiod can be represented by the following equation (Mitsch and Gosselink 1986):

$$\Delta V = P_n + S_i + G_i - E_t - S_o - G_o +/- T$$

where:

$$
\begin{aligned}
V &= \text{Volume of water storage} \\
\Delta V &= \text{Change in volume of water storage} \\
P_n &= \text{Net Precipitation} \\
S_i &= \text{Surface inflow (stream flow and sheet flow)} \\
G_i &= \text{Groundwater inflow} \\
E_t &= \text{Evapotranspiration} \\
S_o &= \text{Surface outflow} \\
G_o &= \text{Groundwater outflow} \\
T &= \text{Tidal inflow (+) or outflow (-)}
\end{aligned}
$$

Wetlands hydrology drives the development and distribution of the other two parameters of wetlands (Tiner 1988). Permanent or periodic inundation or saturation of the root zone during the growing season results in the development of anaerobic conditions, which is one of the chief determinants of hydric soil conditions. Root zone saturation is in turn responsible for the occurrence and distribution of vegetation which can withstand these conditions.

At present, there is a lack of widespread scientific consensus regarding the duration of time required for anaerobic conditions to develop from root zone saturation or inundation. Until recently, the U.S. Army Corps of Engineers required a minimum of two weeks of continuous saturation or inundation within the root zone for an area to be considered to possess wetlands hydrology (U.S. Army Corps of Engineers 1991). Presently, the Corps stipulates that the root zone be saturated or inundated for a consecutive number of days for more than 12.5 percent of the growing season for an area to have wetlands hydrology (U.S. Army Corps of Engineers 1992). Areas inundated or saturated for 5-12.5 percent of the growing season may or may not have wetlands hydrology. Historically, the definition of the root zone and its maximum vertical extent have been critical to determining jurisdictional limits. This vertical extent is a direct function of soil series drainage class and permeability. The *Corps of Engineers Wetlands*

Delineation Manual defines major portions of the root zone to be that area within 30 cm of the soil surface (Environmental Laboratory 1987).

Among the technical parameters, the wetlands hydrology parameter has been the most controversial (U.S. Environmental Protection Agency 1991). A quick reference for side by side comparisons of the various jurisdictional (current and proposed) performance standards for the wetlands hydrology criterion is available in the *National Wetlands Newsletter* (Environmental Law Institute 1991).

The ability to identify and delineate wetlands relative to the hydrology parameter relies on the investigator's ability to recognize field indicators of wetlands hydrology. Generally, quantitative studies and hydrological modeling are not required for onsite wetlands identification and delineation. Although many of the technical aspects of the scientific discipline of hydrology are widely applicable to wetlands formation, only a rudimentary knowledge of the underlying principles are crucial to wetlands delineation. Proficiency in recognizing wetlands hydrological indicators, however, is required.

Wetlands Hydrological Field Indicators

Onsite wetlands inspection and delineation requires the ability to recognize and distinguish wetland hydrological field indicators (Table 1). These indicators provide visual or assumed evidence of soil saturation or surface inundation. Certain indicators provide strong evidence of the frequency and duration of saturation or inundation, and can be interpreted to support wetlands boundary determinations. Field indicators include the direct observation of wetlands hydrological conditions (during the growing season), soil characteristics, morphological plant adaptations and evidence of water movement (Environmental Laboratory 1987, Federal Interagency Committee for Wetland Delineation 1989).

The strongest field indicator of wetlands hydrology is the direct observation of surface inundation or soil saturation within the root zone during the growing season (Figure 1). This indicator represents a real-time visual observation of wetlands diagnostic conditions. Inundation is a valid indicator of wetlands hydrology, however, the boundary or areal extent of a wetland may indeed extend far beyond the limit of the inundated area. Typically, as topographic elevation increases, the areal extent of inundation decreases, with soil saturation within the root zone becoming the dominant defining characteristic. The depth to which soil saturation is present, and its persistence, determines whether the area is a jurisdictional wetland. Depth is also related to the hydric soil criterion, specifically its drainage class and permeability. Generally, the soil should be saturated within 0 to 46 cm of the surface during the growing season. Soil saturation can be field-confirmed through an auger boring to a depth of 46 cm to determine the water table elevation. Saturated soils will occur below this elevation, as well as slightly above, due to capillary action (Environmental Laboratory 1987). Soils can also be squeeze-tested to extract free water to determine saturation.

Wetland drainage patterns occur along riverine, estuarine, palustrine and some lacustrine wetlands. They are typically associated with moving or fluctuating water

Table 1. Wetland hydrological field indicators widely used for identifying wetlands, and delineating the wetland and upland boundary.

Direct observation of inundation

Direct observation of soil saturation

Wetland drainage patterns

Plant morphological adaptations

Adventitious roots	Aerenchyma
Hypertrophied lenticels	Leaf adaptations
Multiple trunks	Oxidized rhizospheres
Pneumatophores	Shallow roots
Stooling	Tree buttressing

Water marks and drift lines

Surface scouring and water borne sediment deposits

Field confirmed hydric soils

Water-stained leaves

Figure 1. As illustrated by this freshwater marsh in New Jersey, the strongest field indicator of wetlands hydrology is the direct observation of surface inundation or soil saturation.

systems such as streams, rivers, creeks, etc. Visual indicators include drainage channels, eroded areas, the absence of litter, litter deposits and characteristic meandering.

Many wetland plants have developed specialized morphological adaptations which enable them to survive and proliferate with their roots in an anoxic environment. These adaptations have developed in response to root zone saturation, and are therefore treated as wetlands hydrology indicators (Federal Interagency Committee for Wetlands Delineation 1989). The adaptations are typically specialized structures enabling the plants to capture molecular oxygen and transport it to stems and roots. This is true of buttressed trees (swollen base of tree trunk), pneumatophores (above-ground root structures), adventitious roots (above ground roots), shallow roots, multiple trunks and stooling, hypertrophied lenticels (exaggerated lenticels on stems), aerenchyma (air-filled tissue) and leaf adaptations (floating and polymorphic leaves). An oxidized rhizosphere is also indicative of wetlands hydrology. Evidenced by channels that have developed along the roots for the transport of oxygen, rhizopheres are difficult to observe and clearly identify unless iron oxide concretions are present.

Water marks are visible indications of inundation on woody vegetation and other permanent structures within wetlands. These are considered strong indicators of the presence of wetlands hydrology due to their confirmation of inundated conditions. Drift lines are visible lines indicating the extent of hydrologically driven actions. Tidal marshes typically display drift lines with debris and driftwood extending up to the spring high tide elevation. Drift lines also accumulate along riverbanks and floodplains.

Surface-scouring occurs in wetlands with widely fluctuating hydroperiods and in areas subject to storm or tidal forces which expose soils, remove leaf litter and cause surficial erosion. Sediment deposits on vegetation also indicate that inundated conditions have occurred in the recent past.

Observations of wetlands hydrology are typically performed at the time of the onsite inspection. However, in cases where long-term observation may be necessary to confirm precise wetlands boundaries, wetlands hydrology should be assessed for longer periods such as entire growing seasons. Long-term hydrological monitoring can be a labor intensive and costly process unless historical groundwater and surface water elevation data is available.

In situations where the jurisdictional limits of a wetland have been debated or questioned, and precise boundaries are required (e.g., real estate transactions), hydrological monitoring can typically be accomplished in a cost effective manner through the installation and monitoring of hand-set groundwater monitoring wells. Typically, 5 cm diameter slotted screen PVC monitoring wells, 60-75 cm long, are installed at random locations along a transect line extending from within the recognizable wetland area out to recognizable upland. Hand augers are sufficient to install wells, and borings should be backfilled with washed sand to allow unrestricted passage of groundwater. Well locations and elevations should be surveyed and plotted onto the site plan, and monitoring of groundwater elevations should be conducted weekly throughout the growing season. Water level recording can be accomplished using a sounder mechanism and an incremented cord or tape rule. Precipitation should be recorded throughout the monitoring period. Regulatory agencies may require several growing seasons worth of data, which can be impractical from the standpoint of project timing and cost. Nevertheless, long-term hydrological information represents the strongest evidence for the extent of wetlands.

HYDRIC SOILS

The soils found in wetlands have unique morphological and other observable properties which differentiate them from upland soils. A hydric soil by definition is a soil which is saturated, flooded, or ponded long enough during the growing season to develop anaerobic conditions in the upper part (U.S. Department of Agriculture Soil Conservation Service 1991). Hydric soil properties are a direct function of the frequency and duration of saturation and inundation, specifically in the root zone. Soils which display flooded and saturated conditions for an extended period (two weeks or more) during the growing season create an environment where free oxygen is deficient, and ultimately unavailable to plants. As a result of this saturation and inundation, hydric soils display observable field indicators which are diagnostic of wetland conditions.

The United States Department of Agriculture Soil Conservation Service has developed a classification system which provides criteria for listing a hydric soil, as well as categorization of listed hydric soils. This list, *Hydric Soils of the United States*, categorizes hydric soils into two major groups: organic soils and mineral soils (U.S. Department of Agriculture Soil Conservation Service 1991). Generally, soils with at least 46 cm of organic matter in the upper part of the soil profile are considered organic soils, or histosols (Tiner and Veneman 1989). Organic soils are divided into groups based on the degree to which plant fibers and material are decomposed. Fibrists (peats), hemists (mucky-peats and peaty-mucks) and saprists (muck) are organic hydric soils listed in increasing order of plant material decomposition. Folists are the fourth group of organic soils, but are not considered hydric soils because the organic component does not derive from saturation or inundation (Tiner and Veneman 1989).

Mineral soils generally have less organic material in the upper part of the profile than organic soils, and have differing field indicators. Mineral hydric soils are also taxonomically arranged, and include soils in Aquic suborders, Aquic subgroups, Albolls suborder, Salorthids great groups and Pell great groups of Vertisols (shrinking or swelling dark clay soils) (Federal Interagency Committee for Wetlands Delineation 1989). Mineral soils are considered hydric soils when any of the following criteria are satisfied (Tiner and Veneman 1989):

- The soils are somewhat poorly drained and have a water table less than 15 cm from the surface for a significant period during the growing season
- The soils are poorly drained or very poorly drained and have either a water table at less than 30 cm from the surface for a significant period during the growing season if permeability is equal to or greater than 15 cm/hr in all layers within the top 50 cm, or a water table less than 46

cm from the surface for a significant period during the growing season if permeability is less than 15 cm/hr in any layer within the top 50 cm

- Water is ponded for a long duration (more than 7 days), or a very long duration (greater than a month), during the growing season
- The soils are frequently flooded for a long duration (more than 7 days), or a very long duration (more than a month), during the growing season

A significant period was most recently defined as at least 15 consecutive days of saturation or 7 days of inundation during the growing season (U.S. Army Corps of Engineers 1992). The 1987 Corps Manual, which is the current jurisdictional document, defines growing season as that portion of the year when soil temperatures at 50 cm below the soil surface are higher than biologic zero (5° C). For ease in determination, the growing period can be estimated to occur when air temperature exceeds 2.2° C (U.S. Army Corps of Engineers 1992).

Drainage classes are a significant criterion when determining the presence of hydric soils, as the soils relate to individual taxonomic groups (Smith 1973, Tiner and Veneman 1989). All very poorly and poorly drained soils are hydric soils, assuming the soils have not been drained. A very poorly drained soil is a soil where water is removed from the soil so slowly that free water remains at or near the surface during most of the growing season. A poorly drained soil is a soil where water is removed so slowly that the soil is saturated periodically during the growing season or remains wet for long period. Many somewhat poorly drained soils are also hydric. A somewhat poorly drained soil is one where water is removed slowly enough that the soil is wet for significant periods during the growing season. Soils mapping provided by the U.S. Department of Agriculture Soil Conservation Service indicates the soil series drainage class, which can be field confirmed using criteria developed by the U.S. Army Corps of Engineers (U.S. Army Corps of Engineers 1991).

Essential to the investigation of soils for hydric properties is an understanding of soil horizons. Soil horizons, or layers, develop in response to localized chemical and physical processes resulting from the activities of soil organisms, the addition of organic matter, precipitation and percolation. Horizons can be distinguished based upon color, texture and composition (Environmental Laboratory 1987). However, the soil horizon is essentially a continuum and there is no clear cut distinction between one horizon and another.

Soils typically have 4 major horizons: an organic layer (O), and 3 mineral layers (A, B and C) (Figure 2). The O horizon is the surface layer and is composed of fresh or partially decomposed organic material. The A horizon is characterized by an accumulation of organic matter, and the loss of clay, iron and aluminum. Together, the O and A horizons constitute the zone of maximum biologic activity. The B horizon is characterized by the accumulation of silicates, clay, iron, aluminum and humus, whereas the C horizon contains weathered material either similar or dissimilar to the parent material.

Soil colors also provide critical information on soil wetness, and the degree of saturation and inundation. Three aspects of color are standardized by the

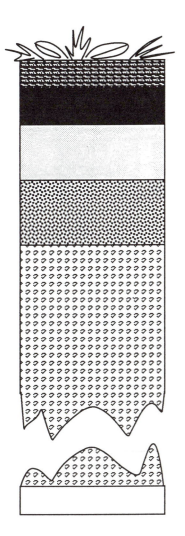

O1 - loose leaves and organic debris

O2 - partly decomposed or matted organic debris

A1 - dark colored horizon with a high content of organic matter mixed with mineral matter

A2 - light colored horizon of maximum leaching

B - A deeper colored horizon of maximum accumulation of clay minerals or of iron and organic matter

C - weathered material, either like or unlike the material from which the soil formed

R - consolidated rock

Figure 2. Generalized soil profile. Actual horizons represent a continuum and will not appear this distinct in the field.

Munsell Color Chart: hue, value and chroma (Kollmorgen Corporation 1975). Hue describes the soil based on its relation to the spectral colors (red, yellow, green, blue, purple, or a mixture of these colors), value describes the degree of lightness, and the chroma indicates the strength or purity of the color. The Munsell Color Chart depicts individual hues on separate pages, with the corresponding value on the vertical axis and the chroma on the horizontal axis. Soil color is reported as hue, value/chroma, for example 7.5 YR 2/1.

Soil colors are an important field indicator for field verification of mapped hydric soils, as well as hydric soil inclusions in mapped nonhydric soils. Both soil matrix, the predominant color derived from the parent material or deposits, and mottles and concretions, contrasting spots within the matrix, should be characterized. Soil colors are diagnostic of wetland conditions when the matrix chroma immediately below the A horizon is 2 or less in mottled soils, or 1 or less in unmottled soils (Environmental Laboratory 1987).

Hydric Soil Field Indicators

Hydric soils display certain field indicators which provide clues to the degree of saturation and inundation. These clues are in turn indicative of the presence or absence of wetlands hydrology. The field indicators presented in Table 2 are nationally accepted, and were developed based on U.S. Department of Agriculture Soil Conservation Service parameters (Environmental Laboratory 1987, Federal Interagency Committee for Wetland Delineation 1989, Tiner and Veneman 1989). Recognizing and applying these field indicators will enable the wetland scientist to field confirm or refute the hydric soils criterion. Most of these field indicators are used when attempting to recognize hydric mineral soils, and some, such as a high organic content at the surface, subsurface streaking and wet spodosols, are particularly useful when attempting to recognize sandy hydric mineral soils.

Table 2. Field indicators of hydric soils. Indicators are applied to the soil profile within the root zone.

Organic soils (histosols)

Histic epipedons

Sulfidic material

Aquic or peraquic moisture regime

Reducing condition

Soil color

Mottling or concretions

High organic matter content in surface horizon

Subsurface streaking

Spodic horizon

As mentioned previously, all organic soils except folists are considered hydric soils. Organic soils are readily identifiable in the field due to the relatively high percentage of poorly decomposed organic matter such as vegetative material, litter, etc. in the soil matrix. Organic soils also often possess dark hues and very dark chromas (7.5 YR 2/0, 10 YR 2/1, etc.).

Histic epipedons are thick (20-40 cm) organic surface layers overlying hydric mineral soils. Histic epipedons develop as a result of prolonged saturation within the root zone, at least 30 consecutive days or more in most years (Environmental Laboratory 1987).

The presence of sulfidic material is a strong indicator of permanently saturated soil conditions. Sulfidic material is evidenced by a rotten egg-like odor, typically following soil disturbance. The odor occurs as a result of sulfates being reduced to sulfides under anaerobic conditions associated with prolonged root zone saturation. This indicator is especially common in tidal marshes.

Aquic and peraquic moisture regimes are characterized by elevated groundwater levels. Aquic moisture regimes are mostly reducing, and are nearly free of dissolved oxygen as a result of saturation due to groundwater or the capillary fringe. A part or all of the root zone is reduced and totally deficient in dissolved oxygen during the growing season. Peraquic moisture regimes occur when groundwater is always at or near the soil surface.

Reducing conditions can be observed in hydric mineral soils of high iron content using a calorimetric test kit. In a reducing environment, ferrous (reduced) ions are detected using a-a-dipyridil (Environmental Laboratory 1987). It should be understood that this test only reveals reducing conditions during the time of the field investigation, and may not reflect conditions typical of a significant part of the growing season. Repeated sampling over an extended temporal scale is required to satisfy the hydric soil criterion.

Soil colors typically provide the most useful and diagnostic field indicators for hydric mineral soils. When water is at or near the soil surface for most or all of the year soils may become gleyed, and take on a dull gray or bluish color. The color arises from the conversion of oxidized (ferric) iron to its reduced (ferrous) state, and the subsequent removal of the latter from the soil. Oxidizing conditions are apparent only along root channels in gleyed soils. Mottling occurs in soils where the water table fluctuates periodically. Saturated, anaerobic conditions cause the solubilization of iron compounds. Subsequent drops in the water table create an aerobic environment, and iron ions are converted to insoluble oxides which appear as orange or red mottles in the soil matrix. Relatively longer periods of soil saturation and anaerobic conditions, followed by aerobic conditions, converts manganese ions to oxides. The manganese oxides are deposited as concretions, and are evidenced by small, dark brown or black nodules interspersed with the soil matrix. Both mottles and concretions persist for long periods after formation and may not reflect current conditions. However, a combination of mottles and a low matrix chroma are generally indicative of hydric conditions.

Soil color is a poor indicator of hydric condition in sandy soils, and other indicators must be used. A high organic matter content in the surface horizon is indicative of hydric conditions, as it is caused by prolonged inundation or saturation. This prolonged inundation or saturation greatly reduces oxidation of organic matter, leading to its accumulation. Subsequent lowering of the water table may cause streaking of the subsurface horizons as organic matter is moved downward through the sand. Eventually, this downward moving organic matter may accumulate at a point coinciding with the most commonly occurring depth to groundwater. The organic matter may become cemented with aluminum, forming a less permeable, hardened spodic horizon. Each of these characteristics, high organic matter content in the surface horizon, streaking and a spodic horizon, are indicative of hydric conditions.

When examining the soils component during wetlands identifications and delineations, minimum equipment needs include a hand auger, Munsell Color Chart, delineation sheets, pencil and camera. Borings need to be hand-augured at locations along the perceived upland and wetland sides of the boundary, keeping hydrological and topographical indicators in mind. Borings should extend to depths between 30-46 cm to adequately characterize O, A and B Horizons. Diagnostic observations include depth to groundwater or depth of inundation, and examination for each of the field indicators for mineral and organic hydric soils. The hydric soils boundary should be identified based upon the wetland side having observable characteristics of hydric soils, and the upland side having a general lack of these characteristics. Visible hydrology will influence the number of borings needed to adequately define the soils boundary.

HYDROPHYTIC VEGETATION

Probably the most visible and easily recognizable diagnostic feature of wetlands is hydrophytic vegetation. Hydrophytic vegetation is the sum total of macrophytic plant life that occurs in areas where the frequency and duration of inundation or soil saturation is of sufficient duration to exert a controlling influence on the plant species present (Environmental Laboratory 1987). Hydrophytes possess anatomical and physiological adaptations which allow them to survive and thrive in saturated or inundated soils, where oxygen depletion is the primary factor limiting vegetational occurrence (Figure 3). Adaptive structures of hydrophytes include aerenchyma, adventitious roots, stooling, hypertrophied lenticels, buttressing, etc.

Plant species that have a demonstrated ability to achieve maturity and reproduce where the root zone is inundated or saturated during the growing season are listed in the *National List of Plant Species that Occur in Wetlands* (Reed 1988). Approximately 7,000 species of hydrophytes are listed. Originally designed as an appendix to *Classification of Wetlands and Deepwater Habitats of the United States* (Cowardin et al. 1979), the list was modified to assist in determining the probability that a vegetation community is a wetland (Wentworth and Johnson 1986). This is accomplished by assigning each plant to an indicator category, which reflects the probability, expressed as a frequency of occurrence, of a species occurring in wetland or nonwetland (Table 3). Modifiers (+ or -) are also assigned to the categories to further refine individual species affinity towards wet conditions. A '+' indicates an increased probability of wetland tolerance, and a '-' indicate an affinity for drier conditions (Reed 1988). The categories should not be equated to degrees of soil wetness. For example, many obligate wetland species occur in permanently flooded wetlands, whereas other obligate plants occur in seasonally or temporarily flooded wetlands.

Figure 3. Hydrophytes, such as this skunk cabbage *(Symplocarpus foetidus)*, possess anatomical and physiological adaptations which allow them to survive in saturated soils.

Indicators of Hydrophytic Vegetation

Dominance

The most reliable indicator for determining whether hydrophytic vegetation is present, dominance by hydrophytes, has its basis in the current federal definition of wetlands, which includes the phrase *a prevalence of vegetation typically adapted for life in saturated soil conditions.* Vegetation is said to be prevalent when it is dominant, i.e. the species contribute more to the character of a plant community than other species (Environmental Laboratory 1987). If the dominant species are hydrophytes, the community consists of hydrophytic vegetation. The standard for determining if the vegetation is hydrophytic is the 50 percent rule: more than 50 percent of the dominant vegetation is OBL, FACW or FAC, excluding FAC-, on the appropriate list of plant species that occur in wetlands (Environmental Laboratory 1987, U.S. Army Corps of Engineers 1991, U.S. Army Corps of Engineers 1992).

Table 3. Indicator category and frequency of occurrence in wetlands for plants listed in *National List of Plant Species that Occur in Wetlands* (Reed 1988). Upland species are not listed.

Indicator Category	Frequency of Occurrence in Wetlands	Representative Species
Obligate Wetland (OBL)	> 99 percent	cattail (*Typha spp.*) button bush (*Cephalanthus occidentalis*)
Facultative Wetland (FACW)	67-99 percent	soft rush (*Juncus effusus*) speckled alder (*Alnus rugosa*)
Facultative (FAC)	34-66 percent	red maple (*Acer rubrum*) sweet pepperbush (*Clethra alnifolia*)
Facultative Upland (FACU)	1-33 percent	American beech (*Fagus grandifolia*) white pine (*Pinus strobus*)
Obligate Upland (UPL)	< 1 percent	black oak (*Quercus veltina*) staghorn sumac (*Rhus typhina*)

Various ecologically based methods are acceptable for determining dominant plant species. Hays et al. (1981) provide operational descriptions of several techniques suitable for quantitatively measuring habitat variables. The Corps of Engineers recognizes various techniques which consider individual vegetation strata (Table 4). Consistent with the Corps of Engineers (Environmental Laboratory 1987, U.S. Army Corps of Engineers 1993), trees and lianas within a 9.1 m (30 ft) radius of a sample point are identified and measured for basal area at breast height (1.4 m or 4.5 ft). The basal area of all individual tree species and all individual liana species are summed, and tree species and liana species separately ranked in descending order based upon total basal area. As an alternative to measuring the basal area of lianas, the number of individual stems of each species may be counted. As many lianas branch, stem counts should be conducted at ground level. Lianas are then ranked in descending order of dominance based upon number of stems.

Table 4. Categories used in the assessment of vegetation communities.

Story	Stratum	Definition
Woody Overstory	Tree	woody, nonclimbing, at least 12.7 cm dbh and at least 6.1 m tall
	Liana	woody vines, climbing
Woody Understory	Sapling	woody, nonclimbing, at least 1 cm dbh but less than 12.7 cm dbh and at least 6.1 m tall
	Shrub	woody, nonclimbing, at least 0.9 m tall but less than 6.1 m tall
Herbaceous Understory	Seedlings and Herbs	woody, less than 0.9 m tall, or nonwoody and any height
	Mosses and Liverworts	small, green, nonflowering

Other strata are typically evaluated based upon percent areal coverage. According to the Corps of Engineers (Environmental Laboratory 1987, U.S. Army Corps of Engineers 1993), saplings and shrubs are assessed within a 4.5-m (15 ft) radius, and seedlings and herbs are assessed within a 1.5-m (5 ft) radius, of a sample point. For each stratum, the species are ranked in descending order of dominance based upon percent cover. Mosses and liverworts are only considered when they constitute an important component of the vegetation community.

In determining whether hydrophytic vegetation is present, the Corps of Engineers (Environmental Laboratory 1987, U.S. Army Corps of Engineers 1992) suggests using the three dominant species from each vegetation stratum, five species if only one or two strata are present. The indicator status of each species is recorded. If the majority of dominant species are OBL, FACW or FAC, excluding FAC-, then the vegetation is said to be hydrophytic, and therefore indicative of wetlands.

A variation of the dominance method requires calculation of a prevalence index (Federal Interagency Committee for Wetland Delineation 1989). Each indicator category is assigned an index value: OBL = 1.0, FACW = 2.0, FAC = 3.0, FACU = 4.0 and UPL = 5.0. Dominant vegetation is identified based upon its relative frequency of occurrence or relative areal coverage. An index value is assigned to each species, and weighted by its relative dominance in the community. If the sum of the index value is less than 3.0, then the vegetation is considered to be hydrophytic.

Other Indicators of Hydrophytic Vegetation

Although dominance by hydrophytes is the most reliable indicator of hydrophytic vegetation, other indicators exist. In general, these other indicators should be applied only after application of the dominant species method. Nevertheless, visual observation of plant species growing in areas of prolonged inundation or soil saturation, particularly if those species have been observed in other wetland areas, suggests hydrophytic vegetation. This approach may be applied with some reliability for OBL, and to a lesser extent FACW, species, but is considerably less reliable for FAC species. The presence of standing water or saturated soil is, in many cases, insufficient evidence that the observed species are capable of tolerating anaerobic conditions for an extended period. When in doubt, determine hydrology and soils characteristics.

Morphological adaptations to inundation or soil saturation such as buttressed tree trunks, adventitious roots and shallow root systems are also suggestive of hydrophytic vegetation (see Table 1). Most individuals of the dominant species should exhibit morphological adaptations for this indicator to be reliable. However, not all hydrophytic species have obvious morphological adaptations. Conversely, apparent morphological adaptations may develop in response to factors other than soil wetness.

IDENTIFYING AND DELINEATING WETLANDS

Undisturbed Areas

When a site is relatively undisturbed, identifying and delineating wetlands is accomplished by simultaneously applying the criteria for wetlands hydrology, hydric soils and hydrophytic vegetation. Typically, the best time to identify or delineate wetlands is during the growing season, when dominant vegetation, especially annuals, are evident, and hydrology can be directly observed. Nevertheless, identification and delineation of wetlands can be reasonably accomplished at any time of the year, although the process is made infinitely more difficult by snow or ice cover.

To facilitate wetland identification and delineation, it is helpful to establish a working baseline. The baseline should extend parallel to any major watercourses or waterbodies, or to any potential wetland areas (Figure 4). It is also helpful to establish the baseline perpendicular to the topographical gradient. Transects are established perpendicular to the baseline, with the number of transects depending upon the length of the baseline and the diversity of plant communities on the site. Minimally, at least one transect should sample each plant community type. The Corps of Engineers (Environmental Laboratory 1987) offers guidelines for establishing transects. A number of observation points should be established along each transect. Again, the Corps of Engineers (Environmental Laboratory 1987) offers guidelines for establishing observation points. An experienced investigator will minimize effort by establishing one observation point in recognizable wetlands, and a second point in recognizable nonwetlands. Using observed hydrological, soil and vegetation characteristics at these two observation points as references, the investigator can then focus his or her efforts on areas that are less readily discernible.

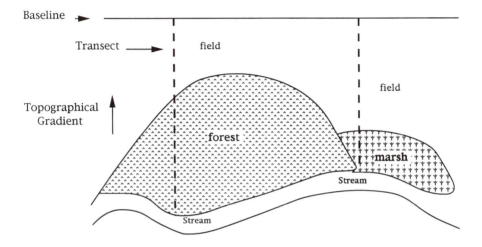

Figure 4. Identification and delineation of wetlands is facilitated by establishment of a baseline parallel to observed watercourses and perpendicular to the topographical gradient. Transects are established perpendicular to the baseline, and at least one transect should sample each plant community type.

At each observation point, hydrology, soil and vegetation are simultaneously characterized, and a determination is made as to whether the area constitutes wetlands or nonwetlands. More precisely, the observation point is inspected for indicators of wetlands hydrology, indicators of hydric soil, and hydrophytic vegetation. Under normal conditions, indicators of wetland hydrology, hydric soil and hydrophytic vegetation must all be present for an area to be considered a wetland. A determination that the observation point constitutes a wetland is sufficient if the purpose of the exercise is only identification of wetlands. For the purpose of delineation, subsequent observation points along the transect, upgradient of the wetland observation point, must be characterized until wetland indicators are absent. Conversely, initial identification of a nonwetland observation point requires subsequent investigation of downgradient observation points until all wetland indicators are present.

Disturbed Areas

When a site is disturbed, identifying and delineating wetlands through application of the criteria for wetlands hydrology, hydric soils and hydrophytic vegetation may be impossible. The disturbance may have been intentional or inadvertent, the result of unauthorized activities or natural events. Site hydrology may be altered due to dam construction, water diversion, channelization or groundwater withdrawal, resulting in a site which is wetter or drier than normal. Soil may be buried, removed or mixed, whereas vegetation may be cut, burned or converted to agricultural crops.

In many cases, normal application of the methods discussed above will lead to the conclusion that the site is not a wetland, as one or more of the wetland indicators are likely to be absent. If one or more wetland indicators cannot be assessed, and the site has clearly been disturbed, it is appropriate to evaluate the remaining indicators and to identify other sources of information regarding the missing indicator(s). In general, as many sources as possible should be considered prior to any determination.

There are several avenues of investigation for determining normal site hydrology. Recent aerial photographs of the site, taken during the growing season, may provide relatively conclusive information about site inundation, and somewhat less conclusive information about soil saturation. In some, albeit few, instances a stream or tidal gauging station may be located in close proximity to the site, allowing calculation of high and low water elevations. Certain field indicators of wetland hydrology, such as plant morphological adaptations or high water marks, may be visible. Less reliable but of some usefulness are historic records, floodplain management maps and the personal knowledge of local officials or residents.

If fill has been placed on the site, buried soils can reasonably be examined for hydric soils indicators. Conversely, the removal of surface soil layers may allow for examination of the exposed subsurface horizons. Plowed soils can be examined immediately beneath the plow zone. Although inconclusive, review of soil survey maps, or examination of adjacent, undisturbed soils of the same series may be useful.

Information relative to the preexisting vegetation community varies in its reliability. The most reliable information comes from examination of vegetation communities only partially removed, or from review of recent documentation of site vegetation. Somewhat less reliable is examination of recent aerial photography, and examination of vegetation in adjacent, undisturbed areas. Least reliable are review of soil survey plant community descriptions, review of National Wetland Inventory Maps, and the recollection of local officials or residents.

Difficult Areas

Some wetlands are inherently more difficult to identify and delineate than other wetlands. The difficulty arises because one or more of the wetland indicators may be impermanent. Slope wetlands in glaciated wetlands have thin soils over relatively impermeable till or varying hydraulic conditions which produce groundwater seepage. Hydric soils and hydrophytic vegetation are generally readily apparent, but wetland hydrology is present only during wetter parts of the growing season. Seasonal wetlands such as vernal pools lack wetland hydrology during all but the wettest part of the growing season also. Prairie potholes have standing water for the majority of the growing season in normal or wet years, but may be dry in years of below average precipitation. Vegetated flats are dominated by annual plant species during the growing season, but are devoid of vegetation outside of the growing season. Almost without exception, identification and delineation for each of these wetland types, as well as other difficult wetlands, can reasonably be accomplished by conducting the evaluation at the appropriate time of the year, or in a year of normal precipitation.

REFERENCES

Cowardin, L.M., V. Carter, F.C. Golet and E.T. LaRoe. 1979. Classification of Wetlands and Deepwater Habitats of the United States. U.S. Department of the Interior, Fish and Wildlife Service Biological Services Program FWS/OBS-79/31. 103 pp.

Environmental Laboratory 1987. Corps of Engineers Wetland Delineation Manual. U.S. Army Engineer Waterways Experiment Station, Vicksburg, MS USA. Tech. Rept. Y-87-1.

Environmental Law Institute. 1991. What is a Jurisdictional Wetland? National Wetlands Newsletter No. 5, September/October.

Federal Interagency Committee for Wetland Delineation. 1989. Federal Manual for Identifying and Delineating Jurisdictional Wetlands. U.S. Army Corps of Engineers, U.S. Environmental Protection Agency, U.S. Fish and Wildlife Service and USDA Soil Conservation Service, Washington, D.C. USA.

Hays, R.L., C. Summers and W. Seitz. 1981. Estimating wildlife habitat variables. USDI Fish and Wildlife Service. FWS/OBS-81/47. 111pp.

Kollmorgen Corporation. 1975. Munsell Soil Color Charts. Munsell Color, Baltimore, MD.

Mitsch, W.J. and J.G. Gosselink. 1986. Wetlands. Van Nostrand Reinhold Co., Inc., New York, NY USA.

Reed, P.B., Jr. 1988. National List of Plant Species that Occur in Wetlands: National Summary. U.S. Fish and Wildlife Service, Washington, D.C., Biol. Rept. 88(24).

Sipple, W.S. 1988. Wetland Identification and Delineation Manual. Volume I: Rational, Wetland Parmaters, and Overview of Jurisdictional Approach. Volume II. Field Methodology. U.S. Environmental Protection Agency, Office of Wetlands Protection, Washington, D.C.

Sipple, W.S. 1992. Time to move on. National Wetlands Newsletter. Environmental Law Institute. 14:4-6.

Smith, G.D. 1973. Soil moisture regimes and their uses in soil taxonomies. In Bruce, R.R., K.W. Flach and H.M. Taylor (eds.) Field Soil Water Regime. Soil Scientist Society of America. Madison, WI.

Tiner, R. W. 1988. Field Guide to Nontidal Wetland Identification. Maryland Department of Natural Resources, Annapolis, Maryland and U.S. Fish and

Wildlife Service, Newton Corner, Massachusetts Cooperative Publication. 283 pp.

Tiner, R.W. and P.L. Veneman. 1989. Hydric soils of New England. University of Massachusetts Cooperative Extension, Revised Bulletin C-183-R, Amherst, MA. 27pp.

Tiner, R.W. 1991. Wetland Delineation. In Aron, G. and E.L. White (eds.) Proceedings of the 1991 Stormwater Management/Wetlands/Floodplain Symposium. Pennsylvania State University, Department of Civil Engineering, University Park, PA.

Tiner, R.W. 1993. The Primary Indicators Method-A Practical Approach to Wetland Recognition and Delineation in the United States. Wetlands 13(1): 50-64.

U.S. Army Corps Of Engineers. 1991. Questions and Answers on the 1987 Corps of Engineers Manual. CECW-OR. 7 October.

U.S. Army Corps of Engineers. 1992. Technical Memorandum: Clarification and Interpretation of the 1987 Manual. USCOE/Washington, D.C. CECW-OR. 6 March 1992.

U.S. Army Corps of Engineers, New England Division. 1993. Performance standards and supplemental definitions for use with the 1987 Corps manual. CENED-OR-R. 9 September.

U.S. Department of Agriculture Soil Conservation Service. 1991. Hydric Soils of the United States. Washington, D.C. Public No.1491.

U.S. Environmental Protection Agency. 1991. Proposed Revisions to the Federal Manual for Delineating Wetlands. Federal Register 58:157. 14 August 1991.

Wentworth, T.R. and G.P. Johnson. 1986. Use of vegetation for the designation of wetlands. U.S. Fish and Wildlife Service, Washington, D.C. 107 pp.

CHAPTER 4

WETLANDS FUNCTIONS AND VALUES

Robert J. Reimold

The physical, chemical and biological interactions within wetlands are often collectively referred to as wetland functions. Similarly, the characteristics of wetlands that are beneficial to society are considered as wetlands values. This chapter examines the basis of wetlands functions and values, and discusses the different approaches used to assess wetlands functions and methods used to ascribe values to wetlands.

Wetland functions involve the performance or execution of changes within the wetlands ecosystem. In conformance with the laws of thermodynamics, these functions include biological, chemical and physical transformations in the diversity of forms and substances that exist within the wetland. Representative biological functions include providing habitat for reproduction, feeding, and resting. Representative physical functions include flood attenuation, groundwater recharge and sediment entrapment. Representative chemical functions include nutrient removal and toxics decontamination. These functions can be measured in quantitative terms, and involve an assessment of change over a finite period of time.

Values on the other hand are sociological, subjective terms, which are particularly malleable. Values of wetlands are based on anthropogenic properties by which the wetlands are determined to be useful, or impart *public good*. The value establishes a worth, excellence, utility or importance of a given wetland function.

Wetlands have historically been considered to be of value based on certain functions that occur in wetlands. For example, the values of different wetlands have been expressed based on their production of cranberries, forest products, peat, etc.

Due to the inextricably linked interactions between wetlands functions, values, and historic uses, it is important to consider both functions and values associated with judgments regarding wetlands uses. Wetlands functions and values have been commingled as the basis for establishment of regulations regarding changes that may be made to wetlands. Wetlands functions and values have been used for making decisions regarding a wide variety of land and water related planning issues. The regulatory framework for wetlands management has evolved because of the significant percentage of the wetlands that have been destroyed and/or altered by development activities (Shaw and Fredine 1956, Walker 1973, Reimold 1977, Dahl 1990).

In addressing the functions and values of wetlands, there are a number of important subcomponents to consider. Some perceive wetlands functions and values from an ecological perspective while others view the functions and values from a regulatory basis. Scale is an important factor to consider, both with respect to direct measurement of wetlands function or value, or indirect assessment based on measurements of a related function or value. For example, measurement of plant species diversity may indicate the quality of a wetlands habitat. Such indirect measures as species presence or site morphology are often used to infer measures of function. These topics are examined in subsequent sections as direct indicators of function. Other measures of structure, when measured at repeated intervals over time (such as diversity, which describes the number of species and number of individuals of each species present at one point in time), and integrated into a dimensionless function associated with value of function, are described as indirectly measured indicators. These too are examined in subsequent sections, as these indirect indicators document the wetland's ability to maintain a given level of the wetland's functions.

Changes in land use (both natural and human-developed environments), whether existent or planned, result in changes in functions and values of wetlands. Whether from a landscape ecology perspective (Westman 1977, Urban et al. 1987, Odum and Turner 1990), an organismal perspective (Loomis and Walsh 1986), or a physical perspective (Kentula, et. al. 1992), changes in wetlands functions and values are concepts used in monitoring wetlands and managing wetlands. With continued efforts to curtail wetlands losses, as well as to monitor the efficacy of wetlands restoration efforts, it is essential to have a clear understanding of the functions and values of wetlands. To be successfully used as indicators however, the functions and values must be able to determine change within a reasonable amount of time, at a reasonable cost, and with minimal negative impact to the wetland.

DEFINITIONS

Ecological Basis

The biological community and the nonliving environment function as an interactive system: an ecosystem (Odum 1989a). While *ecosystem* is an internationally accepted term, some European literature refers to ecosystem using the term *biogeocoenosis*, which connotes the functioning of life and earth together. The ecosystem basis of wetland functions and values are thus based on the living and nonliving components of the wetlands ecosystem. The *biology* (the study of the flora and fauna), the *pedology* (a term commonly used in the agricultural and geological literature meaning the study of soils), and the *hydrology* (the study of water on the surface of the land, in the soil and underlying rock, and in the atmosphere) are the three essential components used in defining wetlands ecosystems (Cowardin et al. 1979).

Early approaches to assessing wetlands functions and values focused primarily on waterfowl and wildlife uses in different types of wetlands (Shaw and Fredine 1956). Golet (1973) established ten criteria related to wildlife production and diversity, but did not consider aquatic vertebrates and invertebrates or plants. Increased scientific interest resulted in the Sharitz and Gibbons (1989) symposium volume on the functions and values of wetlands. Many contemporary approaches to assessing wetlands functions and assigning values are based on ecosystem components (Adamus 1983, Adamus and Stockwell 1983, Adamus et al. 1991, Atkinson 1985, Golet 1976, Greeson et al. 1979, Harker et al. 1981, Montanari and Kusler 1978, Odum, E. 1989b, Odum, H. 1978, Schamberger and Kumpf 1980, Schroeder 1987, Kent et al. 1992).

Biology

Wetlands play a crucial role in the survival of numerous fauna and flora living within wetlands, as well as fauna and flora living adjacent to wetlands. Wetlands functions are frequently based on vegetation (species, coverage, survival), fauna (species, density, and habitat quality), sanctuary refuge value for fish, wildlife and waterfowl, and food chain production and production export to adjacent ecosystems.

The species of plants living in a wetland as well as the area of land covered by the plants can be directly determined. White et al. (1990) developed a classification system for assessing the functions of pre-project wetland mitigation sites and criteria for determining successful replication of forested wetlands ecosystems. This approach was based on the species of plants living within the wetland area. Similarly, Lugo et al. (1990) described the structure and functional components of forested wetlands in the Caribbean.

In addition to consideration of the plant community, the species and number per unit area (density) of animals occupying a given wetland can be directly determined (Hook et al. 1988, Schamberger and O'Neil 1986). The areal extent of a wetland area can be directly quantified in terms of the sanctuary or refuge value it can provide for aquatic species (Darnell 1976). While historically habitat value provided to wildlife and fish has frequently been used as an indicator of wetland functions (Shaw and Fredine 1956), only a paucity of work has been conducted to assess wetlands functions based on faunal diversity (Brooks et al. 1991).

Primary productivity refers to the amount of organic matter converted from solar energy and mineral nutrients by plants. Wetlands vegetation plays a key role in converting nutrients and solar energy into stored energy in plant tissues. This concentrated form of energy is either directly consumed in the wetlands by grazers (such as insects) or exported from the wetlands by flowing water as detritus (Odum, E. 1961).

Hydrology

Water is a major determinant in all wetlands. The hydrological components of wetlands functions and values include water depth, flow rates, flow patterns, flood attenuation for storm or flood waters, natural water quality improvement, groundwater recharge, shielding other areas from wave action, erosion, or storm damage, and storage for storm or flood waters. The changes in water quality within wetlands has also been used as a major function from which values have been assigned (Farnworth et al. 1979, Hammer 1992, Kibby 1979, Larson 1973, and U.S. Army Corps of Engineers 1980).

Pedology

The soils or substrate are the third major factor in determining the quality of the function or value of wetlands. Wetlands pedology (literally, soil science) includes substrate depth, color, texture, source, organic content and sediment flux. Soil color and mottling are indicative of the depth and duration of soil saturation. Mineral soils are considered to be typically hydric when they are predominately neutral gray in color with occasional greenish or bluish gray mottles. These are known as gleyed soils, i.e. hydric soils that are an integral part of the wetlands ecosystem. Organic hydric soils will never take on such a gleyish appearance, but are instead dark brown or nearly black. The soil organic content is also a factor directly influencing the growth of plants, the flux of nutrients, and microbial activity within the wetlands soils (Fried and Broeshart 1967, Black 1968, Segelquist et al. 1990).

The value of wetlands soils is directly impacted by the hydrologic regime. Due to the influence of permanent or periodic inundation or soil saturation, anaerobic conditions become established in wetlands soils. The anaerobic conditions result in chemical reduction of iron and manganese oxides in the soils and thus impart colors characteristic of wetlands soils.

The soils in wetlands serve a function in binding and complexing chemicals for retention within the wetlands soil matrix. Due to the expanding lattice characteristics of wetlands clays, the sediments have a variety of functions in impacting the bioavailability of sediment-bound chemicals to aquatic organisms. Such interactions impact the fate and effects of sediment-bound chemicals in wetlands ecosystems (Dickson et al. 1987). These functions associated with sediment bioavailability have also been valued.

Morphometry

In addition to the three primary determinants affecting the functions and values of wetlands, the morphometry (measurement of external form) also impacts the functions and values of wetlands. Morphometry includes location, drainage area, surrounding land use, area, slope of wetland, slope of surrounding land, and perimeter to slope area. As most wetlands are located in topographic depressions, these external factors have been used to provide indirect determination of wetlands functions, such as their sediment trapping abilities. Reimold (1980) used the functions of sediment trapping and wildlife habitat as the basis of design for a created wetland.

Cultural Values

A variety of individual, societal and global functions of wetlands involve humans (Queen 1977). Collectively termed cultural values, these less tangible wetland values are culturally perceived attributes and include both consumptive and nonconsumptive uses. Consumptive uses (and thus values) are actual or potential uses of wetlands where consumable products are removed from the wetland. These consumptive uses have traditionally been used for economic valuations of wetlands.

Aesthetics, educational, sociocultural and recreational values of wetlands are important determinants associated with wetlands values. Depending on the intensity of destructive use, these factors can also be determinants of wetland function (Reimold et al. 1980). There are a wide variety of recreational uses of wetlands. Some people enjoy sightseeing, swimming, canoeing, kayaking, sailing, and ice skating. Others enjoy power assisted recreation in wetlands using power boats, snowmobiles, etc. Aesthetic uses include photography, painting, musical compositions and poetry. Educational uses include nature study, bird-watching, research, etc. Consumptive values are associated with fur harvest, aquaculture, timber harvest, peat harvest, commercial shellfish and finfish harvests, hunting, etc.

Nonconsumptive societal values have also been attributed to wetlands (Reppert et al. 1979). Preserving open space, local climate amelioration, wilderness, landscape or heritage values, ecological balance, known presence of archaeological resources, and habitat for threatened or endangered fauna and/or flora, are representative of the landscape ecology components of wetlands to which values have been ascribed (Carls 1979, Costanza 1984, Flanagan 1988, Hansen et al. 1980, Queen 1977). Nonconsumptive considerations such as aesthetics have been overlooked in evaluating wetlands partially due to the difficulty in quantifying these qualities, and partly because of a technological bias toward the physical and biological components (Carls 1979). Because worth or value is usually established based on a monetary standard, value determinations often exclude beauty, purity and other intangible attributes. *Such aesthetic components are attributable to the inherent presence of the undisturbed wetlands* (Daiber 1986).

Regulatory Basis

The regulatory basis of wetlands function and value considerations is based on Section 10 of the Federal River and Harbor Act and Section 404 of the Clean Water Act. Section 10 of the River and Harbor Act of 1899 (33 U.S.C. 403) provides for the regulation of dredge and fill work in navigable waters of the U.S. Section 404 of the Clean Water Act regulates the discharge of dredged or fill materials into waters of the United States, including adjacent wetlands. Applications for permits are administered by the U.S. Army Corps of Engineers. Based on the Fish and Wildlife Act of 1956 (16 U.S.C. 742a, et seq.), the Migratory Marine Game-Fish Act (16 U.S.C. 760c-760g), the Fish and Wildlife Coordination Act (16 U.S.C. 661-666c), the Endangered Species Act (16 U.S.C. 1531 et seq.), and the National Environmental Policy Act (42 U.S.C. 4321-4347), the applications for dredge and fill permits are subject to review of wetlands functions and values by the U.S. Environmental Protection Agency, the U.S. Fish and Wildlife Service and the National Marine Fisheries Service.

There has been a dramatic reduction in the areal extent of wetlands in the U.S. in the twentieth century (Reimold 1977, Ledec and Goodland 1988, Dahl 1990). Since these federal regulatory programs are intended to minimize adverse impacts to wetlands by preventing the unnecessary loss of wetlands and other sensitive aquatic areas, the use of wetlands functions and values plays a pivotal role in decision making regarding future wetlands use.

The Clean Water Act Section 404(b)(1) guidelines explicitly call for the consideration of the following functions and values: physical substrate; water circulation; suspended particulates; contaminants; aquatic ecosystems; specific organisms; cumulative impacts on organisms; and secondary impacts. Consequently, functions and values play an important role in the regulatory aspects of wetlands.

METHODS OF FUNCTIONS AND VALUES ASSESSMENTS

As discussed earlier, there are several approaches that have been utilized for assessment of wetlands functions and values. The direct methods are based on quantitative measurements over time and space, whereas indirect methods for assessing functions and values are based on integration of a number of different direct measurements (Table 1). The following sections describe both direct and indirect indicators.

Table 1. Methods of Measurement of Wetlands Functions and Values.

Directly Measured Indicators	Indirectly Measured Indicators
geohydrology	ecologically based models
flora	economically based models
primary productivity	
fauna	
water quality	

Direct Methodologies

Geohydrology

The geohydrological functions of wetlands are attributes related to the presence and disposition of water within wetlands. Wetlands functions are associated with flood storage capacity and the resultant delay in runoff from major precipitation events. Such attenuation of flood peaks reduces potential damage. Consequently, many government agencies now utilize the natural functions of wetlands by constructing engineered wetlands for flood controls instead of constructing engineered flood control structures such as reservoirs.

Floral Components

The species of plants that make up the floral component of wetlands have been valued based on species presence, as well as a number of ecological community analyses. These are essentially ecological accounting methods integrating the number of species and the number of individuals of each species to derive ecological community indices such as diversity, similarity, evenness, etc. There is widespread public concern regarding loss of floral species and/or diversity in wetlands, as such losses are perceived as having a negative value.

The functions of providing wetlands floral diversity as well as mineral cycling have been investigated for a number of wetland plants (Black 1968, Hardisky and Reimold 1977, Nixon and Lee 1986). In the middle of the twentieth century, the floral components of wetlands were used as factors associated with value. The inherent presence of *Phragmites australis* (common reed) was judged to be of low value while the presence of *Typha latifolia* (cattail) was presumed to be of great value, just based on species presence, without explicit consideration of the habitat functions and values afforded by the plants. The values of different wetlands throughout the world have been expressed based on relationships of the floral communities (Chapman 1974, Chapman 1976).

In addition to species presence and ecological community characteristics of wetlands, the percent cover has been used as an assessment of function and value. The amount of cover provided by the vegetation is a measure of the protection the vegetation provides to various faunal populations that may use the wetlands. Consequently, cover is frequently used as an expression of the function of a wetland, and is expressed as a percentage cover.

Relative Primary Productivity

Wetlands are among the most productive ecosystems in the world (National Academy of Sciences 1975, Odum 1989a). The incorporation of sunlight and mineral nutrients by wetland plants results in the tremendous production of plant biomass. The primary consumers (grazers) as well as those organisms that feed on the dead plant material (detritivores) all benefit from the primary productivity of wetlands. This has been a function of wetlands recognized for many years (Shaw and Fredine 1956, Odum 1961, Reimold and Queen 1974, Reimold and Linthurst 1979, Nixon and Lee 1986). Wetlands have also been valued for their potential to support domestic livestock (primary consumers) that consume the primary production (Ranwell 1972, Chapman 1974, Reimold 1976).

Some wetlands' primary production has been deemed to be of comparatively low value however. The abundant primary production of the common reed and water hyacinths has been valued by some as a nuisance associated with wetlands. For purposes of objectivity, one must consider the primary production of all wetland flora when appraising wetlands values with this direct measurement indicator.

Faunal Components

The functions of wetlands in providing habitat for a diversity of life forms has been recognized for a long time. Shaw and Fredine (1956) were among the first to document the value of wetlands to wildlife and waterfowl. Wetland resource management decisions were historically made based on the value of the wetland to waterfowl. Many species of waterfowl use wetlands for feeding and resting areas during their spring and fall migrations. Resident waterfowl also rely on wetlands for nesting and brood rearing areas. Waterfowl broods derive their needed protein for rapid growth by feeding on wetlands insect larvae, diverse aquatic invertebrates, and protein rich wetlands plants. Similarly, many nongame birds (such as egrets, herons, and bitterns) also depend on wetlands for their life support system.

Wetlands have been determined to be among the most productive ecosystems, and because of this primary production, wetlands have functioned by providing habitat for a diversity of other vertebrates as well as invertebrates. For example, Galtsoff (1964) documented the value of wetlands in production of estuarine shellfish, including the American oyster. Wolf et al. (1975) inferred value associated with wetlands based on the abundance of fiddler crabs, and thus the ability of the wetlands to provide necessary habitat values to support the crabs.

Wetlands provide habitat for important furbearing species such as muskrat and beaver that consume primary production from the wetlands. White-tailed deer use wetlands as browse areas, and depend on the wetlands for cover. Many threatened and endangered species such as the wood turtle, common loon, and osprey also depend on wetlands for life support. In addition, a number of other wildlife species benefit from the rich aquatic and terrestrial food source produced in wetlands. A number of wading birds, songbirds, bats, amphibians and reptiles feed in the wetlands-based insect food webs. Many turtles, snakes, salamanders and toads also use the wetlands for their life support system.

Wetlands also serve as habitat for fish communities. Wetlands provide fish breeding, and nursery grounds, food and cover from predators. In coastal areas, it is estimated that over 90% of all species of fish that live in the coastal zone depend on wetlands for feeding and reproductive habitat. The coastal fisheries' commercial harvests of shrimp, oysters and finfish have been demonstrated to be dependent on outwelling associated with wetlands' primary productivity (Odum 1989a).

Wetlands also provide habitat for a variety of public disease vectors. Several decades before present, the mosquito was identified not only as a nuisance, but also as a public health threat. Consequently, wetlands were managed (extensively ditched and drained) based on this single direct indicator. Values such as these, based on wetlands faunal components, have not received much attention in recent years. Yet to remain objective, such values should be considered.

Water Quality

A variety of approaches have been used to assess the functions of wetlands in managing water quality. Nixon and Lee (1986) provided a regional review of the functions of wetlands in serving as sources, sinks and transformers of nutrients and metals. The role of wetlands in terms of water quality functions has become inextricably linked in recent years to the interest in using wetlands for small-scale sewage treatment as well as stormwater retention or treatment basins. Pioneering work by Kadlec (1979), as well as contemporary work by Mitsch et al. (1988), Hammer (1989, 1992), Mitsch (1992), and Rodgers and Dunn (1992), have considered the assimilative capacity of wetlands in terms of water quality improvements.

While some have viewed wetlands as sinks in the biogeochemical cycling of metals, nutrients and carbon, others have identified wetlands as sources that outwell carbon and nutrients (Odum 1989a). The success of employing wetlands functions in improving water quality relate to the metabolism of the wetlands ecosystem. Stems and leaves of wetlands plants provide surface area for microbes, whereas below ground transport of oxygen by these same plants produces an oxidized zone in the rhizosphere where additional microbial populations exist. This complex of wetland plants and microbes has a high efficiency in modifying nutrients, metals and other compounds. The functions of water quality enhancement by wetlands have recently been incorporated into wetland design to control nonpoint source pollution (Mitsch 1992), treat agricultural runoff from nonpoint sources (Hammer 1992), and to remove pesticides (Rodgers and Dunn 1992). Not only have the functions of engineered wetlands been valued in terms of treatment of wastewater, but nutrient removal by natural wetlands has been valued for its role in water quality improvement. As a result, based on landscape ecology functions related to water quality, the U.S. Environmental Protection Agency (1990) has established national guidance on expected water quality in wetlands.

In addition to nutrient removal and metals attenuation, wetlands also are valued for their ability to remove sediments. When sediment laden flood waters flow over a wetland, the dense wetland vegetation slows the water velocity. This in turn results in deposition of the sediment load of the waters. Emergent macrophyte wetlands (such as cattails) can remove up to 95% of the sediments in the water column by such means (Hammer 1989, 1992).

Indirect Methodologies

Indirect methods for assessing function and value have been developed based on both ecological and economic measurements. Since the two words *ecology* and *economy* have a common root (*oikos*, meaning house or place to live), it is not surprising to note that the business of environmental housekeeping and economic housekeeping have both evolved a variety of indirect methodologies for wetlands valuation. The components of the ecological basis and economic basis for assessing wetlands functions and values are described in the following sections (Table 2).

Table 2. Indirect Methods for Assessing Wetlands Functions and Values

Ecological Based	Economic Based
community	contingent valuation
habitat	opportunity-cost
WET[1]	cumulative assessment
risk	

[1] = Wetlands Evaluation Technique (Adamus et al. 1987)

Ecologically Based Indirect Methods

Habitat based methods for assessing wetlands functions and values exhibit a wide variety of complexity and sophistication. The habitat, the place where a certain species can be found, is the address at which the organism lives (Odum 1989a). While certain species of plants and animals may or may not be present due to temporal or seasonal variations, the presence of habitat suitable for their survival has been frequently used as a measure of the function of wetlands. Habitat assessments involve classifying a prospective area (spatial and temporal limits) in terms of a hierarchical system. The most frequently used habitat classification nomenclature, adopted by the National Wetlands Inventory, is that of Cowardin et al. 1979. The classification scheme is most commonly based on biological and physical criteria.

An early habitat evaluation system was developed for use in the Lower Mississippi Valley (U.S. Army Corps of Engineers 1976, 1980). Concurrent with this development, the U.S. Fish and Wildlife Service (1977, 1980) also developed a habitat evaluation based on cover types (i.e. the vegetative cover of the area such as deciduous forest, coniferous forest, residential woodland, etc.). The habitat evaluation concept is based on community models, i.e. literature reviews of wildlife community ecology. Roberts et al. (1987) and Schroeder (1987) discuss the concepts and approaches on which community models are based. Platt et al. (1987) concurrently developed methods for evaluating wetlands habitats. Adamus et al. (1991) integrated many of these approaches into one commonly referred to as WET. WET provides an organized approach for the assessment of wetlands functions and values, including habitat suitability for 14 waterfowl species groups, 4 fish species groups, 120 wetlands dependent birds, 90 species of freshwater fish, and 133 species of marine fish and invertebrates. WET uses a list of predictors to characterize wetland functions and values.

Incorporating a number of similar approaches (Inhaber 1976, U.S. Fish and Wildlife Service 1976, 1977, Ellis et al. 1979), the U.S. Fish and Wildlife Service (1980) developed the Habitat Evaluation Procedure (HEP). While initially developed for both upland and wetlands habitats, the HEP was quickly implemented in numerous wetlands evaluations related to resource management decisions.

The HEP process compares the relative value of wildlife habitat in terms of different areas compared at the same point in time, or the relative value of the same area at two points in time (present and future). HEP utilizes Habitat Suitability Indices (HSI) to make the comparisons. HSI values (ranging from 0.0 to 1.0) are derived using wildlife models which are hypotheses of species-habitat interactions and not necessarily cause and effect, field tested relationships. The wildlife models are based on species distribution, life history, and specific habitat requirements (food and foraging habitats, water, cover, interspersion, and other factors).

HSI models have been developed by U.S. Fish and Wildlife Service for many species. The published models include habitat variables, life requisites, and resultant habitat value. Updated copies of available models are available from the U.S. Fish and Wildlife Service, U.S. Department of Interior. These habitat suitability models are based on another set of technical publications, known as species profiles, which include information regarding the life histories and environmental requirements of given species (these documents are also available from the U.S. Fish and Wildlife Service). The species profiles are designed to give managers, engineers and planners a brief yet comprehensive summary of the biological characteristics and environmental requirements of a given species and to describe how populations of a given species react to environmental perturbations. The species profiles have sections on taxonomy, life history, ecological role, environmental requirements and economic importance (as applicable).

The Habitat Suitability Index (HSI) multiplied by the areal extent of available habitat yields the Habitat Units (HUs), which can be used to make pre- and postactivity comparisons. Guidance on the use of these procedures (U.S. Fish and Wildlife Service 1980) assists the user in determining the applicability of HEP, the study limits, the baseline HUs, the future HUs, and comparisons. A modification of this HEP procedure (Palmer et al. 1985) was developed as a less labor intensive version of the original HEP. With the PAM HEP procedure, one can evaluate baseline habitat conditions, determine direct construction and operational related impacts of a project, and develop a mitigation plan to offset impacts. The process reduces the level of effort needed for the original HEP, and therefore reduces the time for normal environmental review.

In addition, there are a wide variety of existing community models developed in support of many of the following described habitat-based approaches. Using a concept of the potential natural vegetation type, Short (1982) developed the Habitat

Gradient Model.   The system is based on an assumption of what vegetative community would become established over a finite period of time, given satisfactory growing conditions.   Typical categories include grasslands and woodlands.

Specific habitat quality evaluation systems have been developed based on waterfowl (Colwell 1978) where eight different types of land use classes were evaluated. The U.S. Forest Service (1981) wildlife and fish habitat relationships method is a habitat-based system consisting of 24 vegetation types. U.S. Fish and Wildlife Service also developed the rapid assessment methodologies (Asherin et al. 1979) in which there are six primary divisions, including one (native vegetation) with nine secondary classes and 42 subclasses.

While there are a multitude of structured habitat evaluation systems with adequate documentation, there is to date no universally accepted approach. Schamberger and O'Neil (1986) have listed a number of constraints to such attempts to assign functions and values based on habitat considerations. Wakeley and O'Neil (1988) further addressed alternatives to increase efficiency and reduce effort associated with use of the habitat evaluation procedures.

The direct measurement of wetlands evaluation using species lists noting rare, threatened, endangered and common species, has largely been replaced with habitat-based approaches (Atkinson 1985). While a diversity of habitat-based valuation systems exist, they are more standardized than the direct measurement approaches. The habitat-based systems involve classification schemes, multi-variate models of ecosystems, and measure various parameters. From such measurements (primary determinations), inferences are drawn regarding habitat quality. Since habitats are more stable than individual populations (or species) the concept is a useful one for evaluating current conditions and for estimating future conditions within a wetland ecosystem.

An integrated approach to the assessment of wetlands functions has been developed by Adamus et al. (1991). The Wetland Evaluation Technique (WET) includes eleven functions and values commonly attributed to wetland systems. WET includes groundwater recharge and discharge, floodflow alteration, sediment stabilization, sediment/toxicant retention, nutrient removal/transformation, production export, aquatic diversity/abundance, wildlife diversity/abundance, recreation, and uniqueness/heritage functions (Adamus et al. 1991).   This contemporary approach represents a simplification of earlier wetlands evaluation procedures (Adamus 1983, Adamus and Stockwell 1983) that were developed for the Federal Highway Administration.   While WET is a qualitative tool for screening, it can also be used to determine whether or not one of the more quantitative assessment methods, such as the Habitat Evaluation Procedure (U.S. Fish and Wildlife Service 1980), should be used. The WET approach is most accurate where detailed, extensive data sets are available.

Some states have developed their own integrated wetlands value assessment methods (Ammann et al. 1986). These approaches have all been based on the wetlands functions of such factors as agricultural potential, forestry potential, flood control, ecological integrity, wildlife habitat, finfish habitat, educational potential, nutrient removal, sediment trapping, visual or aesthetic quality, water based recreation, groundwater use potential, dissipation of erosion forces, and special values such as historic or archaeological sites, or critical habitat for threatened or endangered species.

Atkinson (1985) developed a summary of a number of habitat evaluation procedures. Sather and Stuber (1984) contains numerous specific technical discussions of habitat-based wetlands evaluation approaches. General observations, research needs, and analysis of keys and predictors are provided for hydrology, water quality, food chains, socioeconomics, and habitat. Their intent was to develop a national approach to wetlands values assessment methods.

Another contemporary approach to an indirect method (Simenstad et al. 1991) uses a systematic, on-site measurement of habitat attributes which are functionally important to fish and wildlife. Integrating biological habitat concepts with physical attributes, Brooks et al. (1991) developed a method to monitor cumulative biological impacts of wetlands, streams and other riparian components of watersheds.

Another family of indirectly measured indicators of wetland functions and values is based on the risk assessment approach. Based on a diversity of approaches for wetlands models (for example, Mitsch et al. 1988, Mitsch 1990, 1992), ecology has advanced from mere natural history to ecological energetics. Scheifele (1987) developed an approach for the assessment of Ontario's wetlands using predictive models for environmentally sensitive areas. He used a multiple regression equation to predict the total wetland value based on the number of vegetation communities, the wetland size, the proximity to urban centers, the number of wetland types (bog, swamp, fen or marsh), and the percentage of organic soil present. Scheifele (1987) recommended that this model be utilized to pre-stratify wetlands so that those of greatest value could be assessed more thoroughly.

Defining the wetland ecosystem in terms of quantitative health factors (relating the wetland ecosystem health to nonhuman animal health as well as human animal health) results in a functional definition of state and condition of the wetlands (Schaeffer et al. 1988). Based on practical and economic considerations, wetland ecosystem management decisions are often made based on limited knowledge. Thus, a small number of diagnostic variables must be selected to accurately reflect the wetland ecosystem. Johnson (1998) developed techniques, including multiple linear regression, discriminant analysis and visual inspection of graphical data, to use for diagnostic criteria. The use of these diagnostic variables as predictors of ecological risk was also evaluated with respect to cost-benefit. While the approach affords a new insight into evaluation, extensive data sets are required to obtain reliable estimates of the risk functions. Johnson (1988) concludes *Therefore, a good deal of data snooping, exploratory analysis, inspection of graphical displays, and ecological judgment will probably always be needed in the search for adequate predictors of ecological risk.*

Economically Based Indirect Methods

In addition to ecosystem based approaches, a variety of economic valuation systems have been developed for wetlands. For example, Shabman and Batie

(1980) stressed the importance of estimating the economic value of coastal wetlands. The three most commonly used methods include contingent valuation (willingness-to-pay), the opportunity cost method, and the cumulative assessment approach.

The contingent valuation method of Mitchell and Carson (1989) has been used by Bardecki et al. (1989) to evaluate the opportunity costs associated with wetland conservation.   In this willingness-to-pay approach, a hypothetical market is established for non-market goods or services. Using a telephone survey, Bardecki et al. (1989) assessed recognition of the term *wetland*, recognition of specific wetland values, opinions concerning wetlands conservation (and associated willingness-to-pay for wetland conservation), and related socio-economic characteristics. The contingent value approach demonstrates economic interest as well as the political will of the respondents.

Another economic approach to wetland valuation is associated with the opportunity cost of conserving the wetland. This is the net monetary benefit derived from the best use of the area had the area not been regulated as a wetland. Costanza (1984), Costanza and Farber (1985), and Raphael and Jaworski (1979) all conducted economic analyses of wetlands based on opportunity costs. While a demand exists for many wetlands related products and services, most of these are not priced in the marketplace. Absent market-oriented transaction data, the opportunity cost approach has general imperfections when applied to wetlands. Data necessary to evaluate wetlands are not usually collected, and the majority of wetland values are not directly priced. Nevertheless, such an approach may be useful for comparing potential impacts with large capital undertaking.

A third method, the cumulative assessment approach, affords another alternative economic based valuation system.   In this method, a particular development scenario for a wetland is evaluated against specific societal goals and objectives which are associated with or are served by wetlands. In such an assessment, persons and agencies are forced to publicly weigh the potential valuation of certain goals and objectives associated with the wetland against the benefits to be derived from implementation of the proposed action impacting the wetland.

As there is no marketplace for such ecological services of wetlands as sediment trapping, nutrient removal, or fish and wildlife habitat, markets for ecological services from wetlands do not function according to established economic criteria. In such economic assessment, activities that eliminate wetlands are overvalued while natural wetlands values are undervalued. Consequently, in any holistic evaluation system, it is essential to recognize the importance of the nonmonetary as well as the monetary values.

VARIABLES AFFECTING THE ASSESSMENT OF FUNCTIONS AND VALUES

Based on the previous description of functions and values of wetlands that may be assessed, there are several important variables that impact the assessment of functions, and the ultimate determination of values. Methods of human thought (individual or group), meteorological events, and field assessment variability all impact the ultimate determination of function and/or value.

Expert Opinion

A variety of individuals with extensive education and experience in a particular discipline may provide an assessment of wetlands functions and values. Specialists in hydrology, pedology, ecology, etc., all have limited abilities to individually assign value to wetlands functions. These empirical methods of assigning value were often the basis of controversy between experts regarding wetlands functions and resultant values. In many instances, it was the scientific, technical debate between wetlands experts that shaped the future of wetlands research. In addition, it was the early expert opinions, often unsubstantiated by fact (sometimes referred to as ecomythology) that spotlighted the importance of wetlands and the need for prudent regulation. As a result of individual expert opinions, another approach was considered to value wetlands, i.e. the building of consensus.

Delphi technique

While different experts often have divergent opinions, the Delhi technique (Pill 1971) has proven to be a successful method for developing consensus among experts. Delphi, the ancient Greek meeting site where Oracles gathered, held discussions and gave wise decisions or opinions, was a term applied by the U.S. Air Force to strategic planning in the early 1950's (Dalkey and Helmer 1963). The technique has subsequently been used in corporate decision-making and renewable resource management (Linstone and Turoff 1975). The Delphi exercise is a discussion by knowledgeable participants in hopes of reaching an agreeable conclusion. *The concept is based on the premise that: (1) opinions of experts are justified as inputs to decision-making where absolute answers are unknown; and (2) a consensus of experts will provide a more accurate response to a question than a single expert* (Crance 1987). He employed the Delphi technique to develop habitat suitability index curves. This technique is especially advantageous to aggregate judgments of experts who may have biases or hostilities, or where individual personalities could distract from the decisions being made. The quality of the Delphi responses is strongly dependent on the interest and strong level of motivation and commitment of the participants (Delbecq et al. 1975).

In the implementation of the various habitat suitability indices and other value systems based on integration of different functions (discussed earlier), the consensus building of experts has been effectively employed. The PAM HEP procedure (Palmer et al. 1985), employs the consensus building approach of experts to reach decisions regarding the value of wetlands (existing and future). Most of the numerical, manipulative methods of assessing functions and integrating them into a value, are based on the generic application of the Delphi approach.

Meteorological Anomalies

In determining the functions of wetlands, be they directly measured functions such as primary production or indirectly measured functions such as habitat suitability, a variety of meteorological events can influence functional assessment and resultant values. Significant differences in standing crop biomass of wetlands primary productivity can be attributed to differences in growing seasons in two different years. The presence of a prolonged period of drought can be a major determinant in the presence of a given species (such as amphibians). A period of drought can also make a significant difference in the water level within a wetland and thus impact the determination of functions. It is important to take into consideration the seasonal aspects of solar radiance, precipitation, etc., in making the functional assessment.

The occurrence of other extreme meteorological events, such as hurricanes, tornadoes, or other acts of God, all can have a significant impact on the outcome of a wetlands value determination based on wetlands functions. In order to accommodate these extremes, it is necessary for the wetlands evaluator to always consider the hydrology in terms of the lowest flow occurring over varying time spans, for example over a one, seven, and ten year period. In addition, one must consider the average as well as the daily maximum and minimum air and water temperature, solar irradiation, etc. Without providing detailed examples of each, the wetlands evaluator must develop a sense of need for a large time series data set in wetland valuation systems. Longer time series data sets of wetlands functions result in a more defensible assignment of wetlands values.

Field Assessment Techniques

The endpoints selected for field assessment techniques and the presence of contaminants are key variables impacting ultimate determination of wetlands functions and values. In ecological surveys, the evaluation of indicator species must include all potential species such that the ecological community structure (diversity, evenness, abundance, and dominance) can be accurately determined. In addition, the guild structure (Landers 1983) which includes the trophic (feeding) structure, must be accurately determined to assess community functions such as primary and secondary productivity. Chemical testing of plant and animal tissue must also be done accurately so as to assess bioaccumulation, and potential partitioning of contaminants between aqueous and organic compartments. The accuracy and thoroughness of the assessment of the health of specific organisms (in terms of histopathology, gross observations, biochemistry/physiological indicators, mutagenicity, sonar/telemetry tracking, etc.), all significantly impact the value assigned to a specific organism's health.

SUMMARY

This chapter has presented a variety of wetlands functions and values. Intensification of development will result in additional demands to use (abuse) of wetlands. The role of wetlands function, when interpreted as values, will become the basis for future decision making. The unknowns associated with the functions of wetlands will form the template for future wetlands research.

Using the important functions from application of the WET, Marble (1990) worked backwards to identify predictors which could result in a *high* rating. One can expect that as new tools are developed to assess the value of wetlands using measurements of functions, more reasonable wetlands permit-decision-making will result regarding wetlands preservation. In addition, using the knowledge of wetlands functions which are rated as *high*, environmental engineers and natural resource decision makers will be able to better achieve wetlands functional designs to replace those wetland areas which are threatened by development related activities.

It is crucial that persons involved in wetlands functional determinations understand the limitations to the state of knowledge regarding individual functions. Available information varies widely between different parts of the country, between ecological functions and between wetland types (Jahn 1981). Perhaps the recent approach of Kentula et al. (1992) affords the most realistic use of wetlands functions. Using a performance curve concept to document the development of ecological functions over time between a reference and a perturbated wetland enables comparison of both the function and the time dependent nature of the function. Based on an evaluation of a national data base, Kentula et al. (1992) determined that there is a reasonably broad data base of measures of wetland structure (discussed earlier as directly measured indicators) which often better *meet the requirements of expedience and economy than do direct measures of function. Therefore, measures of structure are frequently used as indicators of wetland function.*

As wetlands are defined based on soils, vegetation and hydrology, at a minimum any work dealing with structure should include at least one measurement of each. In other words, at least one hydric soil variable, one hydrology variable, and one vegetation variable (at a minimum) should be measured to determine structure and thus infer function.

Whether measuring function, or features of function, (i.e. structure), once the functions have been accurately determined, the assignment of value becomes a matter of choice. The value depends on how humans allocate resources among competing uses to maximize their own well-being (Batie and Shabman 1982). Integration of these various functional determinations facilitates determination of what functions contribute to well-being. Consequently, while precise dollar valuation approaches are not currently available to integrate all the societal well-being factors, it may still be appropriate to continue to convert functions into monetary equivalents as a means of appraising the public interest and thus making conservation-(wise use) based decisions about the future of wetlands.

REFERENCES

Adamus, P.R. 1983. A method for wetlands functions assessment. Vol. II. *FHWA Assessment Model.* Environmental Division. Office of Research. U.S. Department of Transportation. Federal Highway Administration. Washington, DC. FHWA-IP-82-24. 134pp.

Adamus, P.R. and L.T. Stockwell. 1983. A method for wetland functional assessment. Vol. I. *FHWA Assessment Model.* Environmental Division. Office of Research. U.S. Department of Transportation. Federal Highway Administration. Washington, D.C. FHWA-IP-82-23. 176pp.

Adamus, P.R., L.T. Stockwell, W.J. Clairain, Jr., M.E. Morrow, L.P. Rozas, and R.D. Smith. 1991. Wetland Evaluation Technique (WET). Volume 1: Literature Review and Evaluation Rationale. U.S. Army Corps of Engineers. Waterways Experiment Station. Vicksburg, MI. Wetlands Research Program Technical Report WRP-DE-2. 287pp.

Ammann, A.P., R.W. Franzen, and J.L. Johnson. 1986. Method for the evaluation of inland wetlands in Connecticut. Natural Resources Center. Connecticut Department of Environmental Protection. Hartford, CT. 68pp plus appendices.

Asherin, D.A., H.L. Short, and J.E. Roelle. 1979. Regional evaluation of wildlife habitat quality using rapid assessment methodologies. in: *Transactions of the Forty-Fourth North American Wildlife and Natural Resources Conference.* Wildlife Management Institute. Washington, D.C.

Atkinson, S.F. 1985. Habitat-based methods for biological impact assessment. *The Environmental Professional.* 7:265-282.

Batie, S. and L.A. Shabman. 1982. Estimating the economic value of wetlands: principles, methods, and limitations. *Coastal Zone Management Journal.* 10(3): 255-278.

Bardecki, M.J., E.W. Manning, and W.K. Bond. 1989. The reality of valuing wetlands: the case of Greenock Swamp. In: *Wetlands: Concerns and Successes.* American Water Resources Association. pp. 81-90.

Black, C.A. 1968. *Soil-Plant Relationships.* Second Edition. John Wiley & Sons. New York, NY. 792pp.

Brooks, R.P, E.D. Bellis, C.S. Keener, M.J. Croonquist and D.E. Arnold. 1991. A methodology for biological monitoring of cumulative impacts on wetland, stream, and riparian components of watersheds. in: *Proceedings of the International Symposium: Wetlands and River Corridor Management.* J.A. Kusler and S. Daly, eds. Association of State Wetland Managers. Berne, NY. pp. 387-3998.

Carls, E.G. 1979. Coastal Recreation: Esthetics and Ethics. *Coastal Zone Management Journal.* 5:119-130.

Chapman, V.J. 1974. *Salt Marshes and Salt Deserts of the World.* Second, Supplemented Reprint Edition. J. Cramer Publ. Leutershausen, West Germany. 392pp.

Chapman, F.J. 1976. *Mangrove Vegetation.* J. Cramer Publ. Leutershausen, West Germany. 447pp.

Colwell, J.E. 1978. Use of Landsat data to assess waterfowl habitat quality. Environmental Research Institute of Michigan. 120000-15-F. January. Ann Arbor, MI.

Costanza, R. 1984. Natural resource valuation and management: toward an ecological economics. in: *Integration of Economy and Ecology: an Outlook for the Eighties.* A.M. Jansson, ed. University of Stockholm Press. Stockholm, Sweden. pp. 7-18.

Costanza, R. and S.C. Farber. 1985. The economic value of coastal wetlands in Louisiana. Louisiana State University. Baton Rouge, LA. 65pp.

Cowardin, L.M., V. Carter, F.C. Golet and E.T. LaRoe. 1979. Classification of wetlands and deepwater habitats of the United States. U.S. Fish and Wildlife Service. Washington, D.C. FWS/OBS-79/31. 102pp.

Crance, J.H. 1987. Guidelines for using the Delphi technique to develop habitat suitability index curves. National Ecology Center. U.S. Fish and Wildlife Service. Washington, D.C. Biological Report 82(10.134) 21pp.

Dahl, T.E. 1990. Wetlands losses in the United States 1780's to 1980's. U.S. Department of Interior. Fish and Wildlife Service. Washington, D.C.

Daiber, F.C. 1986. *Conservation of Tidal Marshes.* Van Nostrand Reinhold Co. Inc. New York, NY. 341pp.

Dalkey, N.C. and O. Helmer. 1963. An experimental application of the Delphi method to the use of experts. *Management Science.* 9:458-467.

Darnell, R.M. 1976. Impacts of construction activities in wetlands of the United States. Office of Research and Development. U.S. Environmental Protection Agency. Ecological Research Series. EPA-600/3-76-045. 392pp.

Delbecq, A.L., A.H. Van deVen and D.H. Gustafson. 1975. *Group techniques for program planning - a guide to normal group and Delphi processes.* Scott Foresman and Company. Glenview, IL. 194pp.

Dickson, K.L., A.W. Maki and W.A. Grungs, eds. 1987. *Fate and Effects of Sediment Bound Chemicals in Aquatic Systems.* Pergamon Press. Elmsford, NY.

Ellis, J.A., J.N. Burroughs, M.J. Armbruster, D.L. Hallet, P.A. Korte and T.S. Baskett. 1979. Appraising four field methods of terrestrial habitat evaluation. *Transactions North American Wildlife Natural Resources Conference.* 44:369-379.

Farnworth, E.G., M.C. Nichols, C.N. Jann, L.G. Wolfson, R.W. Bosserman, P.R. Hendrix, F.B. Golley, and J.L. Cooley. 1979. Impacts of sediment and nutrients on biota in surface waters of the United States. Office of Research and Development. U.S. Environmental Protection Agency. Ecological Research Series. EPA-600/3-79-105. 314pp.

Flanagan, R.D. 1988. Planning for multi-purpose use of greenway corridors. *National Wetlands Newsletter.* Environmental Law Foundation. Washington, D.C. 10(2): 7-8.

Fried, M. and H. Broeshart. 1967. *The Soil-Plant System.* Academic Press, Inc. New York, NY. 358pp.

Galtsoff, P.S. 1964. *The American Oyster.* U.S. Department of Interior. Washington, D.C. Fishery Bulletin Volume 64. 480pp.

Golet, F.C. 1973. Classification and Evaluation of Freshwater Wetlands as Wildlife Habitat in the Glaciated Northeast. Ph.D. Thesis. University of Massachusetts. Amherst, MA.

Golet, F.C. 1976. Wildlife wetland evaluation model. in: *Models for Assessment of Freshwater Wetlands.* J.S. Larson, ed. Water Resources Research Center. University of Massachusetts. Amherst, MA. Publ. No. 32. pp. 13-34.

Greeson, P.E., J.R. Clark and J.E. Clark, eds. 1979. *Wetland Functions and Values: The State of our Understanding.* American Water Resources Association. Tech. Publ. Ser. No. TPS79-2. 564pp.

Hammer, D.A. 1989. *Constructed Wetlands for Wastewater Treatment.* Lewis Publishers. Chelsea, MI. 831pp.

Hammer, D.A. 1992. Designing constructed wetlands systems to treat agricultural nonpoint source pollution. *Ecological Engineering.* 1(1/2): 49-82.

Hanscn, W.J., S.E. Richardson, R.T. Reppert and G.E. Galloway. 1980. Wetlands' values - contributions to environmental quality or to national economic development? in: *Estuarine Perspectives.* V.S. Kennedy, ed. Academic Press. New York, NY. pp. 17-29.

Hardisky, M.A. and R.J. Reimold. 1977. Salt-marsh plant geratology. *Science.* 198: 612-614.

Harker, D.F., Jr., G.M. Gigante and R.R. Cicerello. 1981. *Evaluation of fish and wildlife habitats: a selected bibliography.* Kentucky Nature Preserves Commission. Frankfort, KY. 105pp.

Hook, D.D., W.H. McKee Jr., H.K. Smith, J. Gregory, V.G. Burrell Jr., M.R. DeVoe, R.E. Sojka, S. Gilbert. R. Banks, L.H. Stolzy, C. Grooks, T.D. Matthews and T.H. Shear. 1988. *The Ecology and Management of Wetlands.* Timber Press. Portland, OR.

Inhaber, H. 1976. *Environmental Indices.* John Wiley and Sons. New York, NY. 178pp.

Jahn, L.R. 1981. Resource Management: Challenge of the Eighties. *Water Spectrum.* 13(3): 1-8.

Johnson, A.R. 1988. Diagnostic variables as predictors of ecological risk. *Environmental Management.* 12(4): 515-523.

Kadlec, R.H. 1979. Wetlands for tertiary treatment. in: P.E. Greeson, J.R. Clark and J.E. Clark, eds. *Wetland Functions and Values: The State of Our Understanding.* American Water Resources Association. Minneapolis, MA. pp. 490-504.

Kent, D.M., R.J. Reimold, J.M. Kelly and C.E. Tammi. 1992. Coupling wetlands structure and function: Developing a condition index for wetlands monitoring. Pp. 159-170 In Ecological Indicators Volume 1. Elsevier Science Publishers Ltd., Essex, England.

Kentula, M.E., R.P. Brooks, S. Gwin, C. Holland, A.D. Sherman and J. Sifneos. 1992. *An approach to decision making in wetlands restoration and creation.* Environmental Research Laboratory. U.S. Environmental Protection Agency. Corvallis, OR. EPA/600/R-92/150. 182pp.

Kibby, H.V. 1979. Effects of wetlands on water quality. in: *Strategies for Protection and Management of Floodplain Wetlands and other Riparian Ecosystems.* Proceedings of a Symposium. 11-13 December 1978. Forest Service. U.S. Dept. of Agriculture. Washington, D.C. GTR-WO-12. pp. 289-298.

Landers, P.B. 1983. Use of guild concept in environmental impact assessment. *Environmental Management.* 7(5): 393-397.

Larson, J.S. 1973. Wetlands and Floods. in: *A Guide to Important Characteristics and Values of Freshwater Wetlands in the Northeast.* J.S. Larson, ed. Water Resources Research Center. University of Massachusetts. Amherst, MA. Publ No. 31. pp. 15-16.

Ledec, G. and R. Goodland. 1988. *Wildlands, Their Protection and Management in Economic Development.* The World Bank. Washington, D.C. 278pp.

Linstone, H.A. and M. Turoff, eds. 1975. *The Delphi method.* Addison-Wesley. Reading, MA. 620pp.

Loomis, J.G. and R.G. Walsh. 1986. Assessing wildlife and environmental values in cost-benefit analysis: State of the art. *Journal of Environmental Management.* 22: 125-131.

Lugo, A.E., S. Brown and M. Brinson. 1990. *Forested Wetlands.* Ecosystems of the World 15. Elsevier Scientific Publishing Co. Amsterdam.

Marble, A.D. 1990. A guide to wetland functional design. Federal Highway Administration. McLean, VA. Report Number FHWA-IP-90-010.

Mitchell, T.C. and R.T. Carson. 1989. *Using surveys to value public goods: the contingent valuation method.* Resources for the Future. Washington, D.C. 482pp.

Mitsch, W.J. 1990. Ecological engineering and ecotechnology with wetlands: applications of system approaches. in: *Advances in Environmental Modelling.* A. Marani, ed. Elsevier Scientific. Amsterdam. pp. 565-580.

Mitsch, W.J. 1992. Landscape design and the role of created, resorted, and natural riparian wetlands in controlling nonpoint source pollution. *Ecological Engineering.* 1(1/2): 27-48.

Mitsch, W.J., M. Straskraba and S.E. Jorgensen, eds. 1988. *Wetland Modeling.* Elsevier Science Publishing Co., Inc. New York, NY.

Montanari, J.H. and J.A. Kusler, eds. 1978. *Proceedings of the National Wetland Protection Symposium.* Office of Biological Series. U.S. Fish and Wildlife Service. Washington, D.C. FWS/OBS-78/97. 255pp.

National Academy of Sciences. 1975. *Productivity of World Ecosystems.* National Academy of Sciences. Washington, D.C. 165pp.

Nixon, S.W. and V. Lee. 1986. Wetlands and Water Quality. Wetlands Research Program. U.S. Army Corps of Engineers. Washington, D.C. Technical Report Y-86-2. 229pp.

Odum, E.P. 1961. The role of tidal marshes in estuarine production. *New York State Conservationist.* 16:12-15, 35.

Odum, E.P. 1989a. *Ecology and Our Endangered Life-Support System.* Sinauer Associates, Inc. Sunderland, MA. 283pp.

Odum, E.P. 1989b. Input management of production systems. *Science.* 243: 177-182.

Odum, H.T. 1978. Value of wetlands as domestic ecosystems. in: *National Wetland Protection Symposium Proceedings.* J.H. Montanari and J.A. Kusler, eds. Biological Services Program. U.S. Fish and Wildlife Series. Washington, D.C. FWS/OBS-78/97. pp. 9-18.

Odum, H.T. and M.G. Turner. 1990. The Georgia Landscape: a changing resource. in: *Changing Landscapes: An Ecological Perspective.* I.S. Zonneveld and R.T.T. Forman, eds. Springer-Verlag. New York, NY. pp. 137-164.

Pill, J. 1971. The Delphi method: substance, context, a critique and an annotated bibliography. *SocioEcon. Plan. Sci.* 5:57-71.

Platt, W.S., C. Armour, G.D. Booth, M. Bryant, J.L. Bufford, P. Cuplin, S. Jensen, G.W. Lienkaemper, G.W. Minshall, S.B. Monsen, R.L. Nelson, J.R. Dedell and J.S. Tuhy. 1987. Methods for evaluating riparian habitats with applications to management. U.S. Department of Agriculture. Forest Service. Ogden, UT. General Technical Report INT-221. 177pp.

Palmer, J.H., M.T. Chezik, R.D. Heaslip, G.A. Rogalsky, D.J. Putnam, R.W. McCoy, and J.A. Arway. 1985. Pennsylvania Modified 1980 Habitat Evaluation Procedure. U.S. Fish and Wildlife Service. Washington, D.C.

Queen, W.H. 1977. Human uses of salt marshes. in: *Wet Coastal Ecosystems.* V.J. Chapman, ed. Elsevier Scientific Publ. Co. Amsterdam. pp. 363-368.

Ranwell, D.S. 1972. *Ecology of Salt Marshes and Sand Dunes.* Chapman and Hall Ltd. London. 258pp.

Raphael, C.N. and E. Jaworski. 1979. Economic value of fish, wildlife, and recreation in Michigan's coastal wetlands. *Coastal Zone Management Journal.* 5:181-194.

Reimold, R.J. 1976. Grazing on wetland meadows. in: M. Wiley, ed. *Estuarine Processes. Vol. I. Uses, Stresses and Adaptation to the Estuary.* Academic Press, New York, NY. pp. 219-225.

Reimold, R.J. 1977. Mangles and salt marshes of eastern United States. in: V.J. Chapman, ed. *Wet Coastal Ecosystems.* (Ecosystems of the World Series No. 1). Elsevier-North Holland Publ. Co. New York, NY. pp. 157-166.

Reimold, R.J. and W.H. Queen. 1974. *Ecology of Halophytes.* Academic Press, Inc. New York, NY. 605pp.

Reimold, R.J. and R.A. Linthurst. 1979. Estimated net aerial primary productivity for selected estuarine angiosperms in Maine, Delaware and Georgia. *Ecology.* 59(5): 945-955.

Reimold, R.J. 1980. Creation of a southeastern United States salt marsh on dredged materials. in: J.C. Lewis and E.W. Bunce, eds. *Rehabilitation and Creation of Selected Coastal Habitats.* U.S. Fish and Wildlife Service. Washington, D.C. pp. 6-22.

Reimold, R.J., J.H. Phillips and M.A. Hardisky. 1980. Sociocultural values of wetlands. in: *Estuarine Perspectives.* V.S. Kennedy, ed. Academic Press. New York, NY. pp. 79-89.

Reppert, R.T., W. Sigleo, E. Stakhiv, L. Messman, and C. Meyers. 1979. Wetlands Values: Concepts and methods for wetlands evaluation. U.S. Army Corps of Engineers. Institute for Water Resources, Research Report 79R1. 109pp.

Roberts, T.H., L.J. O'Neil and W.E. Jabour. 1987. Status and source of habitat models and literature reviews. U.S. Army Engineer. Waterways Experiment Station. Miscellaneous Paper EL-85-1. Vicksburg, MS. 21pp.

Rodgers, J.H. and A. Dunn. 1992. Developing design guidelines for constructed wetlands to remove pesticides from agricultural runoff. *Ecological Engineering.* 1(1/2): 83-96.

Sather, J.S. and P.J.R. Stuber (tech. coords.) 1984. *Proceedings of the National Wetlands Values Assessment Workshop.* U.S. Fish and Wildlife Service. Western Energy and Land Use Team. FWS/OBS-84/12. 100pp.

Schaeffer, D.J., E.E. Herricks and H.W. Kerster. 1988. Ecosystem Health: I. Measuring Ecosystem Health. *Environmental Management.* 12(4): 445-455.

Schamberger, M.L. and H.E. Kumpf. 1980. Wetlands and wildlife values: a practical field approach to quantifying habitat values. in: *Estuarine Perspectives.* V.S. Kennedy, ed. Academic Press. New York, NY. pp. 37-46.

Schamberger, M.L. and L.J. O'Neil. 1986. Concepts and constraints of habitat model testing. in: *Wildlife 2000: modeling habitat relationships of terrestrial vertebrates*. J. Verner, M.L. Morrison and C.J. Falph, eds. University of Wisconsin Press. Madison, WI. pp. 5-10.

Scheifele, G.W. 1987. An assessment of Ontario's wetland evaluation system with reference to predictive models and environmentally sensitive areas studies. M.A. Thesis. University of Waterloo. Waterloo, Ontario, Canada. 166pp.

Schroeder, R.L. 1987. Community models for wildlife impact assessment: a review of concepts and approaches. National Ecology Center. U.S. Fish and Wildlife Service. Washington, D.C. Biological Report 87(2). 41pp.

Segelquist, C.A., W.L. Slauson, M.L. Scott and G.T. Auble. 1990. Synthesis of Soil-Plant Correspondence Data From Twelve Wetland Studies Throughout the United States. U.S. Fish and Wildlife Service. Washington, D.C. Biological Report 90(19).

Shabman, L.A. and S.S. Batie. 1980. Estimating the economic value of coastal wetlands: conceptual issues and research need. in: *Estuarine Perspectives*. V.S. Kennedy, ed. Academic Press. New York, NY. pp. 3-15.

Sharitz, R.R. and J.W. Gibbons. eds. 1989. Freshwater Wetlands and Wildlife. Department of Energy Symposium Series No. 61. Office of Scientific and Technical Information. U.S. Department of Energy. Oak Ridge, TN. CONF-8603101.

Shaw, S.P. and G.C. Fredine. 1956. Wetlands of the United States, Their Extent and Their Value to Waterfowl and Other Wildlife. Circular 39. Superintendent of Documents. U.S. Government Printing Office. Washington, D.C.

Short, H.L. 1982. Development and use of a habitat gradient model to evaluate wildlife habitat. in: *Transactions of the Forty-Seventh North American Wildlife and Natural Resources Conference*. Wildlife Management Institute. Washington, D.C. pp. 57-72.

Simenstad, C.A., C.D. Tanner, R.M. Thom and L.D. Conquest. 1991. *Estuarine Habitat Assessment Protocol. Office of Puget Sound*. U.S. Environmental Protection Agency. Region 10. Seattle, WA. EPA/910/9-91-037.

Urban, D.L., R.V. O'Neill, and H.H. Shugart. 1987. Landscape ecology. *BioScience*. 37: 119-127.

U.S. Army Corps of Engineers. 1976. A tentative habitat evaluation system (HES) for water resource planning. November. Lower Mississippi Valley Division. U.S. Army Engineer. Waterways Experiment Station, Vicksburg, MS.

U.S. Army Corps of Engineers. 1980. A habitat evaluation system (HES) for water resources planning. Lower Mississippi Valley Division. U.S. Army Corps of Engineers. Vicksburg, MS. 89pp.

U.S. Environmental Protection Agency. 1990. *Water Quality Standards for Wetlands: National Guidance*. Office of Water Regulations and Standards. US EPA. Washington, D.C. EPA 440/S-90-011.

U.S. Fish and Wildlife Service. 1976. Habitat Evaluation Procedures. Division of Ecological Services. U.S. Department of Interior. Washington, D.C. 30pp.

U.S. Fish and Wildlife Service. 1977. A Handbook for Habitat Evaluation Procedures. U.S. Department of Interior. Washington, D.C. Resource Publication 132.

U.S. Fish and Wildlife Service. 1980. Habitat Evaluation Procedures. Division of Ecological Services. U.S. Fish and Wildlife Service. U.S. Department of Interior. Washington, D.C. ESM 102.

U.S. Forest Service. 1981. Wildlife and fish habitat relationships: Narratives (Vol. I), and Matrices (Vol. II). Rocky Mountain Region. U.S. Forest Service. Denver, CO.

Wakeley, J.S. 1988. A method to create simplified versions of existing habitat suitability index (HSI) models. *Environmental Management.* 12(1). pp. 79-83.

Wakley, J.S. and L.J. O'Neil. 1988. Alternatives to increase efficiency and reduce effort in application of the habitat evaluation procedures (HEP). U.S. Army Engineer. Waterways Experiment Station. Vicksburg, MI. Technical Report

Walker, R.A. 1973. Wetlands preservation and management on Chesapeake Bay: the role of science in natural resource policy. *Coastal Zone Management Journal.* 1(1):75-101.

Westman, W.E. 1977. How much are nature's services worth? *Science.* 197: 960-963.

White, T.A., R. Lea, R.J. Haynes, W.L. Nutter, J.R. Nawrot, M.M. Brinson and A.F. Clewell. 1990. Development and summary of MiST: a classification system for pre-project mitigation sites and criteria for determining successful replication of forested wetlands. in: *Proceedings of the 1990 Mining and Reclamation Conference and Exhibition.* J. Skousen and J. Sencindiver, eds. American Society of Surface Mining and Reclamation. Charleston, WV. pp. 323-335.

Wolf, P.L., S.L. Shanholtzer and R.J. Reimold. 1975. Population estimates for *Uca pugnax* (Smith, 1870) on the Duplin Estuary marsh, Georgia, U.S.A. *Crustaceana.* 29(1): 79-91.

CHAPTER 5

ECOLOGICAL RISK ASSESSMENT OF WETLANDS

David J. Kent, Kenneth D. Jenkins and James F. Hobson

The term risk assessment historically referred to the estimate of risk to human health, typically from chemical exposure. For example, a cancer risk assessment is an estimate of the risk to humans from carcinogenic compounds. Recently, however, the term risk assessment has been applied to ecological systems. An ecological risk assessment is an estimate of the adverse effect to an ecosystem from chemical, physical or biological stressors resulting from anthropogenic activity. This recent interest in assessing ecological health is evidenced by several recent publications (Bartell et al. 1992, Cairns et al. 1992, Suter 1993) including a document produced by the U.S. Environmental Protection Agency (USEPA) entitled *Framework for Ecological Risk Assessment* (1992). The latter document is intended as the first step in a long term program to develop guidelines for the performance of ecological risk assessments.

The principles of ecological risk assessment can be applied to any ecosystem, although they may be particularly relevant to wetlands. The extent and rate of wetland loss, as well as the biologic, economic and social importance of wetlands, is well documented (Mitsch and Gosselink 1986). Moreover, the transitional nature of wetlands may make them especially sensitive to stress. Despite the uniform application of assessment principles to ecological systems, individual wetlands are sufficiently different in their spatial, temporal and physicochemical characters to warrant site-specific sampling and analysis (Figures 1 and 2). These differences will influence the design and interpretation of the ecological risk assessment.

To understand the use of ecological risk assessment for wetland ecosystems, an introduction to the principles of risk assessment is necessary. The current basis for risk assessment is derived from the National Research Council Risk Assessment paradigm (1983). Following this, an in-depth discussion of the new USEPA Ecological Risk Assessment Framework (USEPA 1992) is presented, as is a discussion of the challenges and strategies associated with carrying out assessments. Finally, examples of specific applications to wetlands ecosystems are provided.

Figures 1 and 2. Risk assessment can be broadly applied to a variety of ecosystems. Nevertheless, individual wetlands, such as this New England saltmarsh (Figure 1) and this Arkansas riverine system (Figure 2), are sufficiently different in their spatial, temporal and physicochemical characters to warrant site-specific sampling and analysis.

THE HUMAN HEALTH RISK ASSESSMENT PARADIGM

The risk assessment paradigm has been used for some time to evaluate the chronic impacts of environmental pollutants on human health. This strategy was initially conceptualized by the National Research Council Risk Assessment Panel (NRC 1983) and formalized by the USEPA in its 1986 Guidelines (USEPA 1986a-e). This risk assessment paradigm consists of several components:

- **Hazard identification:** Does a chemical contaminant represent a specific threat to human health? Establishment of cause-effect relationships are central to this component.
- **Defining dose-response:** What is the relationship between the magnitude of the exposure and the probability of an adverse health effect?
- **Exposure assessment:** What is the potential for human exposure to the chemical of concern?
- **Risk characterization:** What is the potential magnitude of risk to human health given the predicted exposure and dose-response data? What is the uncertainty associated with this risk estimate?

Standard methodologies are employed to evaluate potential threats to human health. The methodologies usually involve determining all relevant effects, and then summing those effects to get a total effect value.

ECOLOGICAL RISK ASSESSMENT

Ecological risk assessments (ERAs) examine the probability that undesirable ecological effects are occurring or may occur as a result of exposure to a stressor or a combination of stressors. The term stressors is used here to reflect the broad range of anthropogenic factors which can result in ecological perturbations. Stressors may include any chemical, physical or biological factor resulting from anthropogenic activities which can cause an ecological disturbance. Most often, however, the term stressor refers to toxic chemicals.

ERAs can be used to address a wide range of issues and are generally classified as predictive or retrospective. Predictive ERAs are designed to assess the risks associated with proposed actions, such as the introduction of new chemicals into the environment and the establishment of new sources of stressors or hypothetical accidents (USEPA 1991). Predictive risk assessments have usually followed the NRC human health paradigm relatively closely while emphasizing the choice of biological endpoints and related stressor-response data.

In contrast, retrospective ERAs address the risks associated with stressors released due to current or previous anthropogenic activities. Examples of retrospective assessments include evaluating the ecological impact of hazardous waste sites and previous releases or spills. The goal of this type of risk assessment is to establish and define the relationship between the pollution source, the distribution of stressors, the exposure of biological endpoints, and the level of effect of this exposure on the ecosystem. Retrospective assessments often take advantage of field data to define contaminant sources and measure adverse biological effects. Various levels of data collection or site specific assessments may be necessary to provide the information required to design and conduct the retrospective ecological risk assessment, and to achieve a given level of confidence. The challenge here is to establish cause-effect relationships between the source of stressors and any observed ecological effects.

Some ERAs, such as those used for wetlands, may involve both predictive and retrospective aspects. For example, in an assessment of a hazardous waste site, the current status of the site may require a retrospective evaluation, but the long-term impact of various remediation scenarios would be addressed in a predictive fashion. As the type of risk assessment to be conducted is dependent upon the ultimate application of the results, a clear understanding of the objectives of the risk assessment is essential.

The move to ERAs by the regulatory community is driven by a number of factors. From a legal standpoint, many of the underlying statutes require some form of evaluation of ecological risk. For example, the Superfund Amendments and Reauthorization Act (1987) specifies that the actual and potential risk to public health, and the environment, must be assessed for each hazardous waste site. The use of a basic risk assessment paradigm would provide a structural framework for ecological assessments and a consistent strategy for managing various types of risk. This issue is particularly important when comparing the sensitivity of ecological endpoints relative to human endpoints. In some cases nonhuman endpoints may prove to be more sensitive than human endpoints, and would thus drive the overall risk assessment. This type of comparison is facilitated if there is consistency in the strategies used to carry out risk assessments for both types of endpoints.

Although human health risk assessment strategies provide a useful model, assessing ecological risk has proven to be more complex. A number of factors contribute to this increased complexity, including:

- **Multiple biological endpoints:** These could include multiple species and various levels of biological organization (e.g. subcellular, individual, population, community, and ecosystem).
- **More complex exposure pathways:** These are determined by the biological endpoints of concern.
- **Indirect effects:** Indirect effects such as habitat impairment or disruption of intertrophic relationships may be more important than direct exposure to chemicals.
- **Evaluating impacts on ecosystems:** Ecosystems are complex, their function is often not tightly coupled to stressor inputs, and they show resilience and recovery to varying degrees of stress.

In spite of these complexities there are some distinct advantages to estimating risk to ecological endpoints. Exposures and hazards can often be estimated directly on the species of concern or a closely related surrogate species. This may result in a more accurate estimate of risk, as there is no need for extrapolation from more distantly related species as is almost always the case in human risk assessment. This is particularly useful when multiple stressors or complex exposure matrices are involved. In ERAs, the biological endpoints of interest can often be tested directly against the specific stressor or mixture of stressors of concern, thus eliminating the need to estimate such factors as stressor interactions, chemical form (speciation) and bioavailability.

Despite the aforementioned advantages, the added complexity associated with ERAs results in a higher degree of uncertainty than is normally associated with human health-based risk assessments. This complexity requires more effort in the initial planning stages so that the final assessment is well focused. Furthermore, some modifications of the basic NRC risk assessment paradigm are required. The Risk Assessment Forum within the USEPA has developed a Framework for Ecological Risk Assessment which addresses these issues, yet is conceptually consistent with human health risk assessment strategies. It is important to note that slight differences in terminology exist between the three main ERA structures currently promoted (USEPA 1992, NRC 1983, Suter 1993). However, the elements are fairly analogous. For the sake of simplicity, the discussions in this chapter will utilize the USEPA Framework terminology, although the concepts may be drawn from multiple sources.

THE USEPA ECOLOGICAL RISK ASSESSMENT FRAMEWORK

Two major elements form the basis of the USEPA framework: 1) the characterization of exposure, and 2) the characterization of ecological effects. Aspects of these two elements are considered in all phases of the framework process. While carrying this common thread throughout the paradigm, the framework is divided into three phases: 1) problem formulation, 2) analysis, and 3) risk characterization (Figure 3).

Problem Formulation Phase

The first step in this process requires defining the specific purpose of the ERA. Although this may seem trivial, many ERAs suffer from lack of clear focus and as a result may be ambiguous and misleading. Once the specific purpose of the assessment is defined, the specific goals which must be met to achieve this purpose are formulated. These goals provide a basis for establishing a precise conceptual study design. In undertaking the study design a number of questions must first be addressed, including:

- Is the ERA to be predictive or retrospective?
- Is the ERA to be site-specific or generic?
- What type of ecosystem(s) is at risk?
- What types of stressors are involved?
- What are the potential source(s) for a given stressor or set of stressors?

- What are the relevant biological receptors for these stressors?
- What are the appropriate biological endpoints for assessing hazard and exposure potential for these receptors?

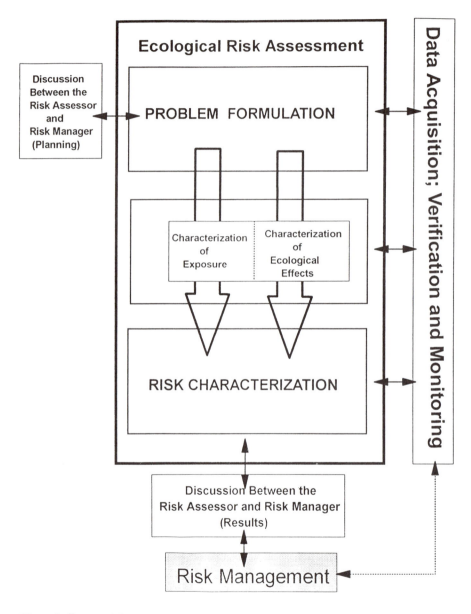

Figure 3. Framework for Ecological Risk Assessment (USEPA 1992)

Addressing these questions requires a rigorous review of available data. Where existing data is unavailable or incomplete, it may be necessary to carry out a preliminary study, particularly to establish the types of potential stressors. Most importantly, ecological risk assessment is often an iterative process. A given level of information is required for developing the design and objectives (i.e. the problem formulation). Additional data may be required for the complete ERA. Ultimately, the level of information needed and the extent of new data collection are dependent upon the objectives, such as the level of certainty desired. As well, the level of information is dependent upon the outcome of the previous iterations. For example, how much ecological impact has occurred at a given site, or how hazardous a new chemical may be based on laboratory hazard assessment.

Choosing Biological Endpoints

An important issue at this stage is determining the appropriate suite of biological endpoints to be used in the evaluation. Biological endpoints should be carefully chosen to specifically address the overall goals of the assessment. The parameters to be considered in choosing these endpoints should include the ecological relevance of the endpoint, and the spatial and temporal occurrence of the endpoint relative to the distribution of the stressors and potential biological receptors.

In ecological assessments the distinction is often made between assessment endpoints and measurement endpoints (Warren-Hicks et al. 1989, Suter 1993). Assessment endpoints represent the ultimate resource(s) or final environmental values that are to be protected. They should have social or biological relevance, be quantifiable, and provide useful information for resource management or regulatory decisions. Successful reproduction of a species in danger of extinction is an appropriate assessment endpoint. For populations that are valued for commercial or sport uses such as anadromous fish (salmon) or estuarine or marine shrimp, the important assessment endpoints could be growth, reproduction, or overall productivity. Other potential assessment endpoints include yield and productivity, market or sport value, recreational quality and reproductive capability (see Warren-Hicks et al. 1989). In general, assessment endpoints focus on population and community parameters, as ecological risk assessments are usually concerned with protecting these higher levels of biological organization rather than individual organisms. However, as when there is concern for endangered species, the assessment endpoints may indeed focus on the individual organism.

In contrast, measurement endpoints represent the specific parameters which are to be measured in a given assessment. They are often chosen based upon practical considerations such as availability and ease of measurement. It is often practically impossible to measure some endpoints (e.g. endangered species) and other more practical measurement endpoints are chosen as surrogates for the actual assessment endpoints of interest. Such measurement endpoints should be well characterized and take into account exposure pathways and temporal factors. Ideally, measurement endpoints should be chosen so that the data from these endpoints can be linked directly or indirectly to appropriate assessment endpoints. This latter issue is often the most difficult to address in ERAs.

At the level of the individual, measurement endpoints may include mortality, growth and fecundity. Abundance and reproductive performance are measurement endpoints on a population level. Other measurement endpoints include species evenness and diversity on a community level, and biomass and productivity on an ecosystem level. Assessment of endangered species poses a special problem. As exposure to an endangered species cannot usually be directly assessed, the residues of a contaminant in a principle food item can be a measurement endpoint. Alternatively, if exposure to a chemical and its potential impact on an endangered species is the assessment endpoint, then a co-existing species with similar life history habits might be used as a surrogate measurement endpoint.

Spatial and Temporal Considerations

Ultimately the problem formulation phase should result in the establishment of the study design, which will form the basis of the assessment. The design must take into account a number of environmental and ecological factors which may effect the stressors or their potential impact on biological systems. Many of these factors have spatial or temporal components which must be taken into account in the design. Spatial factors include potential routes of exposure, other sources of stressors, location of sensitive biological resources and factors which may modify contaminant mobility or availability. The changing composition of sediments is an example of the latter. Temporal factors may include seasonal changes in physical, chemical, or biological aspects of ecosystems which may influence the magnitude or form of the stressor, or its potential to cause a biological effect. For example, increased surface water movement in the wet season in riparian habitats can dramatically affect contaminant migration and the potential for exposure. Also, seasonal variation in physical or chemical parameters such as temperature or pH can modify the bioavailability of contaminants, thus changing the nature of the exposure.

The biological characteristics of the exposure matrix may also change temporally. Species may be present only seasonally, as is typical of migratory water fowl, anadromous fishes, or species which migrate seasonally within a local area. Exposure to a given species, population or community may vary with seasonal changes in life history habits such as seasonal feeding patterns or reproductive cycles. Thus, temporal changes in the biological components of an ecosystem may influence the distribution of the stressor or the availability of the biological end point, within the ecosystem. Temporal variation would affect the ultimate risk assessment. Therefore, these temporal patterns must be addressed in designing the sampling scheme and data collection for the risk assessment.

Temporal considerations may also include long-term historical or predicted trends in stressor influence, and potential seasonal variation in stressor impact. Historical or predicted trends are very important to understanding the overall impact of stress on an ecosystem, and may be important to the application of risk information in risk management decisions such as remediation plans, wetland restoration, registration of a new pesticide, etc.

Other Considerations

Another factor to consider is the presence of biological resources that are either sensitive to stressor impact or may be of particular economic or social importance. Wetlands are areas for many species which inhabit adjacent terrestrial and aquatic communities. Other resources may include populations of economically important species such as anadromous fish populations, which are the basis for economically important fisheries resources, or endangered species, which are protected by law.

In completing the problem formulation phase of the ERA, these issues must be addressed in rigorous and systematic fashion. The ultimate goal is the development of a conceptual model which will serve as a basis for the ERA. The model should address the spatial and temporal distributions of potential stressors, appropriate biological endpoints, as well as probable routes and levels of exposure. Finally, the model should be precisely linked to the goals and purposes of the risk assessment. A well conceived conceptual model is essential to facilitate the implementation of the subsequent components of the risk assessment.

Analysis Phase

The analysis phase of the USEPA framework consists of two activities: characterization of exposure and characterization of ecological effects. It is here that the two main elements of an ecological risk assessment are most prominent and most closely interrelated.

Exposure Characterization

The goal of exposure characterization is to develop an accurate appraisal of the potential for exposure to important biological resources or receptors, as represented by the measurement endpoints, to the identified stressors. Spatial and temporal factors are particularly important in this stage of the ERA. Issues associated with multiple stressors and multiple routes of exposure must also be addressed at this time.

Both the stressor and the ecosystem must be characterized with regard to the distribution and pattern of change. This is accomplished through the use of either modeling or monitoring data or both, depending on whether the ERA is predictive or retrospective. Stressors can be characterized from direct sampling, laboratory or field testing, or through remote sensing. At this stage, it is important to evaluate the means by which stressors can be modified in the ecosystem through biotransformation or other environmental fate processes such as photolysis, hydrolysis and sorption. Transport and fate models are often employed, including interactive mass balance models like the Exposure Analysis Modeling System (EXAMS) developed at the USEPA research laboratory in Athens, Georgia (Burns et al. 1982). Characteristics of ecosystem components that can affect exposure also need to be evaluated. These can include habitat requirements, food preferences, reproductive cycles and seasonally-influenced activities.

After stressor and ecosystem characteristics are defined, the spatial and temporal distributions are combined to evaluate exposure. Finally, the magnitude and distribution patterns are quantified for the scenarios developed during the problem formulation phase of the ERA. All sources of uncertainty should also be quantified to the degree possible for the input into the ERA.

Ecological Effects Characterization

The goal of this portion of the analysis phase is to identify and characterize any adverse ecological effects that are associated with exposure to a particular stressor or stressors. To the extent possible, these effects should be quantified, and any cause and effect relationships evaluated.

Initially, an evaluation must be made of all effects data that are relevant to the stressor. The types of data that are relevant will depend upon the characteristics of both the stressor and the ecological component, and also on whether the ERA is to be predictive or retrospective. Commonly used data types include aquatic toxicity tests, computer models, quantitative structure activity relationships (QSARs), micro- and mesocosms, species diversity analyses, artificial substrate comparisons, the Wetland Evaluation Technique (Adamus et al. 1987), and others (Suter 1993, USEPA 1992, McKim et al. 1987, Barnthouse et al. 1986). Data from laboratory and field observations, as well as from controlled studies, may be used. These data are considered based upon their relevance to the measurement and assessment endpoints. Evaluating the quality of the data, that is the adequacy of sampling and statistical design, is an important component of this stage.

The next step is to quantify the ecological response in terms of the stressor-response relationship. The aim is to describe the relationship between the magnitude, frequency and duration of the stressor, and the magnitude of the response. The dose-response curve in laboratory aquatic toxicity tests is an example of this type of analysis. Determining relative differences in productivity, biomass, species composition and diversity between contaminated and reference sites is also a commonly used means of defining this relationship. Any extrapolations required between the stressor and measurement endpoints must be evaluated, as well as any extrapolations between measurement and assessment endpoints. The strength of the causal relationship must be quantified to the extent possible. Finally, as with the exposure analyses, quantitative estimates of uncertainty such as natural variability in ecological characteristics and responses, should be included in the analyses. It should be emphasized that although the goal is to provide quantification of the stressor-response relationship, often it can only be described qualitatively, and therefore, professional judgment plays an important role in many ERAs.

Risk Characterization Phase

The ultimate purpose of the ERA is to estimate the risk that unacceptable adverse ecological effects will occur due to exposure to anthropogenic stressors. The risk characterization phase provides an estimate of the likelihood that an adverse impact has occurred, or will occur. It should also address the relative ecological consequences which are associated with the various levels of risk.

Ideally, risk characterization should be quantitative and include estimates of uncertainty. However, the complexity of ecological risk assessments may often preclude precise quantification. Under these circumstances, qualitative estimates of risk are often employed.

In the risk characterization phase a hazard-exposure matrix is developed. This matrix represents a fusion of the exposure predictions developed in the exposure characterization phase, and the estimates of effects developed in the ecological effects characterization phase.

Ultimately, the risk characterization phase should evaluate the ecological consequences or ecological significance of predicted or observed effects. This is simpler in some cases for retrospective risk assessments, as the changes can be documented and quantified *in situ*. Also, the ecological effects, that is the resulting changes at higher levels of biological organization or in the ecosystem as a whole, can be monitored directly with the proper sampling design and appropriate reference sites.

In predictive ecological risk assessments these effects or changes in ecosystems must be predicted. The degree of uncertainty for such predictions is relatively large in most situations. As the historical database for ecotoxicology and ecological risk assessment increases, especially field data sets, these predictions will be made with higher levels of confidence. Especially important is the field validation of risk assessments based on laboratory-based hazard assessments and modeled exposure assessments.

As a result of the inherently different goals of predictive and retrospective ecological risk assessments, the remainder of this chapter will provide illustrations of how ERAs may be applied for each of these types with regard to wetland ecosystems. The first type of ERA to be discussed will be the current tiered assessment design, a predictive approach employed by the USEPA in the pesticide regulation program. The second describes a retrospective approach more common in the assessment of hazardous waste sites or spills.

PREDICTIVE ECOLOGICAL ASSESSMENTS

A practical example of a predictive risk assessment can be seen in evaluating the potential ecological impact of the introduction of a new herbicide. Under the Federal Insecticide, Fungicide and Rodenticide Act (FIFRA) and its amendments, all new pesticides and herbicides must be registered for each specific use. The Office of Pesticide Programs of the USEPA prepared the *Standard Evaluation Procedure for Ecological Risk Assessment* (SEP) (Urban and Cook 1986) to define the procedures to be used for the ecological risk assessment process. In short, this document presents a tiered approach to determining the potential for unreasonable adverse effects on the environment as a result of the use of pesticides and herbicides. Additional information for the testing of new pesticides and herbicides can be found in the Pesticide Assessment Guidelines (USEPA 1982a & b).

To illustrate this example, consider a new herbicide for control of weeds in rice paddies adjacent to riparian and emergent wetlands. The goal here is to assess the potential risk of the new herbicide on the wetland ecosystem based on the toxicology and environmental fate of the pesticide under the proposed use pattern.

Problem Formulation

The first activity in this predictive ERA is to define the scope of the problem. The planned use pattern for the herbicide requires its application to fields which have a great dependence on water flow and thus have a large potential for transport of chemicals into adjacent aquatic habitats. It is important, therefore, that the inherent toxicity of herbicides is assessed in relation to the likely exposure to various ecological compartments. In simple terms, the questions to be answered are: 1) how toxic is it? and 2) will there be enough exposure for the toxicity to be realized?

The typical rice field (Figure 4) may be 30-150 acres in size and laser-leveled to control water flow. The fields are not allowed to get too dry and, if sufficient rainfall is lacking, they may be flushed with water to depth of 2.5 cm or so on a fairly regular basis. Once the rice is adequately established, the field is typically flooded to a depth of 7.5-15 cm which is maintained until harvest. The temporary flush and permanent flood waters are drained to an adjacent ditch which runs within 100 m to a small bayou. Emergent and riverine wetlands are associated with the bayou. Approximately 1000 acres of rice drainage enter into the bayou above and below our hypothetical field. For this example, let's assume that the herbicide has a maximum application rate of 2.27 kg (5 lbs) active ingredient per acre, and is applied only once a year just prior to the permanent flood.

Exposure Characterization

Essentially, there are four means by which the herbicide could enter the adjacent wetland habitats. First, exposure may occur if the pilot applying the material oversprays the field and deposits some of the herbicide directly on the wetland. Second, small droplets of the substance have the potential to drift in the wind during application. Third, due to the nature of rice culture, periodic releases of water are made from the fields. And finally, severe storms could potentially create an overflow of paddy water into nearby wetlands. Each of these routes should be examined in the ERA.

In most cases, measured concentrations of the new herbicide in the field will not be available at this point. Therefore, the potential exposure must be estimated. The endpoint of interest in the exposure characterization is the determination of an Estimated Environmental Concentration (EEC). The EEC provides the risk assessor with a concentration of the herbicide that may be present in the environment. A tiered approach is commonly used to determine the EEC, moving from simple to more complex depending on the need for additional information.

The simplest way to determine the EEC is to assume that the herbicide is applied directly to the water in the wetlands adjacent to the rice fields. Although this is unlikely, it is possible if, for example, the pilot applying the herbicide overshoots the field. In any case, it provides an extreme worst-case exposure scenario. At the given application rate, and assuming a direct application to water that is 0.6 m deep, the EEC immediately following application would be 919 ppb. Comparing this EEC to ecological effects would allow a cursory assessment of risk to be made.

Figure 4. The typical rice field may be 30-150 acres in size and laser-leveled to control water flow. Flooding and drainage is accomplished through use of the ditch in the foreground.

However, this simple method of determining the EEC does not consider variables that could modify the actual concentrations in the adjacent wetland system. Normal use practices of the herbicide would preclude direct application to water bodies. Rather, the application would be directly to the field and transport of the chemical would be a function of waterflow off of the field. In addition, physicochemical and environmental fate characteristics of the herbicide may allow it to partition to the sediments or be metabolized or degraded into either more or less persistent compounds.

If a more realistic determination of the EEC is required, additional tiers of evaluation can be performed. For example, a second level may incorporate additional variables such as drainage basin size, surface area and percent runoff into the simple equation. A still more complex EEC determination can be made utilizing computer models such the Pesticide Root Zone Model (PRZM) and EXAMS (Burns et al 1982) to predict herbicide concentrations in aquatic systems resulting from runoff. Finally, the last tier of EEC determination requires an actual field residue monitoring study in which the herbicide is applied at the maximum rate in a test field and residues are measured periodically around the field.

It is important to consider all of the spatial and temporal factors associated with the introduction of the herbicide. For the herbicide example, these can include the application and use rates; water flow dynamics in the rice field (e.g. is water released immediately after application or held for a significant time period before release?), mobility and persistence of the herbicide (e.g. water solubility, hydrolysis/photolysis rates, metabolism rates, etc.), and propensity for bioaccumulation. Other factors include the timing of application with respect to seasonal biological cycles and the type and number of organisms that could potentially be affected.

Ecological Effects Characterization

The purpose of this segment of the ERA is to characterize any adverse impacts that may be associated with the herbicide. Because this is a herbicide, it is designed to be toxic to certain broadleaf weeds. Therefore, the effects characterization should be limited to nontarget flora and fauna that may be exposed to the herbicide either during use or following runoff. To test all of the species that may come in contact with the herbicide at all of the potential sites of application is unrealistic. Typically, a series of standard surrogate species are employed.

As with the determination of the EEC, a tiered approach is usually taken to quantify the toxicity of the herbicide to nontarget organisms. Results from the first tier testing determine the need for second tier testing, and so on. Each tier normally examines increasingly sensitive endpoints.

Tier I addresses acute toxicity to warm water and cold water fish, as well as to aquatic invertebrates and plants (World Wildlife Fund 1992). Acute oral and dietary dosages to game birds such as bobwhites and mallards are also evaluated for toxicity. Additional tests of toxicity to estuarine species, honeybees, etc. may be required depending on the use pattern and environmental fate of the compound. Tier II testing is sometimes required to determine a no-observable effect concentration (NOEC) for effects such as reduced survival, growth and reproduction. Tier II tests for toxicity to fish early-life-stage or invertebrate life-cycle can be triggered by acute toxicity where the median lethal concentration (LC50) is less than 1.0 ppm. These tests can also be triggered by high EEC levels (where EEC is >0.01 of the lowest Tier I LC50 value), chemical persistence in the environment, continuous or recurrent exposures, or high potential for bioaccumulation or reproductive effects. Tier II studies are automatically required for any aquatic use of herbicides. Tier II avian reproduction studies may also be triggered if Tier I avian tests indicate toxicity.

Tier III full life-cycle tests with fish may be required if the EEC is more than 0.1 of the NOEC from the tier II fish early-life-stage study or if data suggest that there is a potential for impairment of fish reproduction. Finally, Tier IV field or mesocosm studies may be invoked if the expected concentrations in the environment exceed the Tier III NOEC determination. Tier IV field testing with avian or mammalian species may also be required. It is important to note that recent guidance by the USEPA suggests that decisions on the reregistration of existing pesticides and herbicides can be made without requiring Tier IV studies (Fisher 1992). However, companies wishing to reregister existing pesticides and herbicides where a hazard for aquatic organisms has been presumed may want to conduct Tier IV testing in an effort to rebut this presumption.

Risk Characterization

By comparing the estimates of environmental concentrations developed in the exposure characterization section and the estimates of toxicity developed in the ecological effects characterization section, an assessment of risk of the rice herbicide to the adjacent wetlands can be generated. There are several methods of examining the relationship between the EEC and environmental effects (Barnthouse et al. 1982a & b). One of these is the Quotient Method of risk analysis and is described below.

The Quotient Method for Risk Characterization

The Quotient Method is often employed for predictive risk characterizations of single compounds such as pesticides and herbicides. This method is based on a simple comparison of the estimated exposure to a stressor response value (SRV) for a specific endpoint. The SRV represents a level of exposure which has been demonstrated to have unacceptable toxicological effects. When the estimated exposure derived from the exposure assessment exceeds the SRV the associated risk is considered unacceptable. Clearly the choice of the SRV is critical to this approach. Usually SRVs are derived from laboratory toxicity studies and are estimates of threshold toxicity for appropriate endpoints, such as one tenth the acute median lethal concentration (LC50) or the chronic NOEC.

The use of the quotient method is straightforward and is useful for single chemicals. However, there are a number of disadvantages to this approach. The estimate of risk from the quotient method is only as good as the hazard and exposure estimates. The latter is often based on cursory environmental fate information and has a low confidence level. The hazard assessment data for ecological endpoints are often based on acute effects (i.e. mortality) rather than subacute effects such as inhibition of growth rates or impaired reproduction. Moreover, these so called Aquatic Hazard Assessments are usually based on clean-water laboratory tests which may not accurately reflect *in situ* exposure conditions. Finally, and most importantly, the quotient method is not useful in estimating risk from multiple stressors.

Most ecological risk assessments, especially retrospective assessments, involve more than one stressor. Risk from multiple stressors cannot be accurately addressed by adding the stressor-response values due to stressor interactions (synergism, antagonism, potentiation, etc.). In these cases, the hazard assessment and the exposure assessment phases of the ERA must be modified and tailored to the suite of stressors. This assumes that the suite of stressors is known, or identified in preliminary characterization carried out as part of the conceptual framework.

A similar methodology is used by the Ecological Effects Branch of the USEPA. However, where the quotient method uses the resultant quotient in a relative ranking to indicate adverse effects, the Ecological Effects Branch compares the EEC and an effect level (e.g. LC50) to regulatory risk criteria. Many of these risk criteria, summarized in Table 1, incorporate safety factors derived from a toxicological model presented in the FIFRA regulations. These safety factors are included to allow for the differential variability and sensitivity among resident fish and wildlife species (Urban and Cook 1986).

Based on the results of the risk characterization, a determination is made as to whether to register the herbicide for use on rice, restrict its use (through required labeling) to certified applicators, restrict its use to areas with limited wetland habitat proximal to the fields, or deny registration of the herbicide (World Wildlife Fund 1992).

RETROSPECTIVE ECOLOGICAL ASSESSMENTS

An example involving an ecological assessment of a wetlands adjacent to a hazardous waste site will demonstrate how the principles of ecological risk assessment are applied in a site-specific retrospective study. The principles developed in the USEPA framework document (USEPA 1992) can be applied to retrospective risk assessments as well as, and in some cases better than, the predictive ERAs (Pascoe et al. 1993). In addition, several other USEPA documents have been produced that provide guidance applicable to ecological assessments at hazardous waste sites (USEPA 1988 and 1989a, Warren-Hicks et al. 1989). Finally, guidance specific to the Superfund program was issued in 1989 (USEPA 1989b & c). This guidance, commonly referred to as RAGS (for Risk Assessment Guidance for Superfund), has become the main source of procedural information for both human and environmental health risk assessments.

Problem Formulation

For the purposes of this example, consider a hazardous waste site that is a long-term chemical manufacturing facility. The facility is located on 5 acres of commercial land which includes a nontidal marsh. In addition, a levee separates the site from an adjacent tidal marsh and a large estuarine bay.

The primary objective is to assess the degree of risk that contamination from the site poses for the adjacent wetland ecosystem. Another objective of this site-specific assessment is to provide information for the development and evaluation of an ecologically sound remedial action plan for the site. In the case of the nontidal wetland, these data can be used to help define the strategy required to address any effects of the contamination from the site on the non-tidal wetland ecosystem. For the tidal marsh these data would provide a basis for determining

Table 1. The regulatory risk criteria listed below can be compared to the estimated environmental concentration (EEC) and the effect level to characterize risk.

	No Risk	Risk Limited by Restricted Use	Unacceptable Risk	
			Nonendangered Species	Endangered Species
Acute Toxicity	Mammals $EEC < 1/5\ LC_{50}$	Mammals $EEC \geq 1/5\ LC_{50}$	$EEC \geq LC_{50}$	$EEC \geq 1/10 LC_{50}$ or $EEC \geq 1/5\ LC_{10}$
	Birds $EEC < 1/5\ LC_{50}$	Birds $1/5\ LC_{50} \leq EEC < LC_{50}$	$EEC \geq LC_{50}$	$EEC \geq 1/10 LC_{50}$ or $EEC \geq 1/5\ LC_{10}$
	Aquatic $EEC < 1/10\ LC_{50}$	Aquatic $1/10\ LC_{50} \leq EEC < 1/2\ LC_{50}$	$EEC \geq 1/2\ LC_{50}$	$EEC > 1/20 LC_{50}$ or $EEC > 1/10\ LC_{10}$
Chronic Toxicity	EEC < Chronic No Effect Level	N/A	$EEC \geq$ Chronic Effect Levels	$EEC \geq$ Chronic Effect Levels

Adapted from Urban and Cook 1986.

if contaminant related impacts are occurring and whether remediation is necessary. Moreover, if it was required, these data could be used to determine the appropriate type and extent of remediation. The following specific goals can be established for performing an assessment of risk to the tidal wetlands.

- Assess the potential for exposure to the contaminants of concern for ecologically and socially important species from the tidal wetlands.
- Determine the potential for acute and chronic toxicological effects from accumulated contaminants on these species.
- Evaluate the structure of the populations and communities of plants, benthic invertebrates and small mammals in the tidal wetlands.
- Relate the exposure, bioaccumulation and toxicity data obtained in this assessment with population, community and ecosystem data to provide an integrated picture of the impact of contamination from the site on the adjacent tidal wetlands.
- Evaluate the source or sources of any stressors that may be identified by examining the distributional patterns of those stressors in the tidal wetlands in relation to identified sources including the site and the bay. That is, do the stressors come from the site or from the other sources via the connection to the bay?

Exposure Characterization

The retrospective approach to exposure characterization is usually more direct than in the predictive approach. In a retrospective ERA stressors are already present in the environment. The goal is to determine which stressors are present and in what concentrations, rather than attempting to predict which stressors might enter the environment. The potential for exposure can be determined by measuring the actual accumulated dose of stressor in environmental matrices, including the tissues of native or transplanted organisms.

The first step in the exposure characterization phase is to determine the nature and distribution of the stressors. As indicated above, this is accomplished by collecting samples of the various environmental matrices and analyzing them for the suspected contaminants. In our hazardous waste site example, sampling should be performed within the site proper, in the onsite nontidal marsh, and at near-field and far-field locations within the adjacent tidal wetland. Additionally, external reference sites should be sampled to provide a comparison to the zones within the site and adjacent wetlands (Figure 5).

The precise sampling strategy employed will be dependent upon the exact goals of the ecological assessment, the structure and function of the ecosystem involved, and the nature of the contaminants of concern. One approach is a random sampling strategy which is designed to evaluate differences over relatively broad areas of the wetland. A grid system comprised of relatively large quadrants is set up within each zone to be examined. The use of randomly selected sampling stations within each grid facilitates statistical comparisons between zones, and the number of samples to be taken per unit area can be calculated (Warren-Hicks et al. 1989). Statistical comparisons of data from the nontidal, the near-field tidal marsh and far-field tidal marsh grids allow for spatial comparisons of the different measurement endpoints relative to their distances from the site. This provides a basis for an internally controlled evaluation of the spatial effects of the site. Data on these three zones can also be compared to the remote reference sites. This strategy provides two independent methods for evaluating the broad scale effects of the site on the adjacent wetlands.

A second approach for evaluating the impact of the site on the adjacent tidal wetlands can be used in parallel with the random sampling approach described above. In this approach, the sampling focuses on the potential route by which onsite stressors may be transported offsite and into the tidal wetlands. In this nonrandom sampling strategy, a series of stations can be established in the major drainage slough to provide a spatial gradient of contamination moving from the site to the bay.

The types of sample measurements that are commonly taken at stations such as those described should include concentrations of contaminants in the sediments, soils and surface waters. Measurements of actual concentrations of the stressors in plants and animals are also helpful. The combined database from all of the sites will provide the basis for interpreting the extent and magnitude of the stressors on the site and in the adjacent wetlands.

The exposure characterization discussion thus far has been limited to determining the nature and distribution of stressors on the site and neighboring wetlands. An equally important component of this phase of the ERA is to evaluate the potential routes by which organisms may be exposed to the stressor. In the example, routes of potential exposure include direct exposure in the sediments and

Figure 5. For hazardous waste sites, initial sampling commonly requires extra protection for team personnel.

surface waters and indirect exposure via the food chain. Soils and sediments commonly serve as sinks for chemical stressors and thus will need to be examined closely. The bioavailability of stressors sequestered in the sediments should also be evaluated, as this could have a profound effect on the actual ecological effects observed. The degree to which the stressor is bioaccumulated or depurated is also very important to the evaluation of actual exposure.

Finally, the temporal aspects of the stressors must be considered. For example, the standing waters of the nontidal wetland would likely recede during the dry season. Marsh plants commonly exhibit a rapid growth period at the end of the wet season and continue this growth into the dry season. As available water diminishes, plant growth subsides, and eventually the plants senesce and die. These and other seasonal changes could modify stressor distribution and bioavailability. Therefore, it is often necessary to perform the ecological assessments during both the growing and nongrowing seasons.

Some inherent complicating factors must be addressed in order to establish cause-effect relationships between stressors in the environment and observed changes in biological endpoints. For example, in highly commercial areas, extensive anthropogenic activities can result in numerous stressors from multiple sources. Moreover, these stressors may have differential mobility within ecosystems and the ratios of one stressor to another may change with distance from the source. This complexity makes it difficult to relate any observed biological effects to specific stressors or to a specific source. The spatial and temporal sampling schemes described above should provide a basis for determining if the concentrations of chemical contaminants and biological parameters in the wetlands show significant correlation with one another, and specific spatial relationships to the site.

Ecological Effects Characterization

To accomplish the retrospective ecological assessment, it is necessary to determine if the stressors from the site have produced adverse effects on the structure and function of the adjacent wetland ecosystems. In the predictive ERA example, the emphasis was on performing laboratory studies to determine potential ecological effects, followed by increasingly higher tier modeling or field dissipation studies if warranted by the lab data. Conversely, with the retrospective ERA, the emphasis is on determining actual ecological effects on the site and in the adjacent wetlands. This is accomplished by emphasizing field evaluations of community structure and the toxicity of soils, sediments and surface waters. It is important to coordinate the ecological effects sampling with the exposure sample collection on both a spatial and temporal scale. In other words, samples for chemical analysis and ecological effects should be taken at the same time and place whenever possible. Examples of the types of ecological effects sample measurements commonly taken are provided in Table 2.

As can be seen in Table 2, the emphasis is on measuring actual effects in the field whenever possible. Nevertheless, laboratory toxicity tests can play an important role in the evaluation of wetlands water and sediments affected by hazardous waste sites (Zimmer et al. 1988, Khan et al. 1993, Woodward et al. 1988) (Figure 6). The toxicity of field collected samples is used to define the extent of contamination with regard to direct biological significance. Onsite studies such as cage studies may also provide valuable information in the delineation of ecological effects.

Table 2. Retrospective ecological risk assessment (ERA) emphasizes field evaluations. Typical ecological effects sample measurements are listed below, and should be collected at the same time and place as chemical measurements.

- Bioaccumulation of contaminants in dominant plant species
- Bioaccumulation of contaminants in dominant benthic invertebrate species
- Bioaccumulation of contaminants in small mammals
- Bioaccumulation of contaminants in dominant fish species
- Sediment toxicity (pore water, elutriates or solid phases)
- Water toxicity
- Plant population and community structure within each grid
- Small mammal population survey for each grid
- Benthic infaunal invertebrate population and community structure
- Histopathological evaluation of native small mammals or fish
- Fish population survey

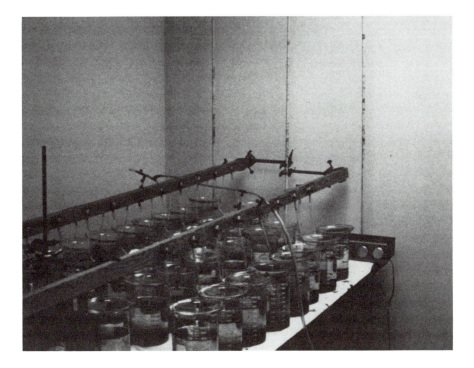

Figure 6. Ecological effects characterization emphasizes measurement of actual effects in the field. Nevertheless, laboratory toxicity tests, such as this sediment test, can play an important role.

Measurement endpoints should be chosen for relevant characteristics. Endpoints may be key ecosystem components which are sensitive to stressors, accumulate the stressor of concern, or are food resources for higher trophic levels. Other endpoints include key indicators of ecosystem structure and function, or surrogates for stressor impacts to endangered species or other assessment endpoints that can not be directly monitored. Species continuously exposed to contaminated sediment or water have the greatest potential for contaminant uptake and serve as excellent indicators of bioaccumulation potential. In wetland sites, marsh plants, benthic invertebrates and filter-feeding mollusks are particularly useful indicators of bioavailability. Transfer of contaminants along the food chain should also be considered.

The biological effects of accumulated contaminants are also complex. Contaminants may be stored in an inactive form, metabolized or excreted. Alternatively, they can interact with cellular macromolecules causing metabolic perturbations and cellular damage that impact the function of the organism. Thus, while bioaccumulation of contaminants reflects the potential for toxicity, supplemental approaches with greater resolution are needed to determine actual toxicity to the organism.

Sublethal impacts on the organism can affect higher levels of biological organization through inhibition of such physiological processes as growth and reproduction. Changes in these parameters can impinge upon populations and communities. In addition, if the marsh plant populations are adversely affected, the ecological impact could result in habitat disturbance for birds and small mammals, including endangered species. As in the previous cases, impact at these higher levels of organization can be inferred or measured directly.

Risk Characterization

Comparison of the exposure and ecological effects characterizations is performed to provide an assessment of risk to both the onsite nontidal wetland and the tidal estuarine wetland adjacent to the site. Because of the complexity of the linkages between contaminant distribution and ecosystem effects, the ecological assessment must be designed to carefully examine the correlations between the concentrations of contaminants in sediment, soil and water and biological function observed at any of several levels of biological organization. The collection of matched samples for measuring contaminant concentrations in the sediments, bioaccumulation, toxicity, and population and community status is important to allow a rigorous examination of the relationships among these endpoints.

The specific procedures used in the risk characterization phase will depend on the design chosen to determine exposure and ecological effects, as well as the overall purpose for performing the ERA. In the retrospective example, the primary objective is to assess the degree of risk that contamination from a hazardous waste site will have on the on-site and adjacent wetland ecosystems. Data collected includes chemical concentrations in the sediments and surface waters, bioaccumulation of the stressors in dominant plant and animal species, sediment and surface water toxicity, fish and wildlife population surveys, and plant and benthic invertebrate population and community structure.

Initially, the stressor concentration data can be tabulated and plotted on both the spatial and temporal scales to evaluate trends in concentration. Are the concentrations higher near the site and lower away from the site? Do concentrations appear to follow a gradient associated with one or more transport routes away from the site? Has the stressor partitioned into the sediments or has it been dispersed for some distance? Furthermore, how do these concentrations relate to water quality standards and criteria established for wetlands (USEPA 1990)? Similarly, the sediment and surface water toxicity data should be tabulated and plotted to show spatial and temporal trends. Are any trends evident and, if so, do they correlate with the stressor concentration trends? Bioavailability can be established by comparison of contaminant levels in the abiotic matrices and biota exposed to these matrices.

Ecological effects on the population and community level can be evaluated using several statistical procedures designed for these types of data (Ludwig and Reynolds 1988). Species diversity, richness, and abundance can be compared using ANOVA and cluster analyses to determine if differences in these parameters can be correlated with stressor concentration.

After comparing all of the collected data, a picture should begin to emerge regarding the levels of exposure and the degree of effect. If the picture is sufficiently muddled because of extensive anthropogenic inputs into the systems under investigation, techniques such as Principal Components Analysis may be used to help define the key contaminants associated with a specific effect. In addition, fate and transport processes can often be modeled based on data collected in the field and the physicochemical characteristics of the stressors of concern. Ultimately, the actual assignment of ecological risk will often depend on the informed opinion of the risk assessor.

SUMMARY

ERA is a rapidly developing and increasingly important discipline which can be used as a management tool in the protection of wetlands. It is a process which estimates the impact of anthropogenic activities on ecosystems. The results of ERAs are used by managers and regulators to determine the appropriate action for the wetland or site in question. The application of the ERA paradigm as described provides a framework for assessing risk to ecological resources for a variety of scientific and regulatory purposes. The principles of ERA can be applied to a variety of situations including retrospective analysis of anthropogenic activities, and predictive assessment of future actions.

Additional definition of the principles and procedures of ERA is needed to help risk assessors standardize their methods as much as possible. With this objective in mind, the American Society for Testing and Materials (ASTM) has undertaken an effort to develop standards for the assessment and valuation of wetlands (Ethier 1993). The use of a standard approach with standard terminology will facilitate communication and interpretation of results and concepts. Furthermore, this standard approach will allow for balanced risk management of human health and environmental concerns under a variety of regulatory programs. However, it is important to understand that even with increased standardization of the ERA framework, ERAs will still need to be designed on a site- or chemical-specific basis to address the specific concerns.

REFERENCES

Adamus, P.R., E.J. Clairain, Jr., R.D. Smith and R.E. Young. 1987. Wetland Evaluation Technique (WET), Volume II: Methodology. Department of the Army, Waterways Experiment Station, Vicksburg, MS.

Barnthouse, L.W., S.M. Bartell, D.L. DeAngelis, R.H. Gardner, R.V. O'Neill, D.D. Powers, G.W. Suter, G.P. Thompson and D.S. Vaughan. 1982a. Preliminary Environmental Risk Analysis for Indirect Coal Liquefaction. Draft Report. Oak Ridge National Laboratory, Oak Ridge, TN.

Barnthouse, L.W., D.L. DeAngelis, R.H. Gardner, R.V. O'Neill, C.D. Powers, G.W. Suter and D.S. Vaughan. 1982b. Methodology for Environmental Risk Analysis. ORNL/TM 8167. Oak Ridge National Laboratory, Oak Ridge, TN.

Barnthouse, L.W., G.W. Suter, S.M. Bartell, J.J. Beauchamp, R.H. Gardner, E. Linder, R.V. O'Neill and A.E. Rosen. 1986. User's Manual for Ecological Risk Assessment. Publication Number 2679, ORNL-6251. Environmental Sciences Division, Oak Ridge National Laboratory, Oak Ridge, TN.

Bartell, S.M., R.H. Gardner and R.V. O'Neill. 1992. Ecological Risk Estimation. Lewis Publishers, Chelsea, Michigan.

Burns, L.A., D.M. Cline and R.R. Lassiter. 1982. Exposure analysis modeling system (EXAMS): User manual and system documentation. EPA/600/3-82/023.

Cairns, J., Jr., B.R. Niederlehner and D.R. Orvos (eds). 1992. Predicting Ecosystem Risk. Princeton Scientific Publishing Co., Inc., Princeton, New Jersey.

Ethier, W.H. 1993. New wetlands standards: Providing great tools for policy makers. Standardization News. 21:26-29.

Fisher, L. 1992. Memo re: Decisions on the Ecological, Fate and Effects Task Force. Program Guidance on Ecological Risk Management.

Khan, A., D.J. Kent and S. Khan. 1993. Freshwater sediment hazard evaluation. In review for Water and Science Technology.

Ludwig, J.A. and J.F. Reynolds. 1988. Statistical Ecology. John Wiley & Sons, Inc., New York. 337 pp.

McKim, J.M., S.P. Bradbury and G.J. Niemi. 1987. Fish acute toxicity syndromes and their use in the QSAR approach to hazard assessment. Environmental Health Perspectives. 71:171-186.

Mitsch, W.J. and J.G. Gosselink. 1986. Wetlands. Van Nostrand Reinhold, New York.

National Research Council. 1983. Risk assessment in the federal government: managing the process. National Research Council. National Academy Press, Washington, DC.

Pascoe, G.A., R.J. Blanchet, G. Linder and C.G. Ingersoll. 1993. Assessment of ecological risks of contaminated wetlands: A Superfund case study. Presented at the First SETAC World Congress, Lisbon, Portugal, March, 1993.

Superfund Amendments and Reauthorization Act. 1987. Federal Register 52: 13378-13410. April 22.

Suter, G.W. 1993. Ecological Risk Assessment. Lewis Publishers, Chelsea, Michigan. 538 pp.

Urban, D.J. and N.J. Cook. 1986. Hazard Evaluation, Standard Evaluation Procedure, Ecological Risk Assessment. EPA 540/9-85-001.

U.S. Environmental Protection Agency. 1982a. Pesticide Assessment Guidelines, Subdivision E, Hazard Evaluation: Wildlife and Aquatic Organisms. EPA 540/9-82-024.

U.S. Environmental Protection Agency. 1982b. Pesticide Assessment Guidelines, Subdivision N, Chemistry: Environmental Fate. EPA 540/9-82-021.

U.S. Environmental Protection Agency (USEPA). 1986a. Guidelines for Carcinogen Risk Assessment. Federal Register. 51:33992-34002. September 24.

U.S. Environmental Protection Agency. 1986b. Guidelines for Mutagenicity Risk Assessment. Federal Register 51:34006-34012. September 24.

U.S. Environmental Protection Agency. 1986c. Guidelines for Health Risk Assessment of Chemical Mixtures. Federal Register. 51:34014-34025. September 24.

U.S. Environmental Protection Agency. 1986d. Guidelines for Health Assessment of Suspect Developmental Toxicants. Federal Register. 51:34028-34040. September 24.

U.S. Environmental Protection Agency. 1986e. Guidelines for Exposure Assessment. Federal Register 51:34042-34054. September 24.

U.S. Environmental Protection Agency. 1988. Review of Ecological Risk Assessment Methods. EPA/230-10-88-041.

U.S. Environmental Protection Agency. 1989a. Rapid Bioassessment Protocols for Use in Streams and Rivers: Benthic Macroinvertebrates and Fish. EPA/444/4-89-001.

U.S. Environmental Protection Agency. 1989b. Risk Assessment Guidance for Superfund, Volume I: Human Health Evaluation Manual (Part A). EPA/540/1-89/002.

U.S. Environmental Protection Agency. 1989c. Risk Assessment Guidance for Superfund, Volume II: Environmental Evaluation Manual.

U.S. Environmental Protection Agency. 1990. Water Quality Standards for Wetlands: National Guidance. EPA 440/S-90-011.

U.S. Environmental Protection Agency. 1991. Summary Report on Issues in Ecological Risk Assessment. EPA/625/3-91/018.

U.S. Environmental Protection Agency. 1992. Framework for Ecological Risk Assessment. EPA/630/R-92/001.

Warren-Hicks, W., B.J. Parkhurst and S.S. Baker, Jr. 1989. Ecological Assessment of Hazardous Waste Sites: A Field And Laboratory Reference Manual. EPA/600/3/89/013.

Woodward, D.F., E. Snyder-Conn, R.G. Riley and T.T. Garland. 1988. Drilling fluids and the arctic tundra of Alaska, U.S.A.: Assessing contamination of wetlands habitat and the toxicity to aquatic invertebrates and fish. Archives of Environmental Contamination and Toxicology 17:683-697.

Zimmer, R.D., G. Buchanun, D. Charters, J. Ferretti and D.J. Kent. 1988. Aquatic toxicological evaluation of soils collected from a mid-western superfund site. Presented at the 61st Annual Conference of the Water Pollution Control Federation, Dallas, TX.

World Wildlife Fund. 1992. Improving aquatic risk assessment under FIFRA: Report of the Aquatic Effects Dialogue Group. Washington, D.C.

CHAPTER 6

WETLANDS AVOIDANCE AND IMPACT MINIMIZATION

Kevin McManus

U.S. Federal Policy toward preservation of remaining wetland habitats has gradually emerged over the past two decades, beginning with the passage of the National Environmental Policy Act of 1969 (U.S. Congress 1969), the Clean Water Act in 1972 (U.S. Congress 1972), and Executive Orders 11998 and 11990 on federal management of floodplains and wetlands (Executive Office of the President). The unprecedented and rapid destruction of wetlands (over 11 million acres lost between the mid-1950s and mid-1970s, Office of Technology Assessment 1984) for land reclamation, flood control, residential and other types of development prompted a number of federal and state initiatives to protect and preserve remaining wetland habitats. At the same time, research on the nation's remaining wetlands habitats provided strong scientific evidence of wetlands ecological importance and the accelerated rates of wetland losses.

The broad federal policy goal expressed within federal executive and legislative actions involves prevention of *unnecessary alteration and destruction of [wetlands]. as these actions are considered ..contrary to the public interest* (U.S. Army Corps of Engineers 1986). This general policy goal requires any project developer (including federal and state agencies) to carefully consider alternative sites, designs, and construction methods which will avoid direct wetlands impacts or minimize the extent of wetland disturbance.

The cornerstone federal legislation for protection of wetlands and other sensitive ecological resources is the National Environmental Policy Act (NEPA) of 1969. This statute requires all federally-proposed, -funded, or -licensed activities which have potentially significant environmental impacts to undergo a thorough environmental evaluation. The guiding principle behind NEPA is to encourage federal agency decision-making, including the U.S. Army Corps of Engineers and Environmental Protection Agency's 404 permit reviews, to consider a reasonable range of alternatives to a proposed action, and avoid, where possible, adverse environmental impacts. Thus, even actions which are not federally funded, but require federal wetlands permits and other approvals, are required to undergo a rigorous alternatives analysis in order to receive necessary authorizations.

NEPA regulations at 40 CFR Part 1508.20 outline the fundamental guidance to be used by federal agencies in considering mitigation for project impacts. The so-called step-down mitigation approach sets out priorities for all project development and planning. In order of importance, the steps to be followed include:

- Avoiding the impact by not taking a certain activity or portion of the proposed action
- Minimizing the impact by limiting the degree or magnitude of the action and its implementation
- Rectifying the impact by repairing, rehabilitating, or restoring the affected environment
- Reducing or eliminating the impact over time by preservation and maintenance operations during the life of the action
- Compensating for the impact by replacing or providing substitute resources or environments

This step-down approach clearly places great emphasis on avoidance and minimization of wetlands impacts. Some of the more commonly cited examples of wetlands mitigation (e.g. wetlands restoration or creation, mitigation banking) are considered by regulatory agencies as last resorts, to be used only if wetlands impacts cannot be avoided or reduced. Inherent in this approach is the belief that a naturally-created, currently functioning wetland is providing a combination of well-known values to the ecosystem, and that creation of wetlands is a relatively new, unproven technique which is only to be used where wetland avoidance is not possible. Thus, project development which does not carefully consider avoidance and minimization, and relies instead upon compensation, will very likely encounter stiff agency resistance.

The Clean Water Act (33 U.S.C. 1125) and its implementing regulations provide additional procedural and technical requirements to justify and document the need for wetlands disturbance. The CWA regulations at 33 CFR Section 320.4 (U.S. Army Corps of Engineers 1986) indicate that one of the key criteria in the Corp's decision to approve a project is ..*the practicability of using reasonable alternative methods and locations to accomplish the objective of the proposed structure or work*. In addition, the Environmental Protection Agency's Section 404(b)(1) guidelines at 40 CFR Part 230 place great weight on the need to ...*examine practicable alternatives to the proposed discharge..[which would result in]..potentially less damaging consequences* (U.S. Environmental Protection Agency 1980). Where wetlands impacts are unavoidable, EPA's 404(b)(1) guidelines (Subpart H) provide specific guidance on actions which can be undertaken to minimize the potential adverse impacts on wetlands. These are discussed in more detail below.

Two executive orders, issued in May 1977, reinforce federal policy toward effective management of wetlands and floodplain resources (Executive Office of the President 1977). Executive Order 11990, Protection of Wetlands, requires all Federal agencies to avoid undertaking or assisting new construction in wetlands, unless no practicable alternatives exist; in addition, all practical measures are to be taken to minimize harm to wetlands. Executive Order 11988, Flood Plain Management, applies similar requirements for avoidance and alternatives analysis for any federally proposed or assisted floodplain development. While these orders do not apply to federal permit decisions, and are considered by some to be of limited effectiveness (Office of Technology Assessment 1984), they do supplement and restate a federally-defined policy to avoid unnecessary development within these resource areas.

In addition to federal mandates, there are a wide variety of state attitudes, and hence regulatory programs, guiding wetlands protection. The Clean Water Act encourages states to develop regulatory programs to assume 404 responsibility for all areas outside traditionally navigable waters, assuming certain minimum Environmental Protection Agency requirements are met. In practice, few states have either the capability or resources to assume control over the 404 program without substantial federal support (Office of Technology Assessment 1984).

Many states, such as Massachusetts, have developed independent and unique wetlands protection statutes which bear little resemblance to the U.S. Army Corps of Engineers' Section 404 program. Others, such as New Jersey and Michigan, have developed programs which closely mirror the Corps' Section 404 program. In general, however, the state programs incorporate many of the key elements of federal wetlands policy. Most contain the strong presumption that wetlands impacts are to be avoided wherever practicable, and that mitigation of unavoidable impacts is a necessary adjunct to project planning.

In addition, several Corps districts specifically include state agencies as part of the joint processing procedures for 404 permit applications, and thus state agencies are closely involved in many Corps permit decisions. Some states, such as North Carolina and New Jersey, have worked closely with the Corps to develop statewide general permit programs, wherein entire categories of actions which meet Corps nationwide permit conditions are jointly processed by the states and the Corps. With this type of federal and state interaction, the federal focus on avoidance and minimization has extended to wetlands permit processing at all levels of government.

The following sections describe in detail some of the key factors used by Corps and state agencies to determine compliance with these broad policy goals, and some recommended techniques for meeting both federal and state agency documentation requirements.

AVOIDANCE OF IMPACTS

In order for a reviewing agency to adopt or concur with a proponent's proposed action and the range of available alternatives, they must first understand the project's intended purpose and need. For example, what functions are being provided by the project, and what private and public benefits are served by the facility? Thus, the determination and effective documentation of a project's main components, and purpose and need, are critical first steps in the application process. For example, a waterfront project whose intended purpose is simply to provide additional commercial, residential and public space is less likely to be viewed favorably than a waterfront project whose stated purpose is to provide public access to XYZ Harbor (including park land and marina space), stabilize an existing eroding bank, and encourage additional water-dependent maritime uses of XYZ Harbor.

The test of a project's water dependency, that is, whether it has to be located on or near water or wetlands to accomplish its stated purpose, is also a critical threshold for any proposed activity. For example a project which proposes to disturb wetlands in order to build a garage is less likely to be permitted than a project which involves construction of a boat launching ramp (even if wetland impacts are greater), because the ramp, unlike the garage, cannot be moved out of jurisdictional areas and accomplish its stated purpose.

Site Selection

Because avoidance is considered by the agencies to be the most effective form of mitigation, virtually all projects requiring wetlands permits must provide documentation of a reasoned, good-faith effort to select a site which meets the Corps definition of the Least Environmentally Damaging Practicable Alternative (LEDPA). Thus, each permit application should provide a detailed description of the site selection process followed in order to select the location for the proposed action.

The level of detail will vary widely for individual projects, depending upon the range of alternatives available to the proponent. For example, a large state- or federally-funded project such as a pipeline or roadway will generally have several alternative alignments or sites under consideration. A private landowner seeking permission to construct a pier or dock facility, has far less flexibility in selecting a site. However, regardless of the geographic extent of the available universe of sites, the proponent must demonstrate to the agencies' satisfaction that a less environmentally damaging feasible alternative was not overlooked.

It is important to consider potential access to the proposed activity as well, as the Corps and state agencies will look at the project as a whole. For example, construction of a road through wetlands, as a means of access to a site which is predominantly in upland areas, must itself be the only practicable alternative means of access. Even in situations like road crossings or utility line installation in wetlands, where the Corps has developed nationwide permits to allow for insignificant project impacts, in most cases individual state water quality certification regulations (under Section 401 of the Clean Water Act) stipulate that certification not be provided if less damaging alternatives exist.

Site selection documentation should include all relevant environmental and other factors which are used to identify the preferred site. These factors or criteria should include both economic and environmental concerns. A central element in the Corps' permit review process, the Public Interest Review, specifically requires the agency to conduct *...a careful weighing of all those factors which become relevant in each particular case* (U.S. Army Corps of Engineers 1986). Typical site selection criteria may include:

- **Site size requirements:** What is the minimum size required to meet the project's purpose and need? If possible, back up selection requirements with an economic analysis demonstrating the financial break-even point for the facility (if applicable).
- **Constructability of sites:** What are the relevant topographic, slope, soil foundation and backfill requirements for the facility? The extent to which additional grading, blasting, or use of imported fill is required can be an important siting consideration in ruling out other potentially suitable sites. Note that these constructability issues all have important environmental implications. For example, additional grading means increased erosion control requirements. Blasting can also have significant short-term environmental and human impacts.
- **Development costs:** What are the total costs of project development at alternative sites (including site acquisition, planning, permitting, design and construction)? It is important to note that the Corps is specifically authorized to include economics as part of their public interest review process (U.S. Army Corps of Engineers 1986). However, this criteria should not be used to automatically screen out all but the least expensive alternative. Only those sites which clearly involve cost-prohibitive development costs should be eliminated from further consideration.
- **Regulatory/Institutional issues:** What other state, federal and local permits and authorizations are required for development of alternative sites? As an example, a site which would require a variance from local floodplain regulations, or a change in zoning, may rank lower than other alternatives. The extent to which existing land use classifications are affected is a legitimate focus of the public interest review process.
- **Proximity to support facilities:** What utilities such as sewer, water, and power, are needed to operate the facilities? Are there other support facilities such as public parking and roadways which are necessary to site the facility?
- **Extent of wetlands impacts:** What are the direct and indirect wetlands impacts at alternative sites? In addition, what types (and functional values) of wetlands would be disturbed at alternative sites?
- **Other environmental impacts:** What other known environmental resources are at or adjacent to the alternative sites? These relevant resources may include fish and wildlife resources (including threatened and endangered species and their habitats), federal or state navigation

channels, public recreation areas, incompatible land uses, sensitive noise or visual impact areas, existing roadways, listed or potentially-significant cultural and historic resources, and designated floodway velocity zones and 100-year floodplains.

For large or potentially controversial projects, it may be useful to summarize the site selection process which was followed in order to select a suitable site. Figure 1 is an example of a matrix which can be used to present and evaluate alternative sites in terms of key selection criteria. Graphic overlays, indicating specific areas to be avoided during site selection (e.g. floodplains, critical wildlife habitats, sensitive receptors) can also be effective means of documenting the efforts employed to avoid environmental impacts.

IMPACT MINIMIZATION

Once the site selection process has been completed, and the LEDPA has been identified, the project focus can shift to design details, site layouts, construction methods, and other specific engineering requirements to minimize, as fully as possible, those unavoidable wetlands impacts associated with project construction. The specific project design elements can be critical in determining whether a project can be permitted and whether it is eligible for permitting under less rigorous requirements. Examples of the latter include the Corps' Nationwide Permit Program state general permit programs. There are no absolute thresholds within the Corps' regulations regarding acceptable versus unacceptable wetlands fills. Site-specific conditions, project purpose and benefit, as well as geographic location all contribute to a project's acceptability. Nevertheless, whenever possible, projects should be designed within certain parameters which allow for permitting under the Nationwide Permit Program (U.S. Army Corps of Engineers 1991). Some examples of key nationwide permit thresholds, which if met will reduce the complexity of permitting, are presented in Table 1. Some of the key design elements which should be addressed in order to minimize potential impacts are described in the following sections.

Project Size and Scope

The project proponent must make a reasoned assessment of the minimum economically and functionally viable size for a proposed structure if wetlands impacts are unavoidable. Particularly for projects which are not water-dependent, such as residential structures or parking lots, the applicant must be prepared to present supporting documentation which demonstrates that the structure or fill cannot be reduced beyond its proposed size and still meet the basic project purpose. Even water-dependent projects, such as marina facilities or dredging projects, must be ready to withstand agency objections and recommendations on reduction of project scope in order to obtain wetlands approvals.

SCREENING CRITERIA	SITE 1	SITE 2	SITE 3 (Preferred Site)
ENVIRONMENTAL			
Aquatic Ecosystem – Physical/Chemical			
Substrate	–	–	0
Water Quality	0	0	0
Water Circulation	0	0	0
Normal water fluctuations	–	0	0
Aquatic Ecosystem – Biological			
Threatened and Endangered Species	0	0	0
Other aquatic organisms	0	0	0
Other wildlife	0	0	0
Special Aquatic Sites			
Sanctuaries/Refuges	0	0	0
Wetlands	–	0	0
Mudflats	–	–	0
Vegetated Shallows	–	–	0
Riffle and Pool Complexes	0	0	0
HUMAN USE			
Municipal and Private Water Supplies	0	0	0
Recreational and Commercial Fisheries	0	0	0
Water–related recreation	0	0	0
Aesthetics	–	–	+
Parks, Preserves, Wilderness Areas	–	0	0
Archaeological or Historical Sites	0	0	0
Compatibility with Adjacent Land Uses	–	0	0
Potential Noise Impacts	–	–	0
Potential Odor Impacts	–	–	0
Property Values	–	0	0
Public Health	0	0	0
Traffic Increase (over existing levels)	–	0	0
TECHNICAL			
Suitable Foundation/Soils Conditions	–	+	+
Adequate Land Area	0	0	+
Access to Existing Roads and Utilities	+	0	o
ECONOMIC			
Land Acquisition	0	0	0
Operation & Maintenance	+	–	+
Capital cost – construction	0	0	0
INSTITUTIONAL			
Public Acceptance	–	–	+
Compliance with Existing Zoning	–	–	+
Need for Permit Variance/Exemption	0	–	o

RATING SYSTEM: + = positive impact
          – = negative impact
          0 = insignificant, or no
              impact perceived

Figure 1. An example of a site selection matrix

Table 1. Examples of Key Nationwide Permit Thresholds

Permit #	Permit	Threshold
3	Maintenance	o Structure or fill not put to different uses
12	Utility Line Backfill/Bedding	o No change in preconstruction contours
13	Bank Stabilization	o Activity less than 500 feet in length
		o Activity will not exceed 1 cubic yard/running foot
14	Road Crossing (1)	o Width of fill limited to minimum necessary
		o Fill in U.S. Waters limited to 1/3 acre
		o Fill in Wetlands less than 200 linear feet
18	Minor Discharges	o Discharge does not exceed 25 cubic yards
		o No more than 1/10 acre loss in wetlands
19	Minor Dredging	o No more than 25 cubic yards below high water mark
26	Headwaters and Isolated Discharges (1)	o No losses of wetlands greater than 10 acres
		o Losses below one acre do not require notification
36	Boat Ramps	o Discharge of fill does not exceed 50 cubic yards
		o Boat ramp does not exceed 20 feet in width

(1) Individual ACOE Districts have discretionary authority to enact more stringent thresholds

Source: U.S. Army Corps of Engineers, Nationwide Permit Program Regulations (33 CFR Part 330)
         Issued November 12, 1991 (effective January 21, 1992)

In the past, developers relied upon the use of pilings, rather than solid fills, in order to reduce the size and scope of wetlands impacts. However, recent guidance from the Army Corps of Engineers indicates that structures which are not, by their basic nature, required to be pile-supported (such as buildings, parking areas, etc.) cannot utilize piles in order to reduce the size and scope of wetlands impacts (U.S. Army Corps of Engineers 1990).

For projects which may involve clearing of trees and other existing vegetation, care should be taken to minimize the limits of clearing to the minimum acreage needed for the project. Maintenance of existing vegetative buffers, particularly within wetlands areas, is not only a valuable means of providing a visual and auditory buffer for the facility, but it also may reduce overall facility wetlands impacts. Because many state regulations specify the importance of setbacks or buffers from delineated wetlands areas, clearing limits should be established to reduce potential permitting issues within this buffer zone. In addition, many state coastal zone management policies specify additional setback requirements from coastal resource areas, which limit the ability to clear and alter existing slope and grade conditions along shorelines. This is particularly true along active coastal shorelines, such as eroding bluffs, beaches and dune environments.

Project Configuration and Layout

The orientation and layout of a project is generally a function of its intended purpose and use. Many projects, such as railways, roads and retaining walls, being linear features, have limited flexibility with regard to basic configuration. However, their actual alignment, relative to wetland areas, can often be optimized to reduce impacts to insignificant levels. Similarly, layouts of buildings and ancillary structures such as garages, walkways, and decks can be adjusted to minimize direct wetlands impacts.

It is important that the project proponent have an accurate wetlands delineation line depicted on site plans prior to development of a final facility configuration in order to evaluate layout options. Developing alternative facility footprints for the project site is often a necessary prerequisite for justification of the final site layout. This is not merely an academic exercise, as regulatory agencies will often take the proponent's facility layout and rotate, shift, and otherwise tweak the proposed location in order to identify the least impacts to wetlands. It is thus in the proponent's best interest to explore these options to site configuration prior to submittal of an application so that agency recommendations or inquiries can be addressed in an expeditious manner.

Project Design Details

The specific design details for a project can also be important factors in reduction of wetlands impacts. For example, the proponent should determine the maximum safe slopes which can be used for site preparation. Because the toe of slope can extend well beyond the area occupied by a facility, an applicant must be able to demonstrate that the slopes have been designed to maximum safety standards, while minimizing incursions into wetland areas. The feasibility of using alternate foundations, such as vertical retaining walls, sheet piling or gabion rock walls in order to reduce impacts should be considered by the design team.

Any proposed crossing of a water body must provide for adequate capacity to withstand 100-year or similar flood events without significant constrictions of flow. While culverts are often used to provide this drainage capacity, their placement and operation can lead to direct losses of additional wetlands, significant effects to existing flow velocities, and increased erosion and scour potential. Thus, agencies often recommend use of bridges to span water bodies in order to reduce direct wetlands impacts and hydrologic changes. The project proponent should evaluate the cost effectiveness of a bridge or span for a particular crossing.

Project Construction Techniques

For many projects, such as construction of subsurface water, sewage, utility and other pipelines, the primary impacts to wetlands are temporary, occurring only during construction of the project. While regulatory agencies generally recognize the difference between short-term, reversible wetlands impacts, and permanent, irreversible impacts, authorization to construct in wetlands will often require specialized construction techniques and restrictions. The use of temporary access materials, specialized construction equipment, and the placement of staging areas can all affect the level of wetlands impacts.

In order to construct through many ecologically-sensitive wetland areas such as estuarine and freshwater marshes, beach or dune environments, and peat bogs, temporary pile-supported construction trestles can be used to significantly reduce direct wetlands impacts (Figure 2). These trestles may be located either directly above the area of work, or directly next to the work area. Equipment can be brought to the work area using rail-mounted transport platforms, and the trestle can be constructed in stages to accommodate the construction schedule. These trestles provide a stable temporary work platform which directly impacts little additional wetland acreage.

Another possible method for construction within these environments involves the use of steel sheet piling to isolate the active work area, and the use of temporary wood decking placed directly on top of the sheet piling, to allow equipment to access the work areas. For work within intertidal or shallow water, this method has the added advantage of isolating the active work area, and when combined with siltation curtains, prevents the release of sediment-laden water outside the work area.

For smaller projects which do not warrant use of sheet piling, geotextile fabric, clean granular material, and wood decking can be placed within the project alignment in order to gain access to work areas and minimize impacts to surrounding wetlands. However, with this method, care should be taken to minimize compaction of wetlands soils, as they can lose their original productivity and hydrologic functions if compacted by construction equipment. For work within intertidal areas, the proponent should consider the use of barge-mounted equipment to access sensitive areas. Work barges can be floated into place on rising tides and grounded out to provide suitable access with minimal or no long-term impacts.

All of these methods allow construction to progress through sensitive habitats at a reasonable rate without additional permanent impacts to surrounding wetlands. They are likely to be suggested by regulatory agencies, and have become standard operating practice for work in wetlands in many parts of the United States (Council on Environmental Quality 1980).

For construction of trenches in wetland areas, utilities have developed specialized tracked trenching vehicles which can operate on soft, unstable soils, and which work directly within the proposed project alignment. Other methods which have been developed involve use of wide, low pressure tires on vehicles to distribute the loads on wetlands soils and vegetation. For dredging within wetlands and U.S. waters, clamshell dredge equipment has been fitted with closeable covers and watertight buckets in order to minimize sediment washout and turbidity impacts during dredging operations. As this equipment is being continually refined and improved, it is important that the project design team remain up to date on recent developments in this field so that they can effectively respond to agency recommendations.

PLAN VIEW

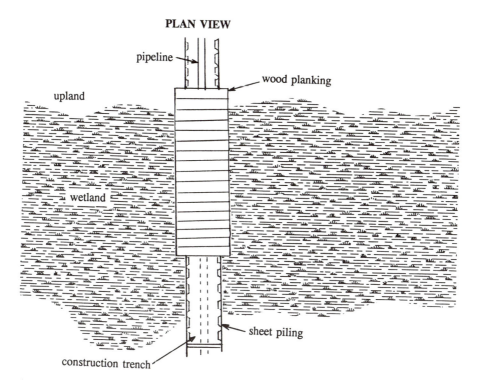

PROFILE

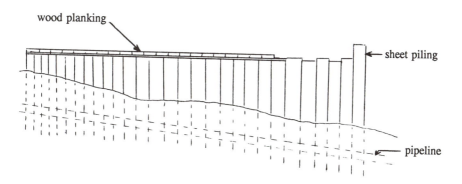

Figure 2. Temporary pile-supported construction trestles can be used to significantly reduce wetland impacts. The trestles may be located either directly above or immediately adjacent to the work area.

Clearly, large construction projects require several different types of equipment, and care must be taken to locate staging areas outside wetlands and their designated buffer zones, and to use paved areas to minimize erosion and groundwater impacts. In addition, they should be sized with adequate stormwater management systems to contain suspended sediments from construction equipment, and to handle accidental releases of fuel oil, lubricants and other potentially hazardous releases from equipment. If discharges are directed into wetland areas, they should first be directed to temporary settling basins (discussed below) to allow physical separation of sediments and contaminants from the collected runoff.

Project Construction Schedule and Sequencing

For projects with temporary, construction-related impacts, proper construction scheduling can help to minimize adverse impacts to wetland habitats and their associated biota. As a general rule, work within wetland habitats should be scheduled in temperate climates during winter and early spring when plants are still dormant and the soils are still frozen or well-consolidated. This allows equipment and temporary structures to reduce the amount of soil compaction, and allows site cleanup and rehabilitation during the coming peak growing season. As an example, in the northeast United States, impacts to estuarine salt marsh areas are generally limited by state and federal permitting agencies to the period of November 1 to April 1 when marsh productivity is lowest and plants are dormant.

Other seasonal restrictions are often applied for work within coastal environments, based upon the expected presence of commercially and recreationally important fish and wildlife species. In the Pacific Northwest, for example, work is often restricted in rivers and streams supporting key anadromous fish species, such as salmon and trout, during their prime spawning and outmigration periods. In the Southeast, work may be restricted in estuarine areas to avoid peak spawning and migratory cycles for several species of shrimp. In the Northeast, work in wetland is generally restricted during the late winter and spring, which are peak periods of migration for anadromous species such as alewife, shad, and smelt. In many areas of New England, work is also restricted during the late-winter and spring periods (March to June) to reduce impacts to overwintering populations of groundfish, such as flounder (pers. comm. National Marine Fisheries Service). In many coastal and inland areas of the U.S., work within or adjacent to wetlands which provide breeding habitat for key waterfowl species such as ducks, swans or geese is often severely restricted to avoid peak migratory periods.

These seasonal restrictions often overlap, depending upon the site-specific conditions of a project location. Thus, it is very important that a proponent contact the relevant federal and state fish and wildlife resource agencies to obtain information on species' use of the project site and potential seasonal work restrictions which may placed upon the applicant. At a minimum, the U.S. Fish and Wildlife Service, and in a coastal or anadromous habitat the National Marine Fisheries Service, should be contacted during early project planning to identify potentially sensitive habitats and work periods for a proposed project site. While these resource agencies have not been provided with the same regulatory clout as the Corps and the Environmental Protection Agency, they are actively involved in review of Section 404 permit applications, and often participate directly in interagency discussions and work groups. As a result, their input and technical expertise is often relied upon by the Corps and the Environmental Protection Agency in the review of applications.

Another method for minimizing the impacts of construction within wetlands is to provide proper sequencing of the work. For example, if a pipeline, roadway, or other linear project is proposed, minimizing the extent of clearing (in front of the active trenching operation) will reduce the potential for soil erosion into adjacent wetlands and reduce impacts to wildlife utilizing the existing vegetative cover. Wherever possible, work which is required within wetland areas should be completed as quickly as possible, without excessive delays between the initial disturbance and rehabilitation. For example, trenching for utility lines within wetland areas should be conducted wherever possible as a single, continuous operation, involving clearing, installation, backfilling and soil restoration, as an open trench can act as a channel to dewater adjacent wetland areas and increase erosion and runoff impacts. Careful attention to construction planning, and development of specific contractor specifications on construction sequencing and scheduling during the permit application process, can provide very visible, demonstrable benefits during agency discussions and project construction.

Project Specifications and Bid Documentation

Environmentally-sound project specifications which still allow for a competitive bidding process and a reasonable construction schedule have become increasingly important over the past decade. This has occurred as a result of financial constraints for project proponents, and strengthened environmental regulatory requirements. The latter have in turn magnified the potential liability issues for construction contractors, engineers, and developers.

The translation of multiple, sometimes vague and contradictory permit requirements into very specific contract specifications requires very close scrutiny during project development. The construction specifications must contractually bind the contractor to meet all applicable commitments agreed to by the project proponent. In addition, the contractor must be informed of the key issues of concern which have been raised by both agencies and the local citizens' groups. The final objectives for all parties involved in project development are to avoid negative local or agency reaction and the issuance of stop-work orders due to unanticipated wetlands or other impacts.

Consultation with potential contractors during the permit application and negotiation process can help to identify whether proposed agency construction restrictions can be implemented, and new or innovative methods which could be used to work within sensitive habitats. Informal discussions with contractors early in the planning phase will often uncover unforeseen constructability issues which can be reviewed with agencies, and resolved prior to preparation of final construction specifications.

The contract specifications should not only append the relevant permit applications, but special notes should be developed to explain, in contractual language, the requirements and intent of each of the permit conditions. In addition, the contractor should be informed directly on the construction drawings of applicable permit conditions, such as the locations of specified erosion and sedimentation control methods, limits of clearing or disturbance, adjacent sensitive resources to be avoided, and designated laydown and transport routes.

To prevent the unforeseen disturbance of wetlands, or incomplete restoration of disturbed wetlands, each construction specification package should require a contractor to post a performance bond to be held in escrow pending completion of all construction and restoration work. The amount and length of this bond requirement should match the worst-case schedule and areal impacts of nonperformance in order to provide a strong financial incentive for the contractor to conduct work in accordance with the specifications.

Each specification should contain a wetlands restoration section which specifies the need for pre- and postconstruction surveys of affected wetland areas. These surveys should include detailed requirements as to the amount of soil compaction, replanting schedule, accuracy of topographic surveys, and other key elements. Monitoring of restored wetlands areas should also be specified, and should be of sufficient duration (2-5 years minimum) to ensure that pre-construction conditions are maintained.

In order to reinforce the environmental requirements of a construction project within wetlands, the potential bidders should be provided with an overview of the applicable permit requirements during a preconstruction bidder conference. This will help them to develop bids without large contingencies or incorrect construction scenarios that could require clarification during the bid process or construction change orders.

Each contractor working on projects within wetland areas should be required to hire a fully qualified wetlands biologist with specific experience in wetlands construction and mitigation to provide onsite support and monitoring of construction activities in wetlands. This person should be provided with sufficient authority to notify construction residents and field supervisors of any observed omissions or deviations from approved permit conditions. Functioning as a compliance supervisor, the biologist would also be responsible for ensuring that wetlands backfilling, restoration, reseeding, replanting and other mitigation requirements are fully achieved and documented during the construction process. To ensure that the person is qualified, their resumé and relevant experience should be reviewed, and their selection approved, by the project proponent or engineer.

Sedimentation and Erosion Control Techniques

Protection of adjacent wetlands from the temporary and permanent impacts due to project construction is a primary consideration in the development of acceptable permit conditions. There are numerous methods for sedimentation and erosion control, all of which seek to isolate and contain, to the maximum extent possible, sediment-laden runoff generated during project construction activities. The performance of these various methods in the field varies considerably depending upon the type of soils, water flows, exposure, and other site-specific factors. Table 2 summarizes some of the more popular sedimentation control methods used, their suitability for use in certain wetland and other jurisdictional areas, and performance limitations.

Sedimentation and erosion control methods can be used singly or in combination. By limiting the amount of incremental and total land clearing, and maintaining existing ground cover to the maximum extent possible, potential runoff, gully creation, rutting, and airborne dust formation can be reduced to acceptable levels. Where feasible, a project site layout should take advantage of existing vegetation between the clearing limits and adjacent wetlands, In addition, vegetative buffers should be preserved for projects with sufficient shoreline frontage in order to meet local zoning requirements and protect structures from wave and flooding impacts.

Installation of hay bales, using wooden stakes, within shallow cut-off trenches upgradient of wetland areas can be an effective and inexpensive method to isolate upland work areas from adjacent wetlands.   Too often, however, the field installation of hay bales is substandard. Bales are not set into in the ground properly, gaps are left between bales, and bales are damaged, moved, or destroyed during construction without proper maintenance and monitoring by field personnel. Construction specifications should provide for regular checks of the condition and effectiveness of the haybale protection systems.

Siltation curtains can be used effectively in both wetlands and open water environments. Curtains can be wrapped around hay bales and staked into the ground to provide an extra measure of protection against the release of fine-grained materials.   Siltation curtains can be used to surround subaqueous dredging operations, particularly those occurring within sheet piling, to isolate trench water from the surrounding environment. Curtains with floatation may also be installed around shoreline construction projects and anchored in place to isolate the work area. However, the effectiveness of these structures decreases significantly in areas of strong river currents, tidal flows, and large tidal ranges, particularly if the curtain is installed perpendicular to the current flow. In these cases, the siltation boom experiences roll-over or submergence from flows, and is susceptible to damage from debris. Thus, they are most effective in ponds, lakes, and other sheltered water bodies with little or no variation in water height.

Geotextile and filter fabrics can be installed on newly graded slopes in conjunction with reseeding efforts to hold soil in place until vegetation is reestablished. They can be used effectively as a barrier between wetlands soils and temporary fills or mats installed to allow access to wetlands work areas, and prevent migration of fine-grained materials down into the wetlands soils. Fabrics are also used in conjunction with staked hay bales to provide a higher level of erosion control between active work areas and adjacent wetlands.

TABLE 2. POTENTIAL SEDIMENTATION AND EROSION CONTROL METHODS
FOR REDUCTION OF WETLANDS IMPACTS

TECHNIQUE	SUITABILITY OF USE						LIMITATIONS
	VEGETATED WETLANDS	INLAND WATERS	RIVERS	ESTUARINE WATERS	COASTAL WATERS	OFFSHORE AREAS	
Minimize Clearing/Disturbance	■						none
Maintain Vegetative Buffers	■						none
Install Staked Hay Bales	■						Requires maintenance and proper installation
Install Siltation Curtains				▨	▨		limited effectiveness in tidal areas, currents > 1 kt.
Use of Geotextile Fabrics		■	■	■			Requires maintenance and proper installation
Construct within Sheet Piling		■	■	■	■		Can be more costly, must avoid shallow bedrock areas
Require Prompt Reseeding/Revegetation	■						none
Specify Strict Dredging Methods		■	■	■	■	■	Dredge bucket modifications may require field testing
Develop Dust Control Procedures	■						none
Use of Proper Material Handling Techniques	■						none
Use of Temporary Settling Basins	■						Space and dewatering flow requirements
Construct Stormwater Detention Basins	■						Requires dedicated open space, maintenance/monitoring
Reinjection of Groundwater	■						Costly, may require pre-treatment
Catch Basin/Storm Drain Protection	■						none

□ = Not applicable

▨ = Suitable (with limitations)

■ = Recommended (where feasible)

As previously discussed, isolation of the active work area in both wetlands and open water areas is an effective method to limit the horizontal extent of disturbance, particularly in areas where significant dredging is required. In such cases, dredging open trenches beyond 1-3 feet in depth requires side slopes which can range from 3:1 to 5:1 or greater, meaning that a 10 foot deep trench would disturb a minimum 60 to 100-foot width of sediments. Clearly, this size dredging operation would require the handling and disposal of large amounts of excess dredged material. Conducting this work within sheet piling allows a vertical sidewall, thereby reducing the volume of material to be handled, and isolating the silt-laden trench water from surrounding marsh and other wetland areas.

In any construction project, regardless of the proximity to wetlands or other adjacent sensitive habitats, construction specifications should require prompt reseeding or replanting to stabilize newly-graded slopes. The construction schedule should allow time for vegetation to become reestablished prior to the end of the growing season. In cases where this is not possible, more expensive but generally less effective measures, such as covering exposed areas with geotextile or other fabric filters may be necessary to prevent significant erosion and runoff during winter construction periods.

Open dredging within or adjacent to wetland areas can produce significant amounts of turbidity. If typical dredging equipment is used, for example a barge-mounted crane with a clamshell dredge bucket, methods are available which can reduce turbidity. These include establishing requirements that all lifts of a clamshell dredge bucket through the water column are vertical (without any horizontal motion), dredge buckets be used with closeable covers and gasket seals to prevent washout of sediments as the bucket is lifted through the water column, and suitable filtering of water released from stockpiled dredged material. While some agencies have required the use of rubber or plastic bucket edges to reduce sediment release as the bucket is closed, care should be taken to ensure that the dredged material can be effectively removed with these attachments. If not, repeated attempts to remove small amounts of sediments will, in the aggregate, result in more turbidity over a longer period of time.

Hydraulic dredging can also be used in certain unconsolidated sediments to reduce turbidity. With this method, sediments are removed and pumped as a slurry to a settling barge or disposal site. While initial turbidity at the point of dredging is minimal, large amounts of water must be filtered and removed from sediments at the disposal site, and pumping limitations require that disposal occur in close proximity to the point of dredging.

Construction will often require temporary stockpiling of soils, and care should be taken to continually spray these piles with water, or cover them, in order to prevent wind erosion and transport of these materials. Similarly, newly graded access roads should be frequently sprayed or treated with water or agency-approved dust suppressants to reduce dust formation. In addition, the construction schedule should attempt to minimize the period of time where exposed stockpiles or unpaved road surfaces are required.

Site grading and excavation activities in areas already served by drainage systems are a potential concern for sedimentation, as many parking lots, roadways, and other facilities use stormwater drainage systems which discharge directly into adjacent wetland areas. In order to minimize the impacts from run-off of sediment-laden water, existing catch basins and storm drains should be completely ringed with staked haybales and a layer of filter fabric. These sediment traps will allow stormwater flow to pass through, but will filter out significant amounts of suspended sediments. These structures also provide protection in the event of an accidental fuel oil spill, hydraulic hose rupture, or other hazardous material release, providing some measure of initial containment upgradient of adjacent wetland areas. As with all hay bale structures, these sediment traps need to be maintained and periodically replaced to ensure their effectiveness.

Excavation for foundations, utility trenches and other facilities will often extend below the existing water table, resulting in collection of groundwater within the excavation. In order to dewater these areas and prevent discharge of sediment-laden water into surrounding areas, various types of settling basins and detention structures can be constructed. These structures allow particulate matter to settle and gradually discharge filtered groundwater runoff. Figure 3 is a schematic representation of a typical settling basin which can be constructed upgradient of a wetland area using filter fabric and clean rip rap material to effectively filter silts and sediments at a construction site. Concrete or fiberglass settling basins are also available for use as sedimentation control structures during dewatering operations, and are often used on barges during dredging operations to filter water discharged from stockpiled dredged materials. Geotextile wetland filter bags have also been developed to serve as sedimentation and erosion control devices on construction sites (Figure 4).

Typical Monitoring Requirements and Procedures

The true test of any sedimentation and erosion control plan will occur during the first significant rainfall event during construction. Thus, it is recommended that the onsite resident inspectors monitor the success of the installed erosion control devices during and immediately after a rainstorm or snowmelt. The hay bales, siltation fences, and other structures should be observed on at least a weekly basis to detect damage from wildlife, machinery or other activities on site.

Equally important, the resident inspectors should conduct frequent visual observations of the adjacent wetlands or open water bodies to detect turbidity plumes resulting from onsite runoff. For certain subaqueous activities, significant short-term increases in turbidity are unavoidable, but attention should be focused on the effectiveness of the silt booms, dredging methods, and dewatering practices to ensure that surrounding background levels of turbidity are not significantly increased. For deepwater areas, the use of a Secchi disk or similar device will provide a qualitative measure of the water clarity and amount of suspended sediments during construction.

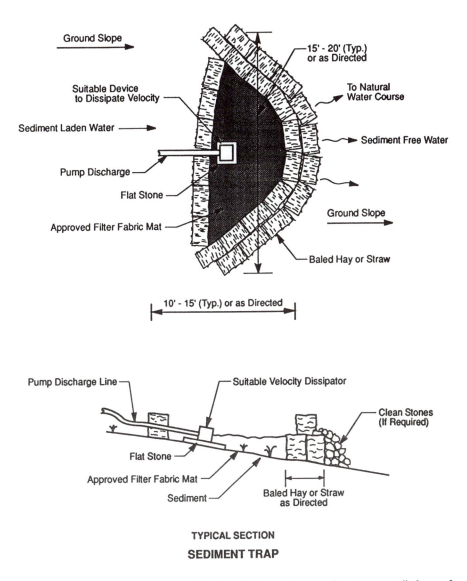

Ground Slope

15' - 20' (Typ.)
or as Directed

To Natural
Water Course

Suitable Device
to Dissipate Velocity

Sediment Laden Water

Sediment Free Water

Pump Discharge

Flat Stone

Ground Slope

Approved Filter Fabric Mat

Baled Hay or Straw

10' - 15' (Typ.) or as Directed

Pump Discharge Line

Suitable Velocity Dissipator

Clean Stones
(If Required)

Flat Stone

Approved Filter Fabric Mat

Sediment

Baled Hay or Straw
as Directed

TYPICAL SECTION

SEDIMENT TRAP

Figure 3. Settling basins are used in conjunction with dewatering operations to prevent discharge of sediment laden water into wetlands. The basins are constructed upgradient of wetlands using filter fabric and clean rip rap material.

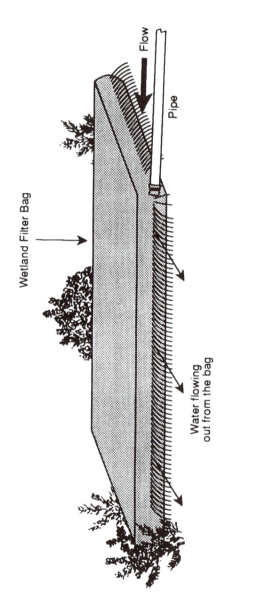

FIGURE 4

WETLAND FILTER BAG

It is recommended that the contractor take readings at regular intervals to provide documentation for agency site inspections or citizens' group meetings. While it may be possible to complete a project within or adjacent to wetlands without any regulatory agency inspections, it is critical that the work always be conducted in accordance with permit conditions. To avoid temporary shut-down of work or a notice of violation, specified monitoring procedures, checklists, and appropriate documentation should be developed during the course of construction in order to address any questions or concerns which might be raised.

SUMMARY

In order to satisfy Corps and other regulatory agency concerns, a project proponent must first demonstrate that all reasonable measures have been taken to avoid impacts. The definition of the project purpose and need is a necessary step toward determining the available universe of alternative sites. Once the purpose is clearly defined, careful documentation of the site selection process is needed, using a clear and comprehensive set of siting criteria. The objectives and goals of this analysis are simple: To demonstrate to the reviewing agencies that the proposed action has emerged as the Least Environmentally-Damaging Practicable Alternative, or LEDPA, and that no less damaging practicable alternatives exist.

The Corps' standard of review for projects will vary depending upon the specific constraints of the applicant and features of the proposed project. The extent to which the project involves permanent wetlands impacts, must be carried out within wetlands or U.S. Waters, and provides both public and private benefits, will all help determine the level of regulatory scrutiny. In all cases, however, avoidance of wetlands impacts must be incorporated as early as practicable into all phases of project planning.

In order to avoid needless additional design and planning expenses, it is recommended that the agencies provide some input during the site selection process. Ideally, a project proponent will hold a series of preapplication meetings with the regulating agencies, particularly the Corps, to obtain their concurrence that the proposed activity is likely to emerge as the LEDPA.

After the LEDPA has been identified and reviewed with the permitting agencies, the details of project design need to undergo a thorough and systematic review to ensure that unavoidable wetlands impacts are minimized to the greatest possible extent. The project's layout, orientation and scale are obviously key elements in minimizing permanent wetlands impacts. However, the construction techniques and contractor specifications are equally important elements to ensure that temporary construction-related impacts are effectively managed. In addition, all staff involved in project development, from engineering and design staff to construction equipment operators, must understand the importance of the wetlands protection measures incorporated into the project design. Adherence to these requirements throughout project development provides suitable ecological protection while also maintaining the project schedule and cost requirements.

REFERENCES

Council on Environmental Quality. 1980. Coastal Environmental Management: Guidelines for Conservation of Resources and Protection Against Storm Hazards. Report prepared by The Conservation Foundation for CEQ, the Department of Commerce, Department of the Interior, the Environmental Protection Agency, and the Federal Emergency Management Agency, Contract #EQ7AC004, Washington, D.C. June, 1980.

Executive Office of the President. 1977. Executive Order 11988, Floodplain Management. Federal Register, Volume 42. U.S. Government Printing Office, Washington, D.C.

Executive Office of the President. 1977. Executive Order 11990, Protection of Wetlands. Federal Register, Volume 42. U.S. Government Printing Office, Washington, DC.

Office of Technology Assessment, U.S. Congress. 1984. Wetlands: Their Use and Regulation. Report #OTA-0-206. U.S. Government Printing Office, Washington, D.C.

U.S. Army Corps of Engineers. 1986. Final Rule for Regulatory Programs of the Corps of Engineers. Federal Register, Volume 51. U.S. Government Printing Office, Washington, DC.

U.S. Army Corps of Engineers. 1990. Regulatory Guidance Letter 90-08 Regarding Applicability of Section 404 to Pilings. Federal Register 14 December, 1990. U.S. Government Printing Office, Washington, D.C.

U.S. Army Corps of Engineers. 1991. Final Rule for Nationwide Permit Program Regulations: Issue, Re-Issue and Modify Nationwide Permits (33 CFR Part 330). Federal Register November 21, 1991. U.S. Government Printing Office, Washington, D.C.

U.S. Congress. 1972. Clean Water Act (as amended by Federal Water Pollution Control Act), PL 92-500. 33 USC 1251 et. seq.

U.S. Congress. 1978. National Environmental Policy Act, PL 91-190. 42 USC 4321. Federal Register, Volume 43. U.S. Government Printing Office, Washington, DC.

U.S. Environmental Protection Agency. 1980. Guidelines for Specification of Disposal Sites for Dredged or Fill Material. Federal Register, Volume 45. U.S. Government Printing Office, Washington, D.C.

CHAPTER 7

ENHANCEMENT, RESTORATION AND CREATION OF FRESHWATER WETLANDS

John Zentner

Frayer et al. (1983) reported that about 95% of the wetlands in the coterminous United States are inland freshwater wetlands. Most of these wetlands are either freshwater marshes or riparian woodlands. Freshwater wetlands develop at elevations above open water aquatic habitats and below uplands. They are found in a wide range of hydrologic conditions, from permanently flooded (to a depth of 1 m) to seasonal saturation of the root zone. Freshwater wetlands occur on a wide variety of soil types including both organic and mineral soils, as well as in nonsoil conditions.

Freshwater marshes are dominated by herbaceous emergents and can be divided into three general categories (Figure 1). From wettest to driest these are: 1) Perennial marshes which are permanently or semipermanently flooded and are dominated by tall emergents such as cattails (*Typha* spp.) or bulrush (*Scirpus* spp.), 2) Seasonal marshes which are seasonally flooded or saturated and are dominated by Cyperaceae and Juncaceae, and 3) Wet meadows which are dominated by graminoids and Juncaceae and which occur in the temporarily or intermittently flooded zones.

By contrast, riparian woodlands are dominated by shrubs and trees. They are characterized by impermanent and varying periods of inundation or root zone saturation during the growing season. Compared to marshes, riparian woodlands occur on relatively permeable and well-oxygenated substrates. Analogous to freshwater marshes, riparian woodlands can be categorized by hydrological regime (Figure 2): 1) Low-terrace woodlands which are semipermanently flooded and are generally dominated by willows (*Salix* spp.), silver maple (*Acer saccharinum*) and similar species with relatively light seeds of limited viability, 2) Mid-terrace woodlands which are seasonally flooded and are generally dominated by green ash (*Fraxinus pennsylvanica*), sycamore (*Platanus occidentalis*) and other species with medium weight seeds, and 3) High-terrace woodlands which are temporarily flooded and are dominated by a variety of species, especially oaks (*Quercus* spp.),

Figure 1. Freshwater marshes are dominated by herbaceous emergents, and can be divided into three hydrological categories: perennial marsh, seasonal marsh and wet meadow.

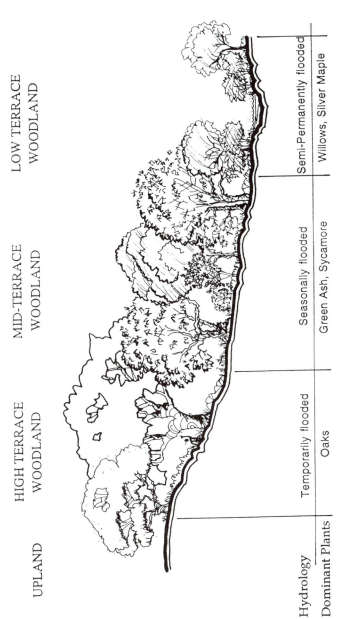

Hydrology	Temporarily flooded	Seasonally flooded	Semi-Permanently flooded
Dominant Plants	Oaks	Green Ash, Sycamore	Willows, Silver Maple

UPLAND — HIGH TERRACE WOODLAND — MID-TERRACE WOODLAND — LOW TERRACE WOODLAND

Figure 2. Riparian woodlands are dominated by shrubs and trees. As with freshwater marshes, riparian woodlands can be categorized by hydrological regime: semipermanently flooded, seasonally flooded and temporarily flooded.

which are typified by heavy seeds. These categories correspond to Categories III (lower hardwood wetlands), IV (medium hardwood wetlands), and V (higher hardwood wetlands) as described by Cowardin et al. (1979), Larson et al. (1981) and Clark and Benforado (1981). Seed weight is included in the above summary because of its importance to natural establishment and resulting growth patterns.

Freshwater wetlands have been the focus of enhancement, restoration and creation efforts. As defined by Lewis (1990), enhancement is an increase in values afforded to a specific vegetation association by construction actions; restoration is the re-creation of a specific vegetation association on a site where that association was once known to occur; and creation is the construction of a wetland on an upland site. These terms are described simply as *construction* in this chapter.

The earliest freshwater wetland construction projects may have occurred many thousands of years ago with the manipulation and creation of wetlands to enhance rice harvests. More recently, waterfowl conservation and hunting organizations have been responsible for numerous wetland construction projects. New York, especially, had numerous small freshwater marshes created in the mid-1950s (Dane 1959). Specific problems associated with wetlands, such as nuisance mosquitoes, have also resulted in a number of useful research and enhancement practices. More recently, interest in freshwater wetland construction is in large part a response to regulatory requirements. Nevertheless, there is also a greater public interest and understanding of the environmental values of wetlands, and a desire for the re-creation of lost or diminished landscapes and values.

This chapter discusses the construction of freshwater marshes and riparian woodlands, and is intended to be a guideline for those seeking to construct freshwater wetlands. The construction process is described in five steps inherent to designing, building and maintaining wetland landscapes. These steps are site analysis, goal setting, construction design, implementation and maintenance. Implicit in these steps are two tenets. First, specification of a target vegetation association (TVA) is the primary goal. A TVA provides the habitat for any plant or wildlife populations that may be desired on the project site and, because of its structural nature, facilitates maintenance and monitoring efforts. Second, there is no substitution for directed observation of the project site and the TVA in construction planning.

SITE ANALYSIS

Site analysis is the first step in a wetland construction project, and may be completed during a wetland delineation effort or as the first steps in a neighborhood restoration program. This step provides the basic framework for goal setting, identification of the TVAs and development of performance standards.

The site analysis typically includes wetland vegetation association mapping and analyses, a review of historic conditions, and analyses of hydrologic, soil, and cultural conditions of the site and any nearby templates, or examples, of TVAs. Small-scale experimental wetland construction efforts should also be initiated in this phase if practicable. Generally, the initial site analysis is completed within a period of three weeks to three months, with relatively intensive effort in defining site vegetation associations, soils, historic and cultural conditions in the first few weeks, and less intensive effort in monitoring hydrology and experimental wetlands over the remaining period.

Vegetation Association Mapping

Existing and proximal wetland vegetation associations are initially identified, and their extent determined, by mapping from an aerial photograph. Aerial photographs are usually available as stock or library film from an aerial survey center. These centers fly important regions every two to three years, providing a backlog or library of film which can be reproduced at a variety of scales. Requesting a halftone mylar as well as a print of the project site is important. The mylar can be used to make blueprints for field use, whereas blackline prints can be reproduced and included in reports.

After initial mapping from an aerial photograph, the borders of the vegetation associations are more precisely defined through onsite analysis. For freshwater marshes, randomly selected 1 m plots representative of each vegetation association can be sampled for species and cover. Randomly selected 0.25 ha polygons (polygons are defined by tree cover and separation among woodland vegetation associations) of species occurrence for all canopy layers are effective in riparian woodlands.

Table 1 provides an example of vegetation association mapping from a hypothetical project site in California characterized by gently sloping hills surrounding a central creek. The result is a table which describes the vegetation associations and the absolute and relative cover of marsh and riparian woodland vegetation. A scan of the table and knowledge of the region reveal that four vegetation associations are present, and that the site is dominated by nonnative species and species representative of disturbed areas. It is also appropriate at this time to identify and survey any upland areas that may be considered for wetland construction to ensure that valuable upland habitats are not inadvertently lost.

Review of Historic Aerials and Maps

Historic aerial and other site photographs, and any available topographic or other maps, are reviewed to define the past conditions and boundaries of the site vegetation associations. Important issues to review with these sources include the location of wetland vegetation associations and their relationship to current vegetation associations, water sources, and cultural elements relative to potential construction sites. Historic aerials are often available from the U.S. Geologic Survey (USGS) and local aerial photography sources. Historic maps are also found at USGS as well as local historical societies. Costs for historic aerials are generally less than current aerials, however reproduction costs for older maps may be significant.

Table 1. Example of vegetation association mapping from a hypothetical project site in California. The table provides absolute and relative information for use in site analysis and target vegetation association selection.

Vegetation Association	Plant Species	Area (ac)	Cover (%)
Wet Meadow		3.5	80
	Lolium perenne *		40
	Hordeum hystrix *		30
	Elymus triticoides		20
	Elymus glaucus		10
Seasonal Marsh		3.4	50
	Lolium perenne *		40
	Juncus balticus		30
	Lotus corniculatus *		20
	Polypogon monspieliensis *		10
Perennial Marsh		2.2	100
	Typha latifolia **		80
	Scirpus acutus		20
Low Terrace Woodland		1.2	100
	Salix lasiolepis **		100

*   non-native species
**   species indicative of disturbance

Hydrological Analysis

The initial site analysis provides an opportunity to define the hydrologic conditions which create the site or template vegetation associations. Generally, hydroperiod, the duration of inundation or saturation, and water depth best define the wetland vegetation association. Accordingly, hydrology stakes and piezometers are placed in wetland and upland areas to define surface and groundwater conditions. Piezometers are essentially shallow, capped observation wells and typically consist of 10-cm diameter PVC pipe inserted 1.2 to 2.4 m into the ground, with a surrounding backfill of gravel or other permeable material. Piezometers and stakes are generally observed weekly during the analysis to define the depth of ponding and the surface approach of groundwater. However, short-term monitoring may not reflect typical groundwater levels. Long-term ground-water level data may be available from local water districts or farmers or may be completely absent.

Soil Analysis

Soil analysis can be one of the most important steps in defining wetland construction opportunities and constraints. The soil analysis can provide clues to the native condition of the vegetation, thereby defining the appropriate TVA, as well as identify any restrictions to construction activities due to soil permeability or soil chemistry. Generally, the best sources for information on soil characteristics for an area are direct site observations supplemented by the U.S. Soil Conservation Service (SCS) soil series maps. The latter are available at no cost from SCS offices. The SCS and associated services such as county agricultural offices often have aerial photographs for review as well. Where soil series maps are unavailable, SCS field personnel may provide onsite assistance.

Site observations typically consist of a series of soil pits, 30 to 50 cm in depth, dug by hand throughout the site for the purpose of defining surface conditions. Use of a backhoe to define conditions at depth is generally helpful, and is essential if significant excavation is required. Test pits (both manual and mechanical) will help identify any significant variation in permeability among soil layers. Where warranted, representative soil samples are analyzed by a lab for soil chemical properties.

Cultural Constraints

Few sites are unconstrained by human artifacts, from prehistoric remains to buried sewer lines. Visible artifacts such as roads, which affect the project site (e.g. drainage impediment), must be identified. Identification of subsurface artifacts is more difficult. Nevertheless, cultural resources can be at least be preliminarily identified through record searches with the State Historic Preservation Office (SHPO) or local universities. Buried infrastructure may be identified through interviews and research at local Public Works Departments.

The Use of Template Associations

Re-creation of an historic vegetation association will in most instances increase the probability of success. In the event that the TVA is missing or poorly represented on the project site, a proximally located vegetation association can be used as a template. The template association should have hydrology and soils similar to existing or intended conditions for the construction site. In addition to determining species composition, the template vegetation association should be studied to determine species evenness and distribution, and identify potential additions to the planting list and likely successional patterns. The specific TVA to be constructed may be difficult to predict at this stage of the project. However, initial review of several different vegetation associations at this stage may have a significant affect on goal setting. For example, participants in a riparian woodland construction project for a degraded waterway in Solano County, California, wished to plant California sycamore (*Platanus racemosa*) extensively, due to its stature and appearance. The project ecologist recognized that the sycamore, adapted to alluvial flats with relatively permeable soils, would not do well on the project site's clay soils. Taking the participants to a regional park which included a well-preserved riparian woodland native to this region and similar to the project site resulted in the selection of several alternative trees, all better adapted to the project site.

Small-Scale Experimental Construction

Where construction planning and the permitting phases take more than six months and occur during a favorable construction season, small-scale experimental projects or pilot studies should be undertaken to identify soil and hydrology constraints, and to refine planning alternatives. Appropriate studies include determining the water-holding capacity of unmodified soils, determining germination and salvage survival rates for target vegetation, refining transplanting techniques and evaluating construction management abilities. Additionally, the importance of small scale pilot projects for building confidence and consensus among the planning group should not be underestimated.

GOAL SETTING

The setting of specific goals is crucial to wetland construction projects. Goals define monitoring elements and protocols, and establish standards for judging success. The latter allows others to objectively evaluate the project, and serves as a basis for practitioner and general public educational efforts.

Elements of a Goal Statement

Construction goals should specifically include a substantive element that is also a measurement parameter. For example, a TVA should include the type of association as well as the dominant species. This is most easily accomplished through comparison with an existing or historically well-described vegetation association from the project site or a nearby site. One of the more difficult tasks is to keep the TVA general enough to be realizable, yet specific enough that the goal is meaningful. Typically, selection of a species-type association, for example a rush-dominated wet meadow, is an effective approach.

Too often, goals are developed without discussion with, or consensus among, all interested parties. Several points should be considered in the development of goals. Construction goals should address the concerns of all those meaningfully affected by the project. Affected parties will include landowners at or adjacent to the project site, regulating agencies, and operation and maintenance entities such as mosquito abatement and flood control districts. Table 2 lists parties potentially affected by wetland construction activities, and goals common to these parties. These parties should all be contacted and their comments solicited on initial goal statements.

In smaller-scale projects where the affected parties' interests are all relatively equal and similar (e.g. neighborhood groups reconstructing a local creek), these comments should be used to develop a consensus on project goals. In most projects, this is not practicable, and the consultative process will involve reviewing each comment and incorporating a response into the construction plan. One of the more frustrating aspects of goal setting in wetland construction is the number of entities with differing mandates and the ability to deny or significantly delay projects.

Experience suggests that a flexible, two-step goal development process is most effective. The initial goals are developed following the initial site analysis and discussions with the affected parties. Final goals are selected at the conclusion of the design analysis process (see below). Even then, the goals may require modification during the construction or even postconstruction monitoring phase as new opportunities and constraints arise.

CONSTRUCTION DESIGN

A substantial body of literature exists on the design of wetland construction projects (Kusler and Kentula, 1989). However, a significant gap exists between those who write plans and those who construct wetlands. This gap mirrors a similar chasm found between landscape architects and landscape contractors, and is likely to exist because of education dissimilarities, professional practices, and even insurance requirements. This gap poses a significant danger to the development of ecological construction as it ensures that important knowledge gained during construction is not incorporated into the design process, and that many projects are not built as designed.

Table 2. Parties potentially affected by wetland construction projects, and goals common to these parties.

Affected Parties	Goals
Landowners	
Site land owner	Minimize costs
Adjacent land owners	Increase aesthetics
Adjacent residents	Eliminate off-site issues
	(mosquitos, odors, floods)
	Increase access
	Reduce vandalism
Permitting and Commenting Agencies	
Corps of Engineers	Provide appropriate data
Wildlife agencies	Respond to all concerns
Public lands agencies	Restore single-species habitat
	Provide appropriate public access
	Ensure success
Operation and Maintenance Entities	
Flood Control Districts	Minimize maintenance costs
Parks Districts	Provide adequate flood
Mosquito Abatement	conveyance
	Ensure compatibility with
	park needs
	Minimize mosquito
	production

Building a wetland shares many basic elements with any landscape construction project, whether it is gardening or reforestation (Table 3). The most important issue in any such project is to determine what conditions create the TVA. Subsequent issues may arise that threaten vegetation association growth, whether one is planting corn, taro, or rushes. Wetlands are among the wettest landscapes and, accordingly, the quantity and quality of the water, as well as the use of soil types which ensure water retention, are extremely important. Also important are those conditions which result from the application of significant amounts of water to the soil, such as erosion and salt build-up, which reduce the viability of the TVA.

Wetland construction strives to achieve ecological function, and presumes that the project, once constructed, will evolve in a natural fashion. Ideally, little or no maintenance will occur following an initial establishment period, making an understanding of the successional aspects of the TVA very important. On the other hand, many wetland construction projects, especially those specifically for waterfowl habitat, will use sources of water which require hands-on operation and maintenance over a significant period of time. Payne (1992) describes in detail the management demands of these types of projects.

Unless the project is built in an extremely remote area, human actions will influence the project. As well, the project may exert some influence over humans. For these reasons, wetland construction must consider the role of people in the design and, in some cases, identify methods for ensuring that potential interactions between the created habitat and people are mutually beneficial.

Table 3. Components, subcomponents, and products of wetlands construction design.

Component	Subcomponents	Products
Hydrologic Analysis	Slope and Basin Models Water Supply	Inundation Depth & Hydroperiod Channel Design
Soils Analysis	Backhoe Tests Soil Lab Tests	Permeability & Texture Chemistry
Vegetation Analysis	Successional Models Planting Design Weed Control	Species Lists Planting Plan Weed Control Program
Cultural Issues	Shape Adjacent Uses Mosquito Control Water Quality	Project Configuration Buffer Requirements Vector Control Program Water Quality Program

Hydrology

As noted earlier, wetland hydroperiod, the duration and frequency of inundation or saturation, and depth of inundation, are generally the major determinants of wetland vegetation associations. However, the manner in which the water is supplied to the construction site may be equally crucial. Artificial supply systems such as drip or spray irrigation and pumps require a significant amount of maintenance and reduce the naturalness of the wetland. Conversely, they eliminate much of the uncertainty involved in water supply planning. More natural water supply systems such as open channels are less manageable but can result in a self-maintaining system which mimics natural systems. Use of natural water supply systems requires a deep understanding of ecology and engineering, and will require more effort in the investigation and planning stages.

Hydroperiod and Depth

Two models are useful in estimating wetland hydroperiod and depth of inundation. Slope models are for use in defining hydroperiod and depth of inundation of wetlands abutting a river or lake where surface water inputs are high. Basin models are for use in modeling most other wetlands.

Slope Wetlands

Slope wetlands are characterized by seasonally varying water levels with no implied storage. The relationship between wetland vegetation associations and water surface elevations is well described (Dickson et al. 1965, Harris et al. 1975). However, water surface elevation analyses of streamflows for storms of various intensity or duration are complex. Nevertheless, channelized streamflow equations have been developed for flood control analyses. These equations factor in the amount of flow, the cross-sectional area and slope of the channel, and a friction element which defines channel roughness. The latter reflects the capability of the streambed and channel vegetation to reduce flow velocity. The HEC-1 model developed by the U.S. Army Corps of Engineers (1981) is an example of this type of model. Used in conjunction with an accurate topographic map of a site, these models can predict the depth of water and duration of flooding at a specific elevation for a specific storm event. However, these models require accurate data which is typically very expensive to generate.

Slope wetlands can often be more simply defined by directed observations of the vegetation associations during specific storm events. The locations of wetland vegetation associations in a slope environment relate to water level, which is related to topographic position. Identifying the elevation of a TVA relative to the channel and then predicting the creation of a new and similar association based on a similar landscape position is crude but relatively effective (Figure 3).

Observations during storm conditions are crucial for determining channel behavior during high flows. Although Figure 3 depicts a classic condition of a channelized stream which might be easily restored by widening it to pre-alteration conditions, other factors may also be critical. Soil or underlying bedrock may be radically different on adjacent reaches. For example, greater permeability in the construction reach might result in less water than predicted. Other factors such as channel roughness in the form of dense vegetation or in-channel features may slow flows and significantly reduce peak flows. A comparison of stormflow peak and duration in the template and construction reaches can eliminate most sources of error.

Basin Wetlands

Basin wetland hydrology is best defined by a water budget. That is, the water available to support the wetland is a function of water inflow minus water outflow. This is illustrated by:

$$S = P + SI + GI - ET - SO - GO$$

where:

S = storage in the wetland basin

P = precipitation

SI = surface inflow

GI = groundwater inflow

ET = evapotranspiration

SO = surface outflow

GO = groundwater outflow or infiltration

Water sources for the wetland include precipitation falling directly on the wetland, surface inflow and groundwater inflow. Precipitation amounts should be identified on at least a monthly basis. Data is usually available from local weather stations or airports. National Oceanic and Atmospheric Administration (NOAA) handbooks are also a good source. Surface inflow is the result of either precipitation within the watershed less the amount lost to evapotranspiration and infiltration, or inflow from adjacent lacustrine or riverine sources. For inflow from adjacent sources, a stage-discharge model may be used to determine surface inflow, or the elevation of the flood flow estimated based upon corresponding vegetation association boundaries. Groundwater inflow is often extremely difficult to determine, but may be approximated through the use of piezometers and well data.

Water is lost from the wetland (and its watershed) through evapotranspiration, surface outflow and groundwater outflow. Evapotranspiration is often described by month in SCS soil surveys, NOAA handbooks and other sources. The low point in the basin will control the basin water surface elevations and surface outflow. The storage capacity is then the remaining volume of water in the basin. Some water flowing into the basin will be lost to outflow through the soil, with the amount of outflow determined by the height of groundwater and the type of soil. Where groundwater levels are high or the soil underneath the basin is at field capacity, no or little loss will occur. Where groundwater levels are low or the soils beneath the basin are not at field capacity, water loss to the soil will occur. The amount of loss during these conditions is dependent upon soil type, with clay soils generally appropriating less water than sandy soils.

The factors used in a water budget calculation are often dependent upon each other, and interaction between the surface and groundwater components can confound the results of the budget analysis (Brown and Stark 1987). Additionally, many other factors can and may affect the parameters or their measurement. In all cases, these models should be verified, validated and calibrated through directed observations of existing or experimental wetlands.

Water Supply

Providing a relatively natural water supply system is often problematic. Much experience, often not positive, has been gained from the design of flood channels that also host wetland creation efforts. Concrete or earth-lined trapezoidal channels have been extremely popular in flood control design because they require little land, are relatively inexpensive to build, keep flood flows below existing ground levels, and are easy to maintain. As these systems were expanded to provide for wetland construction, however, the inherent problems associated with this channel design have become apparent; the perennial saturation of the flat bottom of the channel, which encourages the development of a low terrace riparian woodland dominated by willows and cottonwoods. These species tend to grow densely and rapidly and create a significant impedance to flood flows.

One resolution of these conflicts would be the establishment of a wetland vegetation association with relatively low channel impedance. Riparian woodland associations vary tremendously in their affect on channel impedance but generally follow a pattern of greater impedance with lower elevations. That is, a low-terrace woodland impedes flood flows more than a mid-terrace association, which in turn impedes flow more than a high-terrace association. In a study of riparian woodland establishment patterns at several wetland creation projects in the Central Valley of California, Zentner & Zentner (1992) observed that the low terrace woodland in this region occurs from the upper edge of the summer water level to approximately the mean annual flood line, the mid-terrace zone occurs from the mean annual flood line to about the 10- to 15-year storm line, and the high terrace woodland occurs from the mid-terrace zone to the 100-year flood line. Therefore, to circumvent blockage of flood waters by dense low terrace riparian growth while still providing for wetlands, broad low terraces could be designed which can accommodate flood flows despite the dense nature of the vegetation. Alternatively, the low-terrace zone could be restricted, and terraces would be constructed just above the mean annual flood line where mid-terrace riparian woodland, in friable soils, or marshes in indurated soils, can be maintained (Figures 4 and 5).

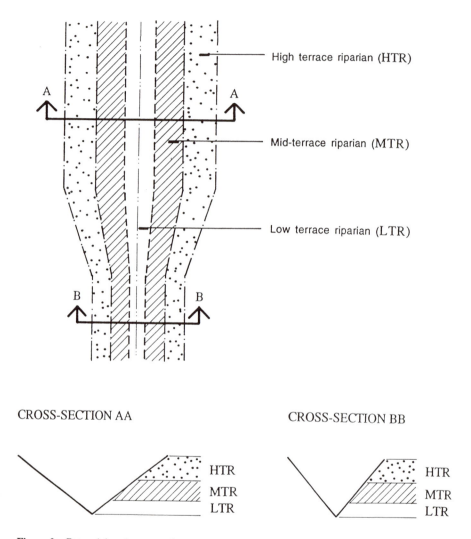

High terrace riparian (HTR)

Mid-terrace riparian (MTR)

Low terrace riparian (LTR)

CROSS-SECTION AA

CROSS-SECTION BB

HTR
MTR
LTR

HTR
MTR
LTR

Figure 3. Determining the appropriate elevation of a constructed slope wetland can sometimes be accomplished through observation of existing vegetation associations and the elevations at which the associations occur.

Additional hydrological factors, such as water velocity, sediment loading and erosion potential can also be important in water supply. Table 4 is a checklist for reviewing channel design issues. The checklist is used to review both upstream and downstream preconstruction and postconstruction conditions.

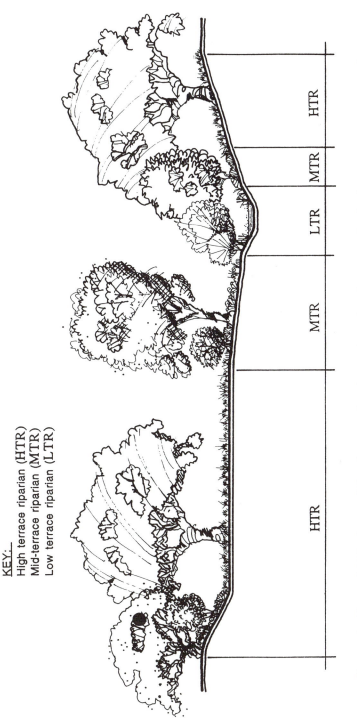

KEY:
High terrace riparian (HTR)
Mid-terrace riparian (MTR)
Low terrace riparian (LTR)

Figure 4. To avoid blockage of flood waters by dense, low riparian growth, the low terrace zone can be restricted, and terraces can be constructed above the mean annual flood line.

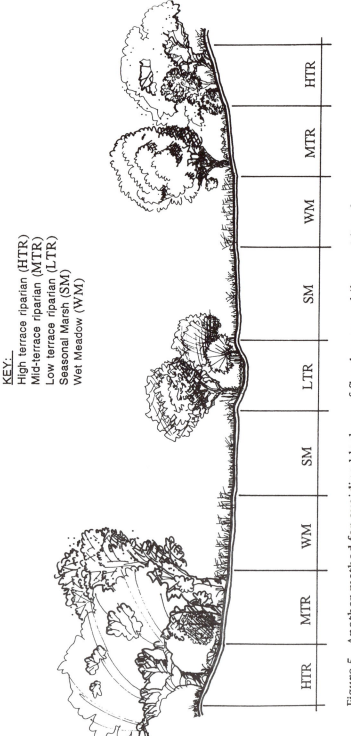

KEY:
High terrace riparian (HTR)
Mid-terrace riparian (MTR)
Low terrace riparian (LTR)
Seasonal Marsh (SM)
Wet Meadow (WM)

Figure 5. Another method for avoiding blockage of flood waters while providing for wetlands is to construct seasonal marsh in low terrace areas.

Table 4. Design checklist for channel wetlands construction. The checklist provides an organized approach for addressing the critical hydrological design components for wetlands construction.

DESIGN COMPONENT	PRECONSTRUCTION CONDITION	POSTCONSTRUCTION CONDITION
Water Surface Elevations		
Seasonal low water level		
Mean annual flood level		
100-year storm level		
Vegetation Association		
High impedance		
Moderate impedance		
Low impedance		
Channel Bottom Slope		
On-site		
Upstream		
Downstream		
Erosion and Erosion Potential		
On-site		
Upstream		
Downstream		
Sedimentation and Sedimentation Potential		
On-site		
Upstream		
Downstream		

Soil

Soil properties have often received little consideration in the design and construction of wetlands. Exceptions include wetlands where soil characteristics determine the wetland form, such as western vernal pools and bogs. In large part, this historic inattention to soils is due to the predominance of perennial marsh construction, which can be supported to some extent on almost any soil as long as an appropriate hydroperiod and water depth are provided. However, constructing the drier wetland types, such as seasonal marshes or wet meadows, has made soil considerations more important.

Soil texture and its corollary soil permeability often strongly influence the type of marsh, or the development of a woodland in place of a marsh. Hydroperiod and inundation depth being equal, finer textured, less permeable soils support a wetter vegetation association and a generally larger wetland than do coarser textured, more permeable soils. More permeable and well-oxygenated soils (including various grades of rock) will support woodlands in addition to, or instead of, marshes.

Habitat Notes

As noted above, perennial marshes can be established on almost any soil type. Soil depth is also an insignificant factor in many instances. Areas with indurated soil layers at the surface have established to perennial marsh within three months of inundation, as have areas of cobble (Zentner & Zentner 1990). Nevertheless, wetland vegetation establishment is easier where significant levels of organic material are present to increase soil water retention capacity. In several cases, peat or other organic mulches have been added to the substrate and have increased the growth of perennial marsh species (Brown et al. 1984).

Seasonal marshes generally require a surface horizon of 10 to 45 cm depth and a soil with some water-holding capacity. This is particularly true where root zone saturation is infrequent. However, if the soil is extremely permeable, seasonal marshes can still be established where water levels are periodically within 15 to 45 cm of the soil surface. Generally 2 weeks per month during the growing season is sufficient. Seasonal marshes can even be established on cobble, but not to the same extent as perennial marshes.

Wet meadows are greatly affected by soil type. Typically, wet meadows occur on clay or clay loam soil which is saturated for 4 to 8 weeks or more in an average year. Generally, inundation is infrequent and soil saturation is the primary source of water. Soil depth may also be important for many of the wet meadow species. Some rushes, for example, have a root depth to 1.8 m.

Riparian woodlands may establish on a broad array of substrates but, as with marshes, wetter vegetation associations are less affected by soil type than drier vegetation associations. Accordingly, low-terrace woodland species may establish on thin cobble layers above mud soils, a useful technique for stabilizing rock weirs in wetlands. However, mid- and high-terrace woodland species require a greater depth of permeable soil (and the concomitant groundwater supplies) as noted above.

In some instances, soil analyses of a potential construction site may show that wetland establishment is not feasible under natural conditions. Godley and Callahan (1984) document the construction of marsh and riparian woodlands in sandhills of Florida using a combination of a clay liner and organic soils to retain water. Pond wetlands with marsh and woodland components have been constructed in the xeric foothills of central California in cobble piles with extreme permeability using a mixture of clay and bentonite to retain water (Zentner & Zentner 1989). Despite the apparent success of these examples, all sites must be examined with an understanding that wetland construction on such sites may best be conducted elsewhere. One of the primary functional objectives of any ecological construction is to minimize or eliminate maintenance after an initial establishment period. Projects which depend on continual maintenance are, at best, unnatural and often unsuccessful in the long term.

Site Testing

The most direct method for handling soil issues related to wetland construction is observation of backhoe pits or trenches excavated in the construction area. The pits or trenches should reach a depth of at least 1.2 m. Where high-terrace woodland species (which generally require permeable soils to significant depth) are proposed, the excavation may necessarily be deeper. Soil texture and variations in texture, especially indurated layers or inclusions of variable material, should be observed and noted for each excavation. Soils are likely to vary considerably on sites of significant size. It is useful to identify similar soils on proximal, undisturbed sites using the SCS soil survey so that natural vegetation associations can be observed and used as a template. The soil survey can also be used to identify engineering constraints.

Soil laboratory analyses are useful for more formally defining texture and soil chemical properties. Salinity and pH, especially, are relatively inexpensive to test for and important for defining planting constraints, although a thorough understanding of the vegetation may also provide a sufficient guide to soil chemistry of a site.

Vegetation

At its most basic, plant ecology is the study of plants and plant associations, and the relationship of plants and plant associations to various edaphic and landscape factors. As wetland construction seeks to place a specific vegetation association in a specific location, it is basically applied plant ecology. Plant ecology encompasses an extremely broad range of subjects. Primary among these for most construction projects are soil and hydrology. Other factors having a significant effect on the establishment and persistence of the TVA are plant succession, planting design and weed control.

Succession

Succession refers to the changes which a vegetation association may go through in response to autogenic or allogenic factors. Two forms of succession are recognized in this chapter. *Phase* succession occurs when the soil and hydrology conditions do not change but the vegetation association changes, usually from a seral to a mature stage. *Type* succession occurs when the soil, hydrology, and vegetation association change. These terms have been used elsewhere for very different purposes, and in many cases are indistinguishable. They are used here to provide a conceptual model of change in the wetland system.

Phase Succession

Generally, most vegetation associations have seral and mature phases. The variation between seral and mature phases of marshes are relatively noticeable. Typically, readily seeded species such as cattails dominate the seral stage of a perennial marsh while species more adapted to vegetative propagation dominate the mature phase. Seed viability for these mature marsh species is often low or noncompetitive on bare ground compared with the more adventitious species. As such, plug planting may be the only practicable means of assuring growth of the mature phase during the early years of a construction project. Conversely, identifying the seral stage species of the TVA allows the ecologist the option of not planting these species, but instead simply ensuring a seed source. Understanding these phases of the TVA is important, and is best accomplished through observation of template sites, or experimental plots.

Seral and mature stages of woodland vegetation associations are often extremely difficult to identify. Successional models for woodlands suggest that the low terrace woodland association is the seral stage for the drier vegetation associations. This may be due to the historic predominance of autogenic succession in woodlands which once assured episodic or periodic pulses of sediment through the system, thereby altering soil and hydrology relations (and leading to type succession as described here). Once these successional forces have been eliminated through flood control, relatively static vegetation associations may result. However, phase succession is still observable in the response of species to canopy openings and, in the propagule stage, to varying degrees of light intensity. Species which, when planted in the field or in nursery conditions, do well in full sun generally represent the seral stage of the TVA, while species which require some shade typically comprise the mature stage. A common problem in woodland construction involves the planting of a full complement of woodland species in a newly constructed site, i.e. a site exposed to full sun, without consideration of the successional status of each species.

Type Change

In wetland construction design, succession and the external forces which seem to often control wetland change are both important parameters. Succession in wetlands often follows the classical autogenic model. Under more or less natural conditions, riparian woodlands become isolated from waterways, or sediment buildup provides for drier conditions, and the wetter association is gradually (or rapidly in some cases) replaced with a less mesic association (McBride and Strahan 1984). Similarly, perennial marshes may gradually increase in surface elevation to become seasonal marshes or wet meadows. However, the allogenic model also often holds true. Marshes may not transition through woodlands. For example, soil conditions adverse to woodlands, especially duripans or heavy clay layers, can favor dense mats of marsh plants that exclude woodlands. Finally, the construction design must be sensitive to those forces that may be periodic or episodic such as beavers, floods, fires, or debris flows.

Cultural factors are also increasingly important in succession. When floods are controlled and the waterway form remains constant, or sediment inputs have been drastically reduced, wetland associations may remain in place without succeeding to more xeric associations. Klimas (1989) found that high terrace, oak-dominated riparian woodlands were significantly underrepresented in the lower Mississippi River valley compared to the more mesic woodland associations. Typically, riparian woodland associations vary in seed size and weight in direct relationship to their successional stage and location relative to seasonal low water levels. That is, low terrace woodland species have the lightest seeds and are the first of the woodland vegetation associations to establish, while the high terrace species have heavy seeds and are a later successional stage. Klimas theorized that levees acted as barriers to the dispersal of the heavier seeds, which were not transported by flood waters as easily as the lighter seeds. Similarly, aggregate-rich lands adjacent to the San Joaquin River, California were mined over an 80-year period, resulting in a series of ponds supported by groundwater in a matrix soil of gravely loam. The soils are very permeable, and little to no flooding occurs. The ponds are all dominated by a low terrace willow-cottonwood association, a typical early sere, regardless of age (Zentner & Zentner, 1989). Without this knowledge, a designer reviewing the woodland stands in both regions would tend to under-represent the upper elevation woodlands (dominated by oaks in both cases) in the planting plan, thereby mimicking an altered vegetation association. The construction design must analyze which set of conditions will govern the construction site and plan for these accordingly.

Planting Design

Construction of a perennial marsh requires only water and at least a thin, preferably organic, soil layer. A more complex task is ensuring that the marsh is not overwhelmed by weedy species such as cattails. Planting marsh species representative of the mature perennial marsh in the proposed establishment zone is cost-effective and provides for rapid establishment of a diverse flora. Marsh species should be salvaged in the form of plugs from nearby sites to supplement on-site salvage, with planting of the marsh plugs to occur on 1-2-m centers.

Seasonal marshes appear to benefit most from salvaging and transplanting of marsh soils, probably due to their relatively high species diversity. At least 20-25 cm of soil and plants should be transplanted into the seasonal marsh zone and immediately supplied with the appropriate hydrologic regime.

Wet meadows are best established using common agricultural practices such as preparation of a seed bed and sowing of seed. In these cases, seed bed preparation (outside of the necessary hydrologic modifications) will consist of weed removal (ensuring the retention of surface soils), discing or otherwise cultivating the soil, possibly a second round of weed removal, seeding, and harrowing to cover the seed. For species which do not produce highly viable seed, plugging rooted plants (similar in size and with similar equipment to those used in tomato planting) or hydromulching with stolons should be considered.

Reviews of woodland plantings have found average densities of mature stands to vary from 500 to 750 trees per acre, roughly equivalent to plantings on 2.5- to 3-m centers. Under natural conditions of course, the initial establishment rate is often much greater, especially for the low terrace woodland species. A sparser planting rate assumes some maintenance or other care in planting that will result in significantly higher survival rates than under natural conditions.

Often neglected in forested wetland construction work, the herb stratum is nevertheless an important component of the riparian woodland. One to two month old plugs of native species should be planted at 1-2 plugs per square meter. Additionally, spreading detritus from an existing forest floor may be very helpful, as is planting stock already inoculated with the appropriate mycorrhizal fungi.

Weed Control

Weeds are nonnative, adventitious plants which replace desired components of the TVA. The term nonnative is used here to describe a species which occurs outside its historic range or landscape position. Weeds include species such as common reed (*Phragmites communis*) which has replaced native species in the New Jersey Meadowlands, or melaleuca (*Melaleuca quinquenervia*) which has invaded south Florida wetlands. Weeds do not include native species, such as cattails, whose growth and spread are promoted by disturbance but which are typically the natural, seral phase of a TVA.

In some instances, weed control or elimination can be the sole action taken in a wetland construction project. The spread of adventitious non-natives throughout the environment threatens many native vegetation associations. However, simply eliminating the exotics will not suffice to restore a TVA. Weed control must take into account the following steps:

1. Identify and address the conditions which allowed or encouraged the exotics to become established.
2. Remove the exotic.
3. Replace the exotic with the appropriate native plant.
4. Ensure that further opportunities for the exotic to become re-established are limited.

Prescriptive steps should be based on the control steps identified above. For example, a noxious pest in California, bristly ox-tongue (*Picris echiodes*) has very light seed which does well in dense annual grasslands which have been heavily grazed as the grazing leaves small openings in the turf which allow the plant to germinate. Mulching bare soil or ensuring a dense carpet of native groundcover and then reducing soil disturbance can effectively eliminate this species.

Cultural Issues

Size and Shape

The size and configuration of the construction site are often determined by market or regulatory forces well in advance of any design analysis. Even where strict limitations exist, the designer must consider the appropriate extent and shape for the different vegetation associations to be constructed. Recent work stemming from MacArthur and Wilson's (1967) research on island biogeography suggests that areas of at least 16 ha contain significantly greater numbers of species of birds, mammals, and invertebrates than do slightly smaller areas (Tilghman 1987, Faeth and Kane 1978). However, configuration may play a major role in these results with greater edge length and habitat diversity often resulting in higher numbers of some species (O'Meara 1984).

Adjacent Uses

Where adjacent uses include natural or restored lands, significant opportunities exist for a larger native landscape or a corridor. MacClintock et al. (1977), Fahrig and Merriam (1985), and Wegner and Merriam (1979) all provide study results which suggest that birds and small mammals actively use natural corridors, and that patches of native landscape connected by such corridors to larger habitats receive greater wildlife use than isolated patches.

Where adjacent uses include built environments, construction projects should consider potential use of the wetland by people and the effects of the human landscape on the natural system. Adams and Dove (1989) document widespread interest in viewing, and interacting with, wildlife within suburban and urban settings. Providing a trail that offers views and interaction with the wetland is often the best means to ensure that visitors do not impact sensitive areas. Leaving the edge of the wetland open so that views of the wetland are created may be the best means to ensure that the wetland is not used for trash and debris dumping. Off-road vehicle access should be eliminated through the use of post and cable barriers between any public roadways and the wetland, unless steep vertical barriers or moats are provided. Domestic and feral cats and dogs are significant predators on ground-nesting wetland birds. Both can be only partially restrained with fences, although open water is an effective barrier to cats.

Mosquitoes

Marshes and mosquitoes, as well as the problems caused by mosquitoes, have been synonymous for millennia. The increase in wetland construction projects has been a cause of some alarm for mosquito abatement officials. Marshes that generate mosquitoes which act as vectors for diseases or which annoy nearby residents will eventually be modified by mosquito abatement officials who, because they operate under a public health and safety mandate, are often immune to wetland protection regulations. Mosquito abatement officials are, therefore, particularly interested in influencing wetland construction design so that postconstruction control is unnecessary.

Mosquitoes have relatively short growth periods, and can fly a significant distance. Generally, optimal conditions for mosquitoes consist of well-vegetated, shallow (less than 20 cm), still water which is isolated from deeper water which would contain predators.

Wetlands can be designed to limit mosquito production (Figure 6). Wet meadow and seasonal marsh areas can be designed to dry by the onset of peak mosquito production periods. This can be accomplished using the basin model described above to compare evapotranspiration rates with the hydrologic requirements of the TVAs to define the weir height which will ensure drawdown at the appropriate time. This in turn determines the water depth. Perennial marshes can be designed as a series of isolated stands adjacent to open water areas that contain mosquito predators. Ponded water areas can include a slide-gated culvert to allow drainage and maintenance of the site should a major outbreak occur.

Water Quality

Untreated stormwater from urban areas is likely to contain pollutants, especially heavy metals, in such concentrations that water quality objectives are exceeded in downstream receiving waters. Many, though not all, pollutants contained in urban runoff are sediment associated. Controls to remove sediments, such as settling or infiltration basins and filtration devices, are recommended in design manuals as the most practical manner to control discharges of pollutants in urban runoff (Stahre and Urbonas, 1989). Settling basins should be designed to retain a specific level of storm event in the basin for a 24-48-hour period. Storms between the mean annual and 10-year storm levels are often used to define the basin size. Infiltration basins should be underlain by a soil that allows percolation at a rate sufficient to trap sediments in the interstitial area for the same storm level. Inflow structures for these basins should be designed so that erosion and short circuiting through the basin are minimized. Outflow structures can be dual low flow and high flow, allowing the passage of extreme events while ensuring that first flush waters receive the extended detention or infiltration time needed to allow settling. Low flow structures used in these basins should include trash racks to prevent clogging.

IMPLEMENTATION

Prior to beginning any construction project, the project designer and the contractors should meet and agree upon a specific order of operations and

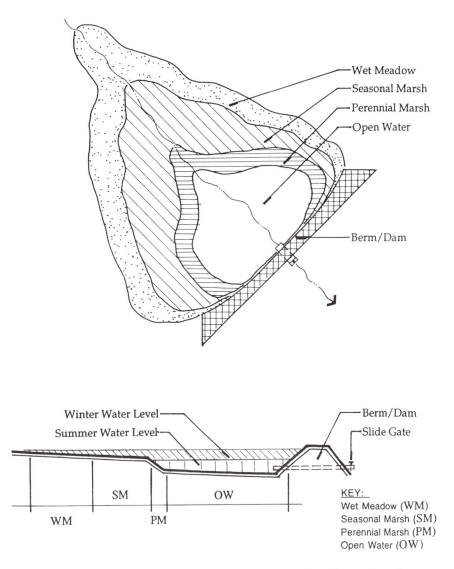

Figure 6. Mosquito production can be limited by designing wet meadow and seasonal marsh areas to periodically dry out, by limiting perennial marsh, and by providing a permanent open water refuge for mosquito predators.

responsibilities. The order of operations set forth in Table 5 is one method of constructing a woodland or marsh project. However, changes in the order of operations may be necessary to allow for weather conditions or other factors. It is imperative that the constructed wetlands, as well as any wetlands to be preserved, are not disturbed by construction activities. This includes using wetlands as right of ways to transport equipment, or as equipment storage areas. Communication and coordination between wetland construction operations and any other construction operations is essential. Access, turning on or off utilities, stockpiling materials on project sites, or any situation which might affect either operation should be resolved in advance.

An Ecological Monitor (EM) should be designated to ensure that the wetland is constructed in accordance with the approved construction plan if portions of the project will be built by non-ecologists. The EM should be responsible for certifying that the project has been built in accordance with the approved construction plan and permits, and should monitor all operations occurring in the wetland areas. The EM should also attend any preconstruction conferences and meetings during construction which are pertinent to the wetland.

Construction Sequencing

For most projects, construction should occur in an orderly sequence (Table 5). This sequence begins with the installation of protective flagging for any areas to be preserved, includes removing invasive species, salvaging desired species, grading, planting, and fencing, and concludes with the production of as-built plans. Construction of a water supply may be included in some instances.

Protective Flagging

Prior to any grading activities, all existing habitats which will remain undisturbed, those which will be salvaged within the general work area, and all travel ways should be staked and flagged. Flagging should be highly visible and replaced as necessary so as to be continuously inplace during construction activities. Stakes should be at least 1 m tall and painted an easily observable color such as fluorescent orange. Flagging should extend around the protected area perimeters or along their entire lengths to prevent any unwanted intrusions or damage. Accidentally damaged plants or areas should be repaired. All fencing and flagging should be removed at the end of the construction project.

Weed Removal

Mowing or scraping requirements for weed removal will vary from site to site depending on the amount and type of weeds and the amount of grading. Typically, areas of noxious weeds should be excavated and the infested soil deposited elsewhere. Alternatively, the weeds should be disked, watered to encourage seed growth, and disked again. The latter cycle should be repeated until weed growth is extremely sparse. Chemical controls may also be used with appropriate controls. Areas with little or no grading will require at least a thorough mowing so that debris will not hamper construction. A flail-type mower should be used, and all debris thoroughly mulched.

Table 5. Recommended sequence of construction events for freshwater wetlands creation.

1.	Install protective flagging and stakes for preserve or salvage areas.
2.	Remove all invasive exotic plants and on-site garbage.
3.	Salvage all specified plants.
4.	Grade construction site to specifications.
5.	Dispose of excavated materials at designated fill sites.
6.	Prepare woodland planting basins, seed beds, and other planting areas.
7.	Install water supply or irrigation system and test.
8.	Plant in designated zones.
9.	Provide protection of construction area.
10.	Prepare as-builts and construction report.

Salvaging

All appropriate plants and soil on site should be salvaged for replanting (Figure 7). Salvaging of wetland soil, and hence the seed bank, can be extremely important for construction success. Dunn and Best (1983) found that representation of plants from all successional stages may only occur where a seed bank has been placed in the marsh. Erwin et al. (1985) found much higher species richness and cover in marshes created with salvaged topsoil than with subsoil. However, seeds of woody species may be absent or underrepresented in seed banks (Leck 1989), requiring collection and broadcasting of woody seeds or the planting of propagules to effect timely development of riparian woodland associations.

If excavation has not proceeded to the point where salvaged plants can be planted immediately (within 48 hours of excavation), they should be kept in the shade and watered. The viability of salvaged plants often declines significantly over time. Additionally, marsh plants may be salvaged from selected locations outside the project site. Salvaging should occur only manually or with a trenching or similar narrow-bucket backhoe, and should remove no more than can be naturally re-established within one year.

Grading

Studies of wetland construction projects show that improper grading and grading management may be one of the most important factors contributing to failure (Garbisch 1977, Zentner 1988). The EM is most useful during this stage and

Figure 7. Early in the construction process, all appropriate plants and soil, such as those of this perennial marsh, should be salvaged.

will often be required to make quick decisions when unanticipated issues develop. For example, while excavating a basin for wetland construction in Folsom, California, the EM noticed a slightly discolored soil area. The discoloration proved to be an unobserved sand lens which, if left untreated, would have resulted in complete drainage of the proposed marsh after inundation. Similarly, large organic debris may be encountered during excavation and should be incorporated by the EM into the project. The EM should recognize that few preconstruction soil surveys are sufficient to precisely define subsurface conditions.

Planting

Plant material may be installed as container stock, bare root seedlings, cuttings, canes, plugs or seed. The first four are typical for woodland plantings; plugs are most often used in marsh projects, while seed is common for both.

Container plants generally range in size from small seedlings to 3.75-liter (1 gallon) or similar-sized stock. Smaller stock is less expensive to grow and plant while larger material beyond the 3.75-liter size can provide an immediate presence to increase buffer capabilities and aesthetics. Wetland or wetland buffer conditions are rarely as protected as nurseries, and larger stock will require additional care during the first year after planting to reach its potential. All plant material used should be collected from within the immediate region. Considerations as to the proximity of the gene pools and planted species vary greatly from one plant species to another. Plant material should be free from disease, insects and weeds, not be rootbound, and identified correctly as to genus and species. All plant species and quantities should correspond to the planting plan, and should be inspected prior to planting.

Container plants should be collected, propagated, and grown for at least one growing season prior to planting. Stock should be planted in a hole twice as deep and wide as the root ball (Figure 8). The root collar of the plants should be level with the surrounding soil level. Slow-release fertilizer, such as agriform or osmocote, may be added to each planting hole prior to planting the seedling. A water-conserving polymer may also be useful in arid regions. In many cases, construction of a basin will help hold water near the plant and ensure adequate irrigation. The basin typically consists of a 0.6-m diameter (or larger) water ring 5 cm deep with a surrounding berm 5 cm above grade centered around the plant. Shredded bark or similar mulch should be placed on watering basins to a depth of 5 cm making sure not to cover the crown of the rootball. A weed mat may be used to further reduce weed growth but should then be covered with mulch to reduce overheating of the near-surface soil. Weed mats generally reduce weed growth, but increase costs significantly. In addition, weed mats can trap moisture, thereby facilitating fungal growth which can lead to reduced plant viability. Where planting basins are eroded, dirt can settle on top of the weed mat and form a substrate for weed growth. The plant should be watered thoroughly after installation, and checked 2-3 days later for settling and stress.

Where woodland planting occurs in low permeability soils, most container plants will benefit from an augured planting hole. Holes should be responsive to the plant form. For example, deep tap root species will use deeper and narrower holes than shallow-rooted species. All holes should be backfilled by the end of that same workday, and should be completely saturated to promote settling of excavated soil.

Bare root seedlings have been a staple of the fruit tree industry for some time and are now commonplace in constructed wetlands. Bare root seedlings are essentially large container stock without the containers. Instructions for container stock noted above are equally applicable to bare-root material. Special care should be taken that the material is planted in a hole deep and wide enough to accommodate the root system without recurving the tap root.

The use of dormant cuttings in low terrace woodland establishment has been well-recognized. Cuttings should generally be 2 to 5 cm in diameter or larger and 1 m in length, although these dimensions will vary by region. At one extreme, construction work in the xeric southwest may involve auguring deep beneath the surface, and the planting of poles up to 6 m long in attempts to reach subsurface water (Anderson et al. 1984). At the other extreme, planting the same species in the more mesic Pacific Northwest may use a pole cutting 30 cm in length or less

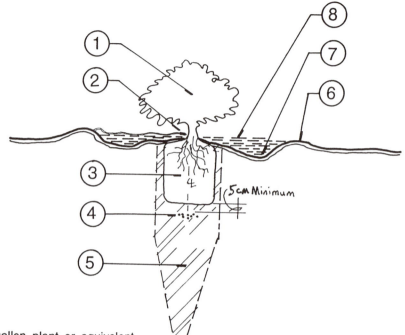

1. One gallon plant or equivalent
2. Drip emitter placed on rootball.
3. Plant rootball.
4. Fertilizer and/or water polymer.
   (See project specifications).
5. Augered plant hole, backfilled
   and watered-in.
6. Berm - 5cm above grade.
7. Water ring - 5cm below grade.
8. Mulch - 5cm deep.
   (See project specifications).

Figure 8. Survival of planted stock can be maximized through careful attention to planting, watering and fertilizing.

(Madrone pers. comm.). All cuttings should be taken from young wood and be free of disease. Two thirds of the cutting should be pounded into the ground ensuring no air pockets are formed near the cutting and soil should be tamped around the cutting. The top should be cut off at a 45-degree angle. Cuttings must be kept moist and can either be planted the same day as they were cut or left in water for 4 to 5 days to harden off. Cuttings are probably the most inexpensive way to quickly establish riparian woodland cover.

Canes are effective planting stock for many vines and resemble cuttings in that they should be nearly dormant and leafless, 60 to 75 cm long and in healthy condition. A shallow trench 3/4 of the cane length should be made and the cane laid in the trench and largely covered, with one end exposed above the soil. The trenches should be oriented on the contour, i.e. parallel to the slope.

Plugs are clumps of herbaceous plant material and associated soil and are easiest to salvage with backhoes and excavators. Manual excavation is used in sensitive areas. Each plug should be about 30 by 30 cm, and should retain as much substrate as possible (at least 10 to 20 cm deep). Plugs should be planted so that the base of the stems are level with the ground surface in inundated to saturated soil.

All seed used on the site should be certified as to germination percentage, and should be weed and disease free. All seeding rates and ratios should follow specifications developed by the grower. The EM should inspect all seed, soil preparation, calibration ratios, and phases of seeding operations. Specific soil preparation and seeding techniques should be adopted for each site and outlined in each site's planting plan. Generally, seeding will work best when it follows local agricultural practices, i.e. ask a nearby farmer. Extensive seedbed preparation (discing, cross-discing, harrowing, and culti-packing) may not be applicable on all sites but will provide good germination at relatively low cost. Hydroseeding or hydromulching with stolons may be applicable where dense plantings or site conditions make tractor- or hand-seeding and seedbed preparation not practicable. The optimal time for all seeding operations is during the dormant period. Seed broadcast early in the dormant season generally results in greater seed emergence than seed broadcast later in the dormant season. In arid regions, one key to success is to establish healthy, growing roots prior to summer drought, so that the roots may follow the subsurface moisture level downward through the soil as summer progresses.

Water Supply

Construction and emplacement of water control structures is relatively simple provided certain issues are resolved. First, review the placement of the structure prior to emplacement. Design analyses are important, and care should be taken to ensure that the location and design make sense in the field before blindly following construction plans. Second, ensure that the structure is firmly sealed into the surrounding soil. Water will tend to erode soil around the structure if proper contact is not provided. Third, check the structure over daily, seasonal and episodic flows. Finally, remember that any structure placed in a waterway will require continual maintenance. Plan ahead to provide the appropriate financing and maintenance reminders.

Enclosed supply systems (irrigation) may be required during the initial establishment period for woodland sites to assure the success of new plantings. Spray irrigation is not recommended due to its propensity to encourage weeds. Where the system is connected to a potable water supply, each system will require a reduced pressure backflow assembly which has been approved by the local regulating agencies. Each backflow assembly will need to be inspected and tested by a certified backflow technician. Where a nonpotable supply is used filtering will be required, usually several large filters in sequence. All irrigation lines should include a tracer line (#14 UF direct burial) for easy identification of line location. Generally, drip irrigation systems consist of a main line of PVC tubing 5-15 cm in diameter running to valves which control the flow. From the valves, secondary or lateral lines, 2.5-5 cm in diameter and also of PVC, transport water to local points where polyethylene (poly) tubing (2.5 cm in diameter) provides water directly to the plants. Before passing into the poly tubing all water should be filtered and the pressure reduced. Smaller poly tubing, known as spaghetti tubing, may be hooked into the larger poly lines for further movement of water. However, spaghetti tubing is difficult to maintain and not recommended for most habitat projects.

In buffer strips along streets and other visible areas all irrigation should be below grade. In more secluded areas, mainlines and valves should be below grade but poly lines may be installed above grade, thereby saving on trenching costs and easing maintenance. Fifteen cm staples or jute hooks should be used to secure poly tubing in place. Staples should be placed every 3 m on a run and 30 cm on either sides of each planting. All lines must be thoroughly flushed before inserting emitters using flushing end caps. Each planting receives its own emitter. It is very important that each emitter is placed directly over the root ball as the density of the root ball and indigenous soil may be quite different. Pressure testing upon completion of the main line and valves should be required. Lateral lines should be left on for a period of at least two hours to visually inspect all leaks. Upon completing the irrigation system, the EM should make a final walk through to verify the system is operating correctly.

Fencing

All habitat areas adjacent to streets should be separated from the roadways by a post and cable or similar barrier. In areas where maintenance access for vehicles is necessary, locking bullards should be installed in accordance with local specifications. Fencing during the construction period may also be necessary.

As-Builts

The EM should be responsible for overseeing preparation of grading, planting and irrigation as-builts when the project is completed. The planting as-builts should include the location, genus and species, and container size of each propagule. The EM should also certify that the project has been built in accordance with the plan, and transmit the as-builts and certification to the appropriate parties.

MAINTENANCE

Popular thought would have it that native plants are preadapted to the growing conditions of a site and should not require substantial maintenance to reach maturity. However, site conditions in the United States and much of the rest of the world have changed so radically over the past two centuries that site conditions often no longer reflect the conditions under which native species developed. A good example are the wet meadows of the Southwest. Prior to European settlement, these meadows were dominated by perennial graminoids. With the Europeans came an invasion of annuals that almost extirpated the native perennials. Over the past 200 years, the annuals and introduced livestock have succeeded in producing a substantial seedbed of annuals, formed thick thatch-dominated near-surface conditions, and fostered the growth of small and large mammal populations. These changes in the western environment have made construction of native meadows much more difficult than simply clearing a patch of ground and spreading some seed. Vigorous maintenance efforts must be carried out to keep the annuals from reinvading the native meadow.

Additionally, construction is not a natural process, but rather entails artificially inseminating a site with plant material. In natural conditions, the plants which reach 10 cm are the few survivors of several thousands of seeds in a favorable microsite at a time when conditions are relatively optimal for their growth. The project proponent rarely has the luxury to wait, or the knowledge to define these conditions.

Consequently, the species to be planted are unlikely to be propagated naturally and may not even be planted in a site for which they are now adapted. Maintenance activities will be needed to ensure that the vegetation association develops as a native vegetation association, without invasion by harmful exotic species. The array of tasks required for any one project may not match the list provided in Table 6, but each project is likely to require some form of maintenance, at least initially. Although maintenance is required, it should be viewed as a short-term measure, not only due to its costs but because long-term maintenance suggests that the TVA is ill-suited for site conditions. Consequently, effective wetland construction will entail: 1) a good design to ensure the best possible match between the selected species and a site, 2) manipulation of the site to improve planting and growth conditions while still recognizing that ecological change which will occur, and 3) maintenance, even intensive maintenance activities, but generally limited to the first three years after the cessation of planting. If a site is not relatively self-sufficient after this time, either additional construction work will be required or the design must be reconsidered.

Weed Control

In most cases, careful design and preplanting weed control procedures will reduce the burden of weed abatement considerably. However, a freshly-graded construction site is a haven for weeds which may be carried in from offsite. The ecologist must make a distinction between ruderal species that are part of natural succession and those species that will impede development of the TVA. The latter, particularly nonnative, aggressive species, can be addressed using manual or chemical controls. The ecologist must also ensure that the appropriate native species replaces the weed, and the conditions which resulted in weed growth will not reoccur.

Table 6. Wetland construction maintenance is an essential step for ensuring project success. Each project should be assessed for the problems listed below, and appropriate remediation initiated.

Problem	Tools	Survey Frequency (First Year)
Weeds		
Annual	Manual	1 to 3 times per year
Perennial	Chemical	
Woody	Hand Removal	
Erosion		
Sheet	Redirect flow	Poststorm and
Scour	Regrade	Annually
Slump	Replant	
Herbivory		
Postplanting	Screens	Monthly
grazing		
Plant Care		
Nutrients	Fertilizer	As needed
Form	Staking	
Water supply	Basin rework	
Pest/diseases	IPM/spray	
Irrigation		
Mortality	Replacement	Monthly
Litter		
Trash	Patrol	Monthly

Manual controls include hand removal of the entire plant or cutting the plant above-ground. Hand removal of the entire weed is labor-intensive and is best suited for woody plants. The advantage of hand removal is that the area of impact is usually quite small. Less specific weed control activities can create the type of disturbance which results in more weed growth. It is often much less expensive to cut the weed and then spot-spray the stump with an herbicide. However, for conditions that militate against the use of chemical controls, an array of tools are available from suppliers that can be used to selectively remove even mid-sized trees without disturbing the adjacent land or plants.

Cutting can include hand-pruning or chopping, weed whips or similar hand-held but power tools, and tractor-driven or floating mowers. Selection of a manual control method will depend on the type and extent of the weed and the project design. Hand-pruning or chopping is labor-intensive and costly unless the personnel consists of volunteers. Prior to selecting this method, the ecologist should personally weed a specific area for at least two hours and develop an estimate of the amount of person-hours then required, multiplying this by the cost of the labor. Where weeding will occur in dense growth, or where only a few plants among many are to be eliminated, this method is the most productive. Weed whips and similar tools are very useful for clearing annual or soft perennial growth over large or small areas where tractors cannot be driven, e.g. slopes. However, these should not be used near planted trees and shrubs or above-ground irrigation lines as even an experienced practitioner can girdle a plant or cut a poly line by mistake. Mechanical mowers are the most efficient means of cutting large areas of vegetation. However, they cannot run on steep slopes (although some have mowing booms which can reach out and cut limited areas) and require a certain distance between trees and shrubs to be preserved. It is important to consider these issues during the design phase. An easily maintained project is generally subject to less disturbance and may prove to be more successful.

Timing is often crucial in manually controlling weeds, especially annuals. Because cutting does not remove the root systems, it is generally important to cut the plant just prior to seed set to maximize the likelihood of complete control. Even this technique may not always work. Mowing of the annual yellow star thistle (*Centaurea solstitialis*) on clay soils just before seed set almost eliminated this noxious weed on one site while mowing of another population on rocky soils (where this species thrives) at the same time resulted in a second flowering several weeks later below the cutting level of the mower.

Chemical controls are often the least expensive means of eradicating or reducing weed growth. However, very few chemicals are licensed for use in wetlands and a licensed technician is often required by local regulations. Further, environmental conditions (e.g. wind speeds) never seem to be as optimal as required and spraying often inflicts damage outside the target area. Chemical controls include pre-emergents that are applied to the ground prior to the emergence of the weeds in the spring, and post-emergents that kill or stunt plants during their active growth stages. Pre-emergents are most useful when applied adjacent to newly planted propagules as they limit the growth of competing weeds. Post-emergent controls are most useful for broad areas completely dominated by weeds, areas where no plants are proposed (firebreaks, for example), or spot-spraying of individual weeds.

Surveys for weed-control needs should start at the beginning of the first growing season following construction. Weeding frequency will depend on the weed species and seasonal growth cycles. Annuals may need controls applied as much as 3 times in a year, whereas woody plants can generally be removed once per year.

Erosion Control

Erosion control is best effected by good design. However, one of the truisms of any construction project is that site conditions are not completely predictable. Unseen channel irregularities, inclusions of different soil types, and similar events will require modifications to the design during construction in order to minimize subsequent maintenance efforts.

Erosion in marshes or the channel banks of riparian woodlands often occurs as sheet erosion, scouring, or slumping. Sheet erosion takes place on relatively flat slopes and generally results in the loss of all or most of the new plantings. This is typically due to poor seedbed preparation, inadequate root penetration prior to the storm, a subsurface soil layer which inhibits root growth, or a combination of the above. The remedy is to try again. If possible, channel the flowing water into a swale rather than across the slope. Scouring occurs on moderate to steep slopes where higher velocities rapidly remove soil to some depth. Scouring is usually the result of modifications which have increased velocities through the site or reoriented high velocity flows. Remedies are cause-specific, but look for forces which have increased upstream flows or reduced downstream elevations. Slumping occurs when the surface soil on a channel bank moves downward as a relatively intact unit, often because it has been undercut. The movement will take anything on top of the soil unit, such as plantings or irrigation lines, with it. Remedies are, again, site specific, but may be corrected by light grading and replanting after the erosive stimulus has been corrected.

It is not atypical to underestimate the flow velocities at specific points in a channel. Water supply structures, especially those perpendicular to the channel, can be undermined or the adjacent side slopes eroded. Where the erosion is relatively mild, for example only a few cm of soil are lost between the structure and the soil, heavy rock interplanted with fast-growing trees may eliminate future erosion. More significant erosion will almost always require replacing the structure, or identifying and rectifying the cause of the erosion.

Reviewing the project site for erosion should occur immediately after the first three storms and each time thereafter for storms of greater intensity. Erosion controls are best applied annually, either immediately following the storm event that caused the erosion or during the appropriate planting or grading period.

Herbivory

Grazing has been and may be used to manage for a TVA. However, the affect of herbivores on target plants is not always beneficial. The most common remedy for deer, rabbit, or beaver grazing is to provide wire screens around the plants or planting area. These screens can be relatively costly to provide, adding as much as $1.00 to the cost of each plant. Consequently, screening might be most efficiently applied after a problem has surfaced. Screens in or adjacent to waterways must be removed prior to high water levels, though, or floating material will catch in the screen, creating enough mass that screen and propagule soon wash away. Monitoring for herbivory should occur monthly during the spring of the first year after planting, and less frequently thereafter.

Plant Care

Plants may be fertilized once each in the spring of the first and second year after planting if nutrient levels are low on the construction site. Fertilizer should be dribbled into the basin around the propagule. Trees may also need to be staked if wind or other forces are warping or bending the tree unnaturally. Two, 2.5-m pressure treated stakes, 5 cm in diameter, should be placed 60 cm into the ground, 30 cm on either side of the tree. The tree should be bound to the stakes using tree tie tape fastened in a figure eight pattern, 30 cm below the top of the stake. Basins around propagules must also be maintained to supply sufficient irrigation water.

Insect or disease infestations are common on habitat construction projects. Generally, all infestations should be treated using Integrated Pest Management (IPM) techniques. Plantings of a single species are especially susceptible to massive outbreaks which are not well treated by IPM, however, and commercial sprays may be required. Monitoring for these problems should be completed at least monthly during the first year of the project with remedial action taken immediately.

Irrigation System Maintenance

The irrigation system should be checked regularly, as much as monthly immediately following planting, for malfunction and vandalism. All drip lines and filters should be flushed as needed. Emitters must be checked regularly for malfunctions, position of emitter over root ball, and vandalism. Flushing end-caps must also be checked regularly. In colder climates, backflow devices and mainline systems should be completely drained at the end of the growing season to prevent rupturing due to freezing.

Irrigation for plantings is a temporary tool to help establish the constructed vegetation association. Water should ideally not be needed by the fourth year. Caution should be exercised not to create lush plants, which would become dependent on regular watering. Deep watering, spaced at the longest time possible before plants show signs of stress, should be used. This type of watering will encourage plant roots to travel downward into the appropriate water zones in the soil.

Litter Removal

Wetlands, because they are usually depressions in the landscape, seem to collect litter. Litter can impede the growth of propagules or seedlings, and increase the cost of plant care maintenance efforts. Therefore, it is recommended that each site be thoroughly policed, and all litter and debris removed from the site at least annually. Construction projects which are near or within residential areas generally collect the most trash. Over 400 pounds of litter was collected from one project in central California in 2 months time, at a cost of 42 hours and $350 in dump fees. It should be assumed that projects in residential areas will require at least one to two hours per month for litter collection.

General Maintenance Frequency

For the first one to two years, most projects will require a walk-through by an ecologist or maintenance technician every one to three weeks. The variation in the number of visits depends on the size of the site, its proximity to disturbance and the time of year. Visits may reasonably be required every week for the first year and twice monthly for the next 2 years. Smaller sites, for example those less than 10 acres, can reasonably be visited once every 2-3 weeks during the first year, and monthly in subsequent years.

Minimizing Maintenance Efforts

Local educational efforts can effectively reduce maintenance efforts associated with vandalism and litter. Efforts should be aimed at informing neighbors as to the project goals, values and needs, and in enlisting their support for the long-term protection of the site. Generally, conducting a replanting project for local school children or similar groups at least once every two years is very helpful. Small brochures for each project which are mailed to local residents can also be beneficial. Preparation of a maintenance manual in the last year of monitoring for long-term maintenance of the project area is also effective.

REFERENCES

Adams, L.W. and L.E. Dove. 1989. Wildlife reserves and corridors in the urban environment. National Institute for Urban Wildlife, Columbia, MD.

Anderson, B.W., J. Disano, D.L. Brooks, and R.D. Ohmart. 1984. Mortality and growth of cottonwood on dredge-spoil. In R.E. Warner and K.M. Hendrix (eds.) California Riparian Systems: Ecology, Conservation and Productive Management. UC Press, Berkeley, CA.

Brown, M.T., F. Gross, and J. Higman. 1984. Studies of a method of wetland reconstruction following phosphate mining. In F.J. Webb (ed.) Proceedings of the Eleventh Annual Conference on Wetland Restoration and Creation. Tampa, FL.

Brown, R.G. and J.R. Stark. 1987. Comparison of ground-water and surface-water interactions in two wetlands. In K.M. Mutz and L.C. Lee(eds.) Wetland and Riparian Ecosystems of the American West. Proceedings of the eighth annual meeting of the Society of Wetland Scientists, Wilmington, NC.

Clark, J.R. and J. Benforado (eds.). 1981. Wetlands of Bottomland Hardwood Forests. Elsevier, Amsterdam.

Cowardin, L. M., V. Carter, F.C. Golet, and E.T. LaRoe. 1979. Classification of wetlands and deepwater habitats of the United States. FWS/OBS-79/31. U.S. Fish and Wildlife Service, Washington, D.C.

Dane, C.W. 1959. Succession of aquatic marsh plants in small artificial marshes in New York State. New York Fish & Game 6(1):57-76.

Dickson, R.E., J.F. Hosner, and N.A. Hosley. 1965. The effects of four water regimes upon the growth of four bottomland tree species. Forest Science 11(3):299-305.

Dunn, W.J. and G.R. Best. 1983. Enhancing ecological succession: seed bank survey of some Florida marshes and role of seed banks in marsh reclamation. In Proceedings of the 1983 Symposium on Surface Mining, Hydrology, Sedimentology, and Reclamation. University of Kentucky, Lexington, KY.

Erwin, K.L., G.R. Best, W.J. Dunn, and P.M. Wallace. 1985. Marsh and forested wetland reclamation of a central Florida phosphate mine. Wetlands 4:87-104.

Faeth, S.H. and T.C. Kane. 1978. Urban biogeography: city parks as islands for Diptera and Coleoptera. Oecologia 32:127-133.

Fahrig, L. and G. Merriam. 1985. Habitat patch connectivity and population survival. Ecology 66:1762-1768.

Foin, T.C. and M.M. Hektner. 1986. Secondary succession and the fate of native species in a California coastal prairie community. Madrono 33:189-206.

Frayer, W.E., T.J. Monahan, D.C. Bowden, and F.A. Graybill. 1983. Status and trends of wetlands and deepwater habitat in the coterminous United States, 1950s to 1970s. Dept. of Forest and Wood Sciences, Colorado State University, Fort Collins, 32 p.

Garbisch, E.W. 1977. Recent and planned marsh establishment work throughout the contiguous United States: a survey and basic guidelines. U.S. Army Corps of Engineers, Waterways Experiment Station, Vicksburg, MS.

Godley, J.S. and R.J. Callahan. 1984. Creation of wetlands in a xeric community. In F.J.Webb (ed.) Proceedings of the Tenth Annual Conference on Wetland Restoration and Creation. Hillsborough Community College, Tampa, FL.

Harris, R.W., A.T. Leiser, and A.T. Fissell. 1975. Plant tolerance to flooding. University of California at Davis, Department of Environmental Horticulture, Davis, CA.

Klimas, Charles V. 1989. Limitations on ecosystem function in the forested corridor along the lower Mississippi River. In Kusler, J. A. and S. Daly (eds.) Wetlands and River Corridor Management. Proceedings of a symposium presented by the Association of Wetland Managers. Charleston, SC, July 5-9.

Kusler, J.A. and M.E. Kentula (eds.). 1989. Wetland Creation and Restoration: The Status of the Science. EPA/600/3-98/0. EPA, Corvqliis, OR.

Larson, J.S., M.S. Bedinger, C.F. Bryan, S. Brown, R.T. Huffman, E.L. Miller, D.G. Rhodes, and B.A. Touchet. 1981. Transition from wetlands to uplands in southeastern bottomland hardwood forests. In Clark, J.R. and J. Benforado (eds.). Wetlands of Bottomland Hardwood Forests. Elsevier, Amsterdam.

Leck, M.A. 1989. Wetland seed banks. Pp. 283-305 In Leck, M.A., V.T. Parker and R.L. Simpson (eds.) Ecology of Soil Seed Banks. Academic Press, Inc., San Diego, CA.

Lewis, R.R. 1990. Wetlands restoration/creation/enhancement terminology: suggestions for standardization. pp. 417-422 in Kusler, J.A. and M.E. Kentula (eds.) *Wetland Creation and Restoration: the Status of the Science.* Island Press, Washington, D.C.

MacArthur, R.H. and E.O. Wilson. 1967. The Theory of Island Biogeography. Princeton University Press, Princeton, NJ. 203 pp.

MacClintock L., R.F. Whitcomb, and B.L. Whitcomb. 1977. Evidence for the value of corridors and the minimization of isolation in preservation of biotic diversity. Am. Birds 31:6-16.

Margules, C., A.J. Higgs, and R.W. Rafe. 1982. Modern biogeographic theory: are there any lessons for nature reserve design? Biol. Conserv. 24:115-128.

McBride, J.R. and J. Strahan. 1984. Establishment and survival of woody riparian species on gravel bars of an intermittent stream. American Midland Naturalist 112(2):235-245.

Mitsch, W.J. and Gosselink J.G. 1986. Wetlands. Van Nostrand Reinhold, NY.

Payne, N.F. 1992. Techniques for Wildlife Habitat Management of Wetlands. McGraw-Hill.

O'Meara, T.E. 1984. Habitat-island effects on the avian community in cypress ponds. Proc. An. Conf. Southeast Assoc. Fish and Wildlife Agencies 38:97-110.

Simberloff, D. and J. Cox. 1987. Consequence and costs of conservation corridors. Conservation Biology 1:63-71.

Stahre, P. and B. Urbonas. 1989. Stormwater Detention. Prentice Hall, Englewood Cliffs, NJ.

Tilghman, N.G. 1987. Characteristics of urban woodlands affecting breeding bird diversity and abundance. Landscape and Urban Planning 14:481-495.

U.S. Army Corps of Engineers. 1981. HEC-1 Flood Hydrograph Package: User's Manual. Water Resources Support Center, Hydrologic Engineering Center, Davis, CA.

Wegner, J.F. and G. Merriam. 1979. Movements by birds and small mammals between and wood and adjoining farmland habitats. Applied Ecology 16:349-357.

Zentner, J. 1988. Wetland restoration success in coastal California. In Zelazny, T. and T.S. Feierabend (eds.). Increasing Our Wetland Resources. Proceedings of a 1987 National Wildlife Federation Conference. Washington, DC.

Zentner & Zentner. 1989. Ball Ranch Mitigation program. Report prepared for the Sienna Corporation. Walnut Creek, CA 150 pp.

Zentner & Zentner. 1990. Natoma Station Monitoring Report. Report prepared for the U.S. Army Corps of Engineers, Sacramento District. Walnut Creek, CA 150 pp.

Zentner & Zentner. 1992. Lower Laguna Creek drainage master plan: Mitigation Program. Report prepared for Sacramento County. Walnut Creek, CA.

CHAPTER 8

ENHANCEMENT, RESTORATION AND CREATION OF COASTAL WETLANDS

Roy R. Lewis

Wetland management does not have to be reinvented, just fine-tuned and applied with definite objectives and goals (Shisler 1990).

Coastal wetlands are generally defined as those wetlands that lie within the realm and effects of salt water. As such, coastal wetlands include seagrass meadows, mangrove forests, and saltmarshes.

Although they are not a traditional wetland type, seagrass meadows (Figure 1) do meet the criteria for protection of aquatic habitat under Section 404 of the Clean Water Act (Fonseca 1990). Seagrass meadows are the only type of coastal wetlands subject to nearly constant submergence. They occur from the subtidal zone to depths of up to 30 meters, depending on light penetration. Seagrass beds are vegetated by true flowering plants. Typical dominant species include eel grass (*Zostera marina*) at temperate latitudes, and turtle grass (*Thalassia testudinum*), shoal grass (*Halodule wrightii*), and manatee grass (*Syringodium filiforme*) at subtropical and tropical latitudes. In shallow areas exposed to occasional freshwater inflow, such as along the Gulf coast and in eastern Florida, widgeon grass (*Ruppia maritima*) is common.

Mangrove forests are limited in distribution to subtropical and tropical zones (Figure 2). In the U.S., they occur predominantly in southern Florida, sparsely along the Gulf coast to the Laguna Madre of Texas, and extensively in Puerto Rico (Kuenzler 1974). Many species of trees of different families are called mangroves, with black mangrove (*Avicennia germinans*), red mangrove (*Rhizophora mangle*) and white mangrove (*Laguncularia racemosa*) common to the United States. There are an additional 31 species worldwide (Tomlinson 1986). All mangroves have features in common that are adaptations for survival in the saline conditions of the intertidal zone (Tomlinson 1986). Morphological adaptations include the aerial roots (pneumatophores) of black mangroves which provide for gas exchange, and the viviparous, floating propagules of red mangroves. In fact, nearly all species of mangroves are viviparous, with the white mangrove being an exception. Physiological mechanisms for dealing with salt excretion or exclusion are also characteristic of mangroves.

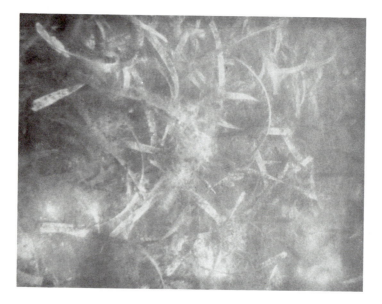

Figure 1. Seagrass meadows satisfy the criteria established under Section 404 of the Clean Water Act and are considered wetlands. Subject to nearly constant inundation, seagrass meadows occur from the subtidal zone to depths of nearly 30 meters.

Saltmarshes are the most ubiquitous type of coastal wetland, occurring on all coasts where appropriate substrate and tidal regimes are present (Figure 3). They are distributed over a relatively broad salinity range, from the high intertidal zone to the oligohaline habitats upstream on tidal tributaries where salinity may never exceed 10 parts per thousand. Various species of cordgrass (*Spartina* spp.) are dominant, with rushes (*Juncus* spp.) also common. On the Pacific coast, *Spartina foliosa* and pickleweed (*Salicornia virginica*) are common. On the Gulf coast and in the southeast, other species of cordgrass (*Spartina alterniflora, S. patens*), saltgrass (*Distichlis spicata*), and black needlerush (*Juncus roemerianus*) are typical. Sawgrass (*Cladium jamaicense*) marshes, such as the Everglades, occur in more brackish water. On the Atlantic coast and to the northeast, *Spartina cynosuroides, S. alterniflora, J. roemerianus*, and some other species are common. Wax myrtle (*Myrica cerifera*) and groundsel tree (*Baccharis halimifolia*) are typical shrubs associated with saltmarshes from the Gulf coast to New England.

Oligohaline (salinity of 0.5 to 5.0 ppt) and tidal freshwater marshes (salinity is less than 0.5 ppt) are herbaceous wetlands located in tidally influenced rivers or streams where the plant community exhibits a diverse mixture of true marine species and typical freshwater taxa that tolerate low salinities (Cowardin et al.

Figure 2. Mangrove forests are found throughout the subtropics and tropics. The tree species which comprise mangrove forests have morphological and physiological adaptations for surviving in intertidal, saline conditions.

1979, Lewis 1990). Tidal freshwater marshes of the mid-Atlantic and Georgia Bight regions may contain as many as 50-60 plant species at a single location (Broome 1990). Ecologically, these marshes provide critical nursery habitat for blue crabs, snook, tarpon, and ladyfish. Because their significance has only recently been recognized, much of this habitat has already been lost. Methods and efforts to restore oligohaline and tidal freshwater marshes lag behind those of other systems.

PRELIMINARY DESIGN CONSIDERATIONS

As defined by Lewis (1990), a created wetland is one that has been converted from a persistent upland or shallow water area into a wetland through some activity of man. An enhanced wetland is an existing one where some activity of man increases one or more values, often with the accompanying decline in other wetland values. A restored wetland has been returned from an altered or disturbed condition to a previously existing natural condition or altered condition by some action of man (e.g., fill removal). Habitat improvement, as used here, may be taken to mean any of the three alternatives: restored, created, or enhanced.

Figure 3. Saltmarshes are the most widely distributed of coastal wetlands. They occur over a relatively broad salinity range, from the high intertidal zone to oligohaline habitats on tidal tributaries.

Establishing Goals

The goals of habitat improvement are frequently constrained by the purpose of the project. Because such projects are so frequently conducted for the purposes of mitigation (i.e, to compensate for some kind of adverse habitat impacts), it is almost a foregone conclusion that the project goals must include replacement of the same type of habitat, in the same area. In other words, the mitigation must be in-kind and on-site. These requirements in turn are based on the assumption that the functional value of the damaged wetland can and should be replaced. This goal is typically evaluated by comparing the flora and fauna of the improved community with a natural community as reference. In this manner, the more similar the communities, the more successful the mitigation project.

However, in-kind, on-site mitigation may not always be ecologically sound. Essentially all of the existing ecosystems of the United States (and most of the rest of the world) have been disturbed to some extent. Therefore, in-kind, on-site mitigation would maintain an artifact of the historical natural community, due to the differential loss or conversion of various plant and animal components of the community. To illustrate this point, consider an actual project involving construction of a seawall adjacent to a causeway across Tampa Bay, Florida. A narrow fringe of mangroves and saltmarsh about four meters wide borders the causeway, and would be impacted by seawall construction (Figure 4). This fringe wetland occupies isolated areas less than 0.5 hectare in size, and its ability to support clapper rails (*Rallus longirostris*) and other coastal species, or to provide other wetland functions, is non-existent due to its narrow width and fragmentation (Roberts 1989). As mitigation, the regulatory agencies required excavation of the narrow upland transition zone fringe (a rare habitat in its own right) upgradient of unimpacted wetlands, and planting of intertidal vegetation. This mitigation attempted to recreate a poorly functioning wetland habitat at the cost of a second habitat. From an ecosystem standpoint, it may have been more prudent to have created a coastal wetland large enough for potential use by clapper rails and other species, and to have created it in a place where it would not be disturbed in the future. Functional equivalency should be a goal of habitat improvement projects conducted for mitigation, but only if the function of the altered wetland is worth replacing.

The goal of replacing damaged or altered wetlands may be becoming obsolete for practical as well as ecological reasons. With the increasing acceptability of mitigation banking, as well as the limited land area available for providing in-kind, on-site mitigation, broader goals should be considered. The alternative provided by mitigation banking and other out-of-kind, off-site mitigation options may in many instances be more realistic and ecologically sound. Josselyn et al. (1990) have suggested that agency staff must be prepared to manage for regional habitat goals, not on a project-by-project basis, and point out that regional habitat goals would not necessarily call for in-kind habitat replacement. In concurrence, Shisler (1990) states that goals should be based upon a wetland system's requirements within a watershed or region.

For habitat improvement outside of the mitigation process, functional equivalency may be an unrealistic requirement. The implementation of large-scale coastal wetland restoration is one of the goals stated in the recently released National Research Council's (1992) report, *Restoration of Aquatic Ecosystems: Science, Technology and Public Policy*, which recommends that *inland and coastal wetlands be restored at a rate that offsets any further loss and contributes to an overall gain of 10 million wetland acres by the year 2010*. Habitat improvement for its own sake allows a much broader range of goals. Outside the mitigation process, it is unnecessary to maintain the same system. In fact, it is under these circumstances that habitat enhancement is most likely to be conducted, altering existing conditions to favor a very specific goal. Often, habitat enhancement is disadvantageous for nontarget species. For example, creating feeding areas for wood storks is necessarily detrimental to fish.

With respect to coastal wetlands in particular, some goals may be more realistic than others. Despite intensive and repeated efforts, the success rate of restoration or creation of seagrass beds through planting is limited. Regardless of whether seedlings, long shoots, or mature plants are used, seagrass beds can rarely be established, maintained, or enlarged through planting (Figure 5). The reasons for this lack of success are not completely known, but existing wave and current patterns, poor water quality, and bioturbation have all been shown to be factors

Figure 4. This narrow fringe of mangroves and saltmarsh along a causeway in Tampa, Florida has a relatively low functional value. In such instances, mitigation for impacts to the system should consider out-of-kind, off-site options.

contributing to failure (Lewis 1987). Nevertheless, increased success of planting efforts and natural expansion of seagrass meadows in Tampa Bay have occurred simultaneously with water quality improvements. Protecting existing seagrass beds from excessive wave action can also keep them from becoming patchy and perhaps disappearing forever.

Saltmarsh and mangrove habitats are more amenable to direct improvement techniques. Success in establishing marsh vegetation is quite high if appropriate technology is applied (Figures 6-8). Kusler and Kentula (1990) have stated, that *in general, the ease with which a project can be constructed and the probability of its success are ... greatest overall for estuarine marshes.* Coastal marsh restoration can be relatively inexpensive and rapidly achieved. In some cases, all that is required is to re-establish a tidal connection. Renewed inundation by saltwater stresses exotic vegetation and allows native species to recover. In other cases such as in Louisiana, life-giving sediments trapped within the levees can be easily transferred into the sediment-starved marshes through manmade gaps (crevasses) in the levees.

For mangrove forests, although the area of application in the U.S. is small (500,000 acres total), the general techniques for successful restoration or creation

Figure 5. Enhancement, restoration or creation of seagrass meadows requires intensive effort, and is rarely successful. Goal establishment should recognize the chances of success, as well as the need to address ancillary issues such as water quality and wave action.

of the plant community have been demonstrated (Figures 9-10). As a result of stricter regulations in the past decade, mitigation for damage to mangrove forests has become a rare occurrence, at least in Florida. Habitat improvement projects for mangroves are likely to specify the initial planting of cordgrass, which serves as a pioneer stage in the successional process leading to a mature mangrove community (Lewis 1982). This approach is also relatively cost-effective.

Other goals of coastal habitat improvement may include optimizing the use of dredged material, creation of increased wetland acreage, erosion control, and water quality improvement (Chabrek 1990, Josselyn et al. 1990). Fonseca (1990) suggests five specific goals for seagrass restoration, which may be applicable to coastal ecosystems in general:

- Vegetative cover should develop and persist
- Equivalent acreage should be attained
- Increased acreage should be achieved where possible
- The species lost should eventually be replaced by the same species
- There should be development of a faunal population equivalent to reference wetlands

Figure 6. The chances for successful habitat improvement are greatest for estuarine marshes. Pictured above is a newly planted saltmarsh.

For seagrass species, Fonseca (1990) notes the difficulty of achieving these goals, and states that *success in recovering climax species is so rare that out-of-kind replacement with another seagrass species is encouraged.*

General Site Selection

The ideal site for proposed habitat improvement will be that which most closely meets the requirements of the community to be improved. The site selection process also depends on whether the improvement project is undertaken for reasons of restoration or for habitat creation. The goals of the project will have substantial bearing on final site selection. Obviously, there will be certain prerequisites to obtaining desired results that are linked to proper site selection. Restoration projects inherently require understanding the initial cause of habitat degradation, and determining the probability that the habitat can be restored and maintained. Wetland creation is generally more difficult because major site alterations are usually involved. These alterations may or may not be economically and environmentally acceptable (Shisler 1990), especially when one habitat must be destroyed to create another. The question of habitat displacement inevitably needs to be addressed in the site selection process.

Figure 7. One year later, the planted cordgrass can be seen to have vegetatively reproduced, and plant coverage has increased.

Figure 8. Eight years later the saltmarsh is entirely covered with vegetation, and is essentially indistinguishable to the eye from a natural system.

Figure 9. Success has also been demonstrated for enhancement, restoration and creation of mangrove forests. Pictured above are newly planted mangrove seedlings.

Two principal criteria that have been identified for the site selection process are: 1) the land should have low fish and wildlife resource value in its present state, and 2) an adequate, natural water supply should be available, such as a river, stream, tidal source, or groundwater (Garbisch 1986). The following discussion summarizes essential factors to be considered with respect to existing conditions, and assumes that these criteria are met.

Site History and Current Status

Shisler (1990) states that *if wetlands are not present in an area that is normally inundated by tides, there has to be a reason.* Determining this reason is fundamental to reviewing a site for potential wetland habitat improvement. In the case of habitat restoration, the causes of wetland degradation must be determined and corrected before the wetland can be restored. If possible, past functioning of the wetland should be evaluated, and all past uses of the site should be reviewed prior to habitat restoration or creation. Current and future land use, zoning regulations and projected sea-level rise may also have a bearing on site selection. Uplands with steep slopes or seawalls should receive higher priority than salt flats and transitional uplands for coastal habitat improvement (Crewz and Lewis 1991).

Figure 10. Five years later, it can be seen that many of the mangrove seedlings have survived and grown. Barring any further perturbations to the system, a mature mangrove forest will eventually occupy the site.

Topography

Elevation and slope are critical factors in determining the success of intertidal zone habitat improvement projects. The optimum elevation for marsh or mangrove plantings is ideally determined by observing and measuring the lower and upper elevation limits of a nearby natural marsh (Broome 1990). It is important to obtain elevations locally because there may be geographic variation of optimal elevation within species. Lewis (1990) further recommends that the lowest and highest points should be disregarded, and only the middle range used for planting. If such reference information is unavailable, an adequate test program should be conducted prior to initiating actual habitat improvement. It is difficult to recommend an absolute planting elevation for a given species because of local and geographic variability in tidal range, but in general, it is the upper one-third of the normal tidal range. Synergistic effects may also alter growth at given elevations. For example, reduced salinity may permit growth of smooth cordgrass at higher and lower elevations than those typically recommended for the species.

Some degree of slope is essential for the proper drainage of intertidal projects. A gradual slope will increase the area available for planting and will dissipate wave energy over a greater area, thereby reducing the possibility of erosion (Broome

1990). Slope should be towards tidal sources to minimize ponding as substrates settle following site preparation. Gentle slopes do not drain as extensively as steeper ones, which is generally beneficial to vegetation, but they are more susceptible to ponding.

Hydrology

The quantity, quality and timing of water entering the wetland is of critical importance to its survival. If the correct hydrology is not present or is insufficient, the desired project will fail (Shisler 1990). Shisler (1990) recommends using adjacent wetlands as a model in the assessment of project design.

Hydrologic factors that need to be evaluated during the site selection process include drainage features, such as channels and ponds, the tidal regime and range, wave intensity, and salinity. Obviously, some of these can be manipulated through design while others cannot. It is sometimes necessary to base site selection upon the most practical trade-off that can be made. In general, wetland habitat improvement projects are more likely to be successful in sheltered locations where wave energy and current velocity are minimal. Salinity variation at the site should be monitored so that appropriate plant species can be selected. Adequate drainage is another requirement, as standing water is typically detrimental to the establishment of wetland vegetation. Wide, shallow channels that retain water at low tide maximize flushing and plant health. Water quality (dissolved oxygen, turbidity, etc.) is an especially important factor determining site selection for seagrass restoration projects. Other factors affecting hydrology are precipitation, surface flows, ground water, and evapotranspiration. Each of these factors affects habitat productivity and species diversity.

Substrate

Soil conditions at the potential site must be appropriate for planting or modification, and are easily checked with soil-auger core samples. Sandy soils are generally more amenable to grading and planting than are silt and clay. However, some degree of organic content is recommended as a nutrient source (Broome 1990). The soil must be of sufficient depth to support planting. A minimum depth of 0.3 m is recommended by Garbisch (1986). Based on literature review and field inspections of created wetlands in New Jersey, substrate preferences are ranked as follows: 1) natural marsh peat, 2) clay and silty clay, 3) estuarine sediments (dredge and fill), and 4) sand (Shisler 1990). If the site requires the addition of soil, or if the site has been filled in the past, the possibility of contamination should be investigated. Chemical analysis of the soil may be necessary in some situations. Some anoxic soils can become highly acidic upon exposure to air, such as during earthwork, and may warrant the addition of a calcareous material, such as crushed shell, as a buffering agent. At sites subject to regular inundation, sedimentation processes should be carefully studied. Some degree of sedimentation may be desirable to stabilize initial plantings; however, burial of seedlings should be avoided.

Light

Light characteristics of potential habitat improvement sites are especially critical for submerged seagrass meadows, but should also be a consideration with respect to other coastal wetland types as well. Shading by nearby vegetation can be a deterrent to project success.

Existing Wetlands, Wildlife and Vegetation

Existing conditions at the proposed site should be carefully evaluated prior to choosing the site for habitat improvement. Potential sites frequently have some degree of ecological value which may be compromised in the process of habitat improvement. Wetlands already present on the site, although disturbed, may have more functional value than would result from a poorly planned and executed improvement project. Potential sites should be carefully surveyed for current wildlife usage to ensure that no disruptions to local populations, especially of protected species, occur.

Adjacent Site Conditions

Conditions near the proposed site for habitat improvement should also be considered. Sites adjacent to existing functioning wetlands offer the greatest chance for success. However, increasing wetland acreage in the watershed may alter hydrologic regimes.

The presence of aggressive exotic species close to the proposed site may also be a problem. In many cases, a buffer zone should be included in the project design. A buffer zone may increase the diversity of flora and fauna present (Josselyn et al. 1990), and will protect the site from human and vehicular traffic. If present, the buffer zone must be kept clear of exotics and other nuisance species that may invade the project site. Native shrubs such as wax myrtle and marsh elder are recommended for planting in buffer zones. Wide buffer zones with tall vegetation provide animals, especially birds and mammals, with refuge from cold winds, high temperatures, high tides and storms.

The proximity or attractiveness of the potential restoration site to predation should be considered. Geese, muskrats, and nutria have all been identified as voracious consumers of newly planted marsh vegetation (Shisler 1990, Broome 1990). Conversely, placement of a well-designed created wetland near a disturbed wetland may offer refuge to dwindling populations of desirable species. Human access should be discouraged. It may sometimes be necessary to select a more remote, less accessible, site simply to provide a degree of protection from anthropogenic disturbance.

Open water near potential sites should be assessed for boat traffic and offshore depth, as both can affect wave energy. Shallow offshore water reduces the severity of waves reaching the shore (Broome 1990). The degree of wave energy can be modified to a certain extent (see below), but sites facing more than

5.5 nautical miles of open water are generally not recommended for planting (Crewz and Lewis 1991).

Special Considerations for Seagrass Meadows

A conservative approach to site selection for seagrass projects is advocated by Fonseca (1990), in that his recommendations assume that in-kind, on-site habitat improvement is required or desired. In order of preference, Fonseca recommends: 1) sites previously impacted by poor water quality, and the water quality has improved; 2) restoration of filled or dredged areas that once supported seagrasses; 3) restoration of filled or dredged areas, irrespective of their previous plant community; and 4) conversion of uplands to seagrass habitat. Fonseca (1990) notes that options 3 and 4 may not always be ecologically sound.

DESIGN CONSIDERATIONS

The degree of design required depends on how close the site is to desired conditions, and how well the site meets the standards established by the project requirements.

Site Location and Structure

Designing a habitat improvement project to be conducted at a given site is dependent on the structure and location of the site. Exposure to wave energy is particularly important. The relationship between wave energy and effects on cordgrass planting can be related to fetch (the distance traveled by waves). At less than one nautical mile of fetch, plantings will be unaffected. Between one and 3.5 nautical miles, some replanting is necessary, and over 3.5 nautical miles, some kind of wave-barrier structure is necessary to protect plantings (Crewz and Lewis 1991).

Orientation of the site with respect to colonizing (or invading) plants and animals is another factor influencing design. Access to desirable flora and fauna should be provided, while exposure to exotic and nuisance species is restricted. Prevailing winds and currents may enhance or deter establishment of volunteers. The presence and condition of adjacent wetlands may also affect design specifics. For example, preserving isolated patches of desirable vegetation may inhibit water circulation by creating berms. Sound ecological judgment must be used in assessing whether existing vegetation impairs or contributes to developing the functional wetlands attributes at the site (Crewz and Lewis 1991).

Site Size

The size of the site selected for habitat improvement can have some bearing on the ultimate design of the project. Larger sites allow for more diversity of species and elevations. In larger projects, complex transitional margins should be an integral part of the design, to allow for inland migration of wetlands as sea level rises. Size may also be a factor in site design when out-of-kind mitigation is being considered. Regulatory agencies can require compensation at ratios greater than 1:1, particularly if habitat enhancement is offered as mitigation for loss of undisturbed natural wetlands. A higher ratio is also justified by the net decline in habitat availability to wildlife (Josselyn et al. 1990).

Wetland loss involves time as well as space. Therefore, to realistically compensate for losses, mitigation trade-offs should be expressed with a temporal as well as a spatial component, for example acre-years (Crewz and Lewis 1991). Because of the different amounts of time required for wetlands of different types to mature, the compensation ratios involved should be adjusted accordingly. Increasing the area of mitigative habitat improvement required may be used to offset discrepancies in the time component.

Slope

Perhaps the most critical aspect in successful habitat improvement design is grading the soil surface to the elevation that provides the optimal hydrologic regime for the desired plant species (Broome 1990). Slopes should be as gradual as possible, while still allowing good surface drainage at low tide.

Gentle slopes at the perimeters of the project sites are recommended to reduce erosion and filter runoff reaching the site. If steep slopes are necessary, they should be stabilized with sod, which will also serve as a deterrent to invasion by exotic or nuisance species. Gradual slopes at the perimeter of coastal habitat improvement sites also serve to prevent compression of vegetative zones due to sea-level rise, the rate of which is predicted to increase rapidly after the year 2025 (Crewz and Lewis 1991).

Hydrology

Improper hydrology often leads to colonization by non-target species. While this in itself may not be cause for concern, it does allow the possibility of invasion by nuisance or undesirable exotic species. For small, fringe marshes, elevation and wave climate are probably the most important design considerations. Larger, broad expanses of marsh require construction of drainage channels simulating natural tidal creeks, to provide tidal exchange and access by wildlife (Figure 11). Distinguishing between the low marsh zone (regularly flooded) and high marsh zone (irregularly flooded) is necessary for proper design of this habitat type.

Wave Barriers

Some kind of proximal offshore structure is recommended for many sites exposed to moderate wave energy. The protection offered by berms or breakwaters can be substantial and critical factor in successful establishment of newly planted vegetation. At some sites, construction of a breakwater may be all that is necessary to stabilize a site, both subtidally and intertidally, and consequently facilitate significant habitat improvement. For example, a small spoil island in

Figure 11. Channelization modifies the hydrological regime and salinity gradient of coastal marshes, in many cases facilitating the establishment of exotic and nuisance plant species. Creation of tidal channels is an essential step in restoring the natural community.

Clearwater Harbor that provided good wildlife habitat for local populations (especially birds) was experiencing severe shoreline erosion. Offshore seagrass beds were also showing signs of stress related to excess wave energy. In 1989, a 140 m long breakwater of large rocks was constructed approximately 25 m offshore of the island, seaward of the grass beds. Following breakwater placement, the shoreline stabilized through formation of a small beach, and the seagrasses appear healthy and may be expanding in area.

Large rocks and rubble are the material recommended for construction of breakwaters in most circumstances. Smaller material may be less stable, or can consolidate excessively, prohibiting adequate tidal flushing. Constructed properly, rock berms used in conjunction with mangrove restoration projects aid in trapping propagules, thereby reducing or eliminating the need for planting.

Vegetation

If a site is ideally suited to a particular type of habitat improvement, the decision whether or not to install plants may present itself. Suitable tidal freshwater sites can rapidly become vegetated on their own in as little as six months (Broome 1990). On the north-central Gulf coast, two general types of marsh habitat improvement projects are common. The first is the use of dredged material, which may be confined by temporary or permanent dikes. Alternatively, dredged material may be deposited unrestricted in open water. Such projects are rarely or only experimentally planted, and are usually allowed to vegetate naturally. The second less common approach to marsh habitat improvement is diversion of river flow. This process provides substrate, and the introduction of fresh water counteracts the effects of saltwater intrusion and rising sea level. Again, natural colonization is allowed to occur in projects of this type (Chabrek 1990). The major benefit of natural colonization is the automatic use of local gene stock. However, the time required for natural colonization varies considerably. Good plant establishment can occur within several months in freshwater tidal wetlands, but may require several years in more saline conditions (Shisler 1990). The presence of an adequate seed or propagule source should be verified if natural colonization is anticipated.

Another option is the use of wetland soils salvaged from another site. The topsoil (sometimes referred to as mulch) removed from a wetland scheduled for destruction can be stockpiled and later used as a top layer in a wetland improvement project. The effectiveness of this soil as a seed bank depends on the plant species present (exotics and nuisance species may be excessive), and the viability of the seeds. The latter depends on the species, the length of time the topsoil has been stockpiled, and the conditions under which the topsoil was stockpiled. This technique is most effective if the soil is in place at the beginning of the growing season.

Plant Selection

The species selected for planting will depend on the type of wetland desired. In Florida, one or two species are commonly used for coastal marsh habitat projects, with the emphasis on smooth cordgrass. However, with the increasing emphasis on wetland function, complex plantings of multiple species may become more common. For specific guidelines on plant species selection, refer to the regional reviews in Kusler and Kentula (1990). Crewz and Lewis (1991) provide specific guidelines for species in Florida.

In addition to identifying species to be planted, the form of the plants needs to be specified. Depending on the species and the location, seeds, seedlings, vegetative shoots, or transplants may be the most desirable planting unit. In the case of mangroves, propagules are recommended for red mangroves, but seedlings of black and white mangroves are recommended in favor of seeds. Smooth cordgrass can be grown from seed, but not throughout its range. However, plugs and bare-root culms are easily transplanted, and nursery-grown units from field-harvested stock are easy to cultivate, handle, and install. Note that permits may be required to remove material from coastal wetlands. Seedlings grown from seeds of cordgrass and black needlerush are used in the southeast (Broome 1990); transplants of black needlerush are generally unsuccessful. Saltgrass can be grown from transplants, rhizomes, seeds, or plugs. Most seagrass species are difficult to cultivate and transplant, although turtle grass has been grown from seeds and seedlings. Other forms of this and other species used in seagrass restoration and mitigation projects, with varying degrees of success, include transplants and vegetative shoots.

In general, plants grown locally should be used because adaptations to local conditions may create ecotypes. This is especially important with regard to species with wide geographic ranges such as smooth cordgrass.

Planting Density

Installation on 1 m centers is the standard for most coastal marsh, mangrove, and seagrass restoration projects (Figure 12). However, this is probably not a realistic specification. Naturally colonizing species, mangroves for example, tend to occur at much higher initial densities. Density decreases with differential survival of the hardiest plants. Crewz and Lewis (1991) recommend 25 cm centers for red mangrove propagules, and 50 cm centers for 1-year old seedlings. As increasing planting density also increases the cost of a project, there should be a distinction made for mitigation versus restoration. The importance of obtaining functional equivalency in mitigation should require higher planting density, while restoration can be more economical, relying on natural colonization and a longer period of time allowed to achieve stated objectives. Planting density should be determined according to the desired rate of plant coverage, which should be rapid in the case of mitigation for short-term wetland losses. The size of the site may also affect planting density, as large sites are often planted at a lower density than a smaller one of the same wetland type, simply for reasons of economics.

EXECUTION

Logistics

Among the factors that expedite project implementation is obtaining all necessary permits prior to scheduled site preparation and planting. There can be several regulatory agencies involved in a habitat improvement project, and the specific requirements of each must be met. These obligations typically include assurance that nearby habitats will not be disturbed, and provision of means to control siltation during construction.

Scheduling of actual planting needs to be coordinated with plant availability, regardless of whether plants are field-collected or nursery-grown. This also needs to be scheduled for the optimal time of year for planting the species involved. Construction and final grading should take place well in advance of the optimum planting dates as several weeks are required for the settling of fill material (Broome 1990). Generally, planting should occur when temperature, rainfall, and salinity are moderate. In the southeast, planting should occur between the first of April and mid-June (Broome 1990). In central and south Florida, this period can be extended to mid-March through September. Predicted tides should be checked to avoid excessively high or low water conditions. Construction should also be timed to avoid disturbance to seasonally nesting or migratory wildlife.

Elevation and Grading

Measurement of elevations at a site requires reference to established benchmarks. Nearby structures may have recorded elevations that can be obtained

Figure 12. Most coastal species are typically planted on one meter centers, but, other planting densities may be more appropriate in some circumstances. Natural plant establishment patterns, cost and purpose of project should be considered when determining planting density.

from local government agencies. At least two benchmarks should be used when establishing a site-reference elevation. As an alternative, local plant populations can be used as a guide to proper elevations. Juveniles or seedlings of the desired species provide the most accurate information because they are more sensitive to elevational differences than adult specimens. If earthwork is required to attain desired slope and elevations, the site should be surveyed prior to plant installation to ensure that the specified characteristics are attained. Topographic surveys should be completed and reviewed before construction equipment is removed from the site, so that slope and elevation can be efficiently adjusted. Shisler (1990) recommends preparing a bathymetric or topographic contour map on 15 cm (0.5 ft) contours prior to planting, but after fill has settled. The map also provides a base for as-built surveys and monitoring efforts.

If the site has been filled or excavated, sufficient time for the substrate to settle should be allowed prior to determining and adjusting final elevations. If planting occurs too soon after earthwork, the site may be susceptible to ponding and resuspension (turbidity), and the substrate will not provide adequate support to young plants.

Drainage

Site drainage should be checked by observing a range of tidal cycles over the site, and observing runoff during and after storm events. If necessary, channels can be created to eliminate undesirable ponding areas. However, some shallow ponded areas can provide refuge for marsh-resident fish and crustaceans during low tide, and feeding areas for wading birds.

Plant Installation

The actual means of plant installation depend on the habitat type, the planting units used, and the substrate. On intertidal sandy substrates, mechanical means (e.g., tobacco planter pulled by a tractor) can be used while on soft or muddy substrate, planting must often be conducted by hand, working off of wooden mats or rafts. Planting submerged seagrasses typically requires the use of snorkels, SCUBA, or hookah equipment.

If using transplanted donor material from another site, it should be removed from the donor site so as to not create conditions attractive to potentially destructive fauna. For example, large open patches are attractive to Canada geese, and invite root predation by avian and mammal populations (Shisler 1990).

Josselyn et al. (1990) offer general guidelines for planting on the Pacific coast. Prior to planting, site soils should be monitored for salinity, moisture and pH. Plants should be native species from the local region, and commercial cultivars should not be used. If transplants are used, the donor site should be monitored to ensure that no permanent damage has occurred. High marsh and transition zones should be planted in the fall, prior to the rainy season, unless irrigation is available.

Fertilizing

The effects of fertilizer on mangroves are unclear, but fertilizer has been shown to accelerate the growth of cordgrass (Broome 1990). Fertilizer and lime can be used to provide short term adjustments to substrate nutrient status and pH. Although plants may respond positively to small amounts of fertilizer at planting, accommodation to predominant long-term substrate conditions is probably desirable (Crewz and Lewis 1991). When used, only a time-release fertilizer incorporated into the substrate at the time of planting is recommended. Broadcasting fertilizer over the site is expensive, increases eutrophication in the water column, and only delays the achievement of nutritional equilibrium.

Protection of Plantings

Some projects will require temporary protection of newly installed plants from excessive wave action during a preliminary establishment period. This can be accomplished through the use of such methods as floating tire breakwaters, earthen dikes, sandbags, and erosion control fabric. Although unattractive, floating tire breakwaters have the advantage of being able to be used at other sites once plant establishment has occurred.

Construction Management

Every habitat improvement project should be overseen by a team composed of, minimally, a qualified wetland scientist, an engineer, and a hydrologist. Problems are to be expected, and are best addressed by an interdisciplinary team. For mitigation projects, permit conditions should include on-site review by experts (Josselyn et al. 1990).

If corrections become necessary during the course of construction, the problem should be corrected at its source, or the decision to accept the change should be made. In permit situations, the regulatory agency involved may be held responsible for making corrections to a plan it has already approved. Therefore, permit conditions should allow for the possibility of plan modification during construction. If plant survival and growth are adequate but coverage is poor, the time allowed to attain specific coverage should be extended. The system is performing, but at a slower rate than predicted. If replanting is necessary during the monitoring period, the monitoring clock should be set back to zero (Fonseca 1990).

Monitoring and Maintenance

Monitoring the habitat improvement project is an integral part of its design and execution. Monitoring needs to be conducted at two levels, one for enforcement and the other for effectiveness. Compliance monitoring is necessary to assure regulatory personnel that permit specifications have been satisfied, and is relatively straightforward. As what is specified and what is actually constructed may differ, a time zero report and documentation of as-built conditions are essential (Lewis 1990). Flexibility may be necessary when monitoring for effectiveness. Many habitat improvement projects may fail to meet all of the specified objectives, yet they may provide unanticipated benefits. As an example, poor substrate at a restoration site may prohibit establishment of the desired plant species, but may allow rapid and healthy colonization by an endangered species. Judgment of success or failure should be based on ecological function rather than on predetermined objectives. Furthermore, the degree of success reported is likely to affect the permitting and acceptability of similar projects in the future (Broome 1990). If possible, monitoring should be conducted by a multidisciplinary team, including an ecologist, botanist, zoologist, hydrologist/engineer, and perhaps a soil scientist.

Monitoring methods vary from simple qualitative observations to highly technical sampling regimes. Photographing the site over time from an array of fixed reference points is probably the most effective way of conveying information about the development of the project. Aerial photography, although expensive, can also be effective, especially for seagrass projects.

Sampling regimes should be designed with subsequent statistical analysis in mind. Consultation with a statistician during monitoring plan preparation can be of assistance. If zonation is likely to occur at the site, stratified random sampling should be used (Broome 1990). Obtaining data adequate for statistical analysis must sometimes be tempered with economics and practicality. Actual methods of sampling can include percent cover, percent cover by species, and enumerating plant colonizers, as well as destructive methods involving actual removal of plant material. While most accurate, methods such as clipping are frequently damaging, especially for seagrasses and needlerush. Smooth cordgrass can be effectively sampled by removal methods if conducted at the end of the growing season (Broome 1990). Additional general guidelines for monitoring programs are given by Lewis (1990).

The time frame over which monitoring is conducted is usually specified in permit conditions, with 3-5 years a reasonable minimum for most project types. Realistically, ten years is necessary in order to draw conclusions regarding community productivity and succession. In either case, the final report should summarize all the data and compare the results with the project goals and literature values of the parameters measured.

Interpreting the sampling results requires comparison with a natural reference marsh, or with literature values. In many cases, comparison with a natural reference wetland is mandated. However, variances as great as 50 percent have been observed in animal populations associated with natural wetlands that appear identical.

Maintenance of the site should be conducted before, during, and after planting. Litter, and nuisance or exotic species, must be removed at all times. Invasion of undesirable plant species should be controlled through physical removal, or alteration of drainage or salinity characteristics. Herbicides should be used only as a last resort, and then only by qualified personnel.

ADDITIONAL CONSIDERATIONS

Permitting and Monitoring

Contingency plans should be included in the permit conditions in case the proposed mitigation project fails. There needs to be some measure of agency responsibility in the site selection and design approval process, in the event the permittee acts in accordance with the permit conditions, yet the project does not attain stated goals. Construction of mitigation projects should be complete before development begins, because of the time required for the constructed wetland to become functional. Funding for restoration or mitigation projects should be guaranteed prior to, or as a condition of, permit issuance (Fonseca 1990). Regional databases of permit requirements and monitoring results, including those of older (10 years and more) projects should be established, so that long-term data becomes available and useful.

Education

There is a wealth of literature now available for many kinds of coastal habitat improvement projects. Educational opportunities for both consultants and regulatory personnel are increasing. Courses in wetland habitat design and construction are available through several commercial sources, and at some academic institutions.

Research Needs

Many issues related to the enhancement, restoration and creation of coastal wetlands would benefit from basic and applied research. Many wetlands are restored and managed with just a few target faunal species in mind and would benefit from increased knowledge about species' life requisites. The physical and chemical properties of wetlands, including topography, soil properties, subsidence rates (which vary locally) and hydraulic circulation are poorly understood. For wetlands designed to improve local water quality, information is needed about the effects of treatment on other wetland functions.

Methods of creating and improving oligohaline and tidal freshwater habitats would benefit from additional study, as would development of methods for cultivating the diverse plant species associated with these habitats. More research is needed on the effects of wave energy, and means of ameliorating this energy. Additional information on the nutrient requirements of marsh plant species, the effects of fertilizer, and nutrient flux in coastal systems is required. The limited availability of funds for habitat improvement projects suggests a need for studies of natural recruitment rates. More information on the use of stockpiled wetland soils is needed to develop methods which minimize problems associated with oxygen exposure and acidification. Some degree of experimentation can be incorporated into both project designs and monitoring plans.

Wetland enhancement, restoration and creation should be examined in the context of the regional ecosystem, with the possible outcome that out-of-kind or off-site creation and restoration may be deemed acceptable due to the regional loss and scarcity of a distinct habitat type such as oligohaline marshes (Lewis 1990). Watersheds should be inventoried to determine what systems are present, where loss and damage have occurred, which are the most effective projects, and which are the most suitable (Shisler 1990). Furthermore, knowledge of the natural wetland development process is a fundamental prerequisite to understanding the creation and restoration process (Shisler 1990). Because coastal wetland systems are destined to change over time, the design and goals of habitat improvement projects should take this dynamic nature into account (Shisler 1990).

REFERENCES

Broome, S.W. 1990. Creation and restoration of tidal wetlands of the Southeastern United States. Pp. 37-72 in Kusler, J.A. and Kentula, M.E. (eds.)

Wetland Creation and Restoration: the Status of the Science. Island Press, Washington, D.C.

Chabrek, R.H. 1990. Creation, restoration, and enhancement of marshes of the north central Gulf Coast. Pp. 125-142 in Kusler, J.A. and Kentula, M.E. (eds.) *Wetland Creation and Restoration: the Status of the Science.* Island Press, Washington, D.C.

Cowardin, L.M., V. Carter, F.C. Golet and E.T. LaRoe. 1979. *Classification of Wetlands and Deepwater Habitats of the United States.* Office of Biological Services, Fish and Wildlife Service, U.S. Dept. of the Interior, Washington, D.C.

Crewz, D.W. and R.R. Lewis. 1991. *An Evaluation of Historical Attempts to Establish Emergent Vegetation in Marine Wetlands in Florida.* Florida Sea Grant College Technical Paper No. 60, Florida Sea Grant College, University of Florida, Gainesville, Fla.

Fonseca, M.S. 1990. Regional analysis of the creation and restoration of seagrass systems. Pp. 171-193 in Kusler, J.A. and Kentula, M.E. (eds.) *Wetland Creation and Restoration: the Status of the Science.* Island Press, Washington, D.C.

Garbisch, E.W. 1986. *Highway and Wetlands: Compensating Wetland Losses.* U.S. Dept. of Transportation, Federal Highway Administration Report No. FHWA-IP-86-22.

Josselyn, M., J. Zedler and T. Griswold. 1990. Wetland mitigation along the Pacific coast of the United States. Pp. 3-36 in Kusler, J.A. and Kentula, M.E. (eds.) *Wetland Creation and Restoration: the Status of the Science.* Island Press, Washington, D.C.

Kuenzler, E.J. 1974. Mangrove swamp systems. Pp. 346-371 in Odum, H.T., B.J. Copeland, and E.A. McMahan (eds.), *Coastal Ecosystems of the United States,* Vol. I, The Conservation Foundation, Washington, D.C. 533 pp.

Kusler, J.A. and M.E. Kentula (eds.). 1990. *Wetland Creation and Restoration: the Status of the Science.* Island Press, Washington, D.C. xxv + 591 pp.

Lewis, R.R. 1982. Mangrove forests. Pp. 153-171 in Lewis, R.R. (ed.) *Creation and Restoration of Coastal Plant Communities.* CRC Press, Boca Raton, Fla. 219 pp.

Lewis, R.R. 1987. The restoration and creation of seagrass meadows in the Southeastern United States. Pp. 159-174 in Durako, M. J., Phillips, R. C. and Lewis, R. R. (eds.) *Proceedings of the Symposium on Subtropical Seagrasses of the Southeastern United States.* Fla. Dept. of Natural Resources Mar. Res. Pub. No. 42, St. Petersburg, Fla.

Lewis, R.R. 1990. Wetlands restoration/creation/enhancement terminology: suggestions for standardization. Pp. 417-422 in Kusler, J.A. and Kentula, M.E. (eds.) *Wetland Creation and Restoration: the Status of the Science.* Island Press, Washington, D.C.

National Research Council. 1992. *Restoration of Aquatic Ecosystems: Science, Technology, and Public Policy.* National Academy Press, Washington, D.C. xxi + 552 pp.

Roberts, T. 1989. Habitat value of man-made coastal marshes in Florida. Pp. 157-179 in Webb, F.J., Jr. (ed.) *Proceedings of the 16th Annual Conference on Wetlands Restoration and Creation.* Hillsborough Community College, Tampa, Fla.

Shisler, J.K. 1990. Creation and restoration of coastal wetlands of the Northeastern United States. Pp. 143-170 in Kusler, J.A. and Kentula, M.E. (eds.) *Wetland Creation and Restoration: the Status of the Science.* Island Press, Washington, D.C.

Tomlinson, P.B. 1986. *The Botany of Mangroves.* Cambridge University Press, Cambridge, U.K.

CHAPTER 9

MONITORING OF WETLANDS

Donald M. Kent

Wetlands monitoring is the checking, watching or tracking of wetlands for the purpose of collecting and interpreting data, which is then used to record or control the wetland, or processes affecting the wetland. Not to be confused with wetlands assessment or evaluation, which is the valuation of wetlands, monitoring of wetlands is a component of mitigation efforts (Kusler and Kentula 1990, U.S. Army Corps of Engineers 1989), the Environmental Protection Agency's Environmental Monitoring and Assessment Program (Paul et al. 1990, Leibowitz et al. 1991), and other programs designed to protect, conserve and understand wetland resources (New Hampshire Water Supply and Pollution Control Commission 1989, Haddad 1990, Walker 1991). These efforts are conducted for a multitude of reasons, using a variety of techniques to measure and assess an array of structural and functional parameters.

The process of developing and implementing a monitoring program can be reduced to four basic steps. First, and foremost, the reason for monitoring must be identified and clearly stated. Second, a determination of the measures appropriate for achieving the stated objective(s) must be made. Third, an approach commensurate with the level of investment and the required return must be selected. The size of the area to be monitored, as well as the length of time the area will be monitored, will affect selection of an approach. Finally, the information gathered from the monitoring effort must be analyzed and interpreted.

REASONS FOR MONITORING

For the most part, wetland monitoring is conducted for a relatively few, discrete reasons. Habitat mapping and trend analysis monitoring is conducted to identify wetlands resources, and to detect changes in these resources over time. Examples of mapping and trend analysis monitoring include efforts in coastal and seaway Canada (Rump 1987, Martin and Bouchard 1991), coastal India (Nayak et al. 1989), migratory bird habitat in central California (Peters 1989), and the National Wetlands Inventory project (Dahl and Pywell 1989).

The largest monitoring effort of this type is that of the Environmental Monitoring and Assessment Program (Paul et al. 1990, Liebowitz 1991) which has stimulated mapping and trend analysis monitoring throughout the United States (Haddad 1990, Johnston and Handley 1990, Orth et al. 1990).   Designed to monitor the condition of wetlands, initial aspects of the wetland ecosystems component of the Environmental Monitoring and Assessment Program focus on determining the sensitivity of various metrics for detecting known levels of stress, and determining the spatial and temporal variability of proposed wetland indicators of condition (U.S. Environmental Protection Agency 1990).

Wildlife and fisheries management monitoring is a type of habitat mapping and trend analysis monitoring. It is conducted to provide information as to species richness and species abundance over time, and to assess the effects of management strategies. Monitoring is either of the wildlife or fisheries population (Henny et al. 1972, Neilson and Green 1981, Hink and Ohmart 1984, Young 1987, Molini 1989), of habitat indicators of wildlife richness and abundance (Weller and Fredrickson 1974, Koeln et al. 1988), or of both the wildlife population and the habitat (Weller 1979, Weller and Voigts 1983).

Enhanced, restored and created wetlands monitoring is conducted to determine the achievement of project goals related to the production of functional wetlands. Examples include evaluation of habitat created using dredge spoil (Newling and Landin 1985, Landin et al. 1989), and restoration of degraded habitats (Pacific Estuarine Research Laboratory 1990). As well, there are numerous monitoring efforts associated with Section 404, state and local wetland fill permits (Kusler and Kentula 1990, U.S. Army Corps of Engineers 1989, Erwin 1991).

Impact analysis monitoring is conducted to determine the response of wetlands to identified direct and indirect impacts.   This includes monitoring of impacts to wetlands on and adjacent to hazardous waste sites (Bosserman and Hill 1985, Watson et al. 1985, Hebert et al. 1990), as well as impacts from discrete and continuous chemical contamination events (McFarlane and Watson 1977, Woodward et al. 1988). Other examples of impact analysis monitoring include studies of the effects of highway construction (Cramer and Hopkins 1981), siting impacts from generating station construction and operation (Wynn and Kiefer 1977), effects on wetland flora from exposure to electromagnetic fields (Guntenspergen et al. 1989) and impacts from agricultural practices (Hawkins and Stewart 1990, Walker 1991).

Treatment monitoring is conducted to determine the potential for, or effectiveness of, wetlands for treating point source or nonpoint source discharges. Treatment monitoring has been applied to studies of the effectiveness of constructed wetlands for wastewater treatment (Hardy 1988, Choate et al. 1990, Tennessee Valley Authority 1990), mine drainage (Eger and Lapakko 1988, Stark et al. 1988, Stillings et al. 1988), stormwater runoff (Meiorin 1991) and agricultural runoff (Costello 1991).

MEASURES

A large number of measures have been applied, or potentially can be applied, to monitoring of wetland structure and function (Table 1). Commonly used measures include measures of the properties of individual plants and animals, measures of the properties of vegetation and wildlife communities, measures of aquatic physical and chemical properties, and measures of soil properties. Less commonly used are measures of hydrologic and hydraulic properties such as flood frequency and groundwater depth. Generally unused are potentially useful measures of landform properties such as heterogeneity and patch characteristics (Forman and Godron 1986). The latter properties are particularly important in the preservation and creation of wetlands for wildlife, and are likely to be useful for other aspects of habitat mapping and trend analysis monitoring. Measures of organismal properties are typical of impact analysis monitoring programs.

Properties of Individual Plants

Measures of the properties of individual plants are of use in assessing the condition of natural plants and propagules. In theory, the properties of a plant are affected by any factor which alters the growth and maintenance of the plant. Factors which affect plant growth and maintenance include soil nutrients, soil moisture, disease, pest infestations, and anthropogenic and other disturbances. Information obtained from measurements of the properties of individual plants can be applied to trend analysis monitoring; enhanced, restored and created wetlands monitoring; impact analysis monitoring; and treatment monitoring.

The simplest measure of an individual plant is survival, i.e. whether the plant is dead or alive. For living plants, measures include basal area, which is the area of exposed stem if the plant was cut horizontally, and stem diameter which is the maximum width of the area of exposed stem if the plant was cut horizontally. Basal area and stem diameter are usually measured 2.5 cm (one inch) above the ground by ecologists and range managers, and 1.4 m (4.5 feet) above the ground by foresters. Plant height is the mean vertical distance from the ground at the base of a plant to the uppermost level of a plant. Cover, including ground cover (herbaceous plants and low growing shrubs) and canopy cover (other shrubs and trees), is that part of the ground area covered by the vertical projection downward of the aerial part of the plant. Typically, the vertical projection downward is viewed as a polygon drawn around the plant's perimeter, and ignores small gaps between branches. Canopy diameter is the average maximum width of the polygon used for canopy cover. Basal area, stem diameter, plant height, cover and canopy diameter, if repeatedly measured over time, can be used as indicators of plant growth rate.

Plants allocate net production to leaves, twigs, stem, bark, roots, flowers and seeds. The accumulated living organic matter is the biomass, and is usually expressed as the dry weight per unit of area. Determining the allocation to each part is generally invasive in that the parts must be removed from the plant and either weighed, or analyzed for energy or nutrient content. Nevertheless, individual plant productivity can be estimated by sampling of leaves, flowers or seeds (Figure 1).

Table 1. Measures of wetland structure and function.

Properties of Individual Plants
 Basal Area
 Biomass
 Canopy Diameter
 Cover

  Growth Rate
  Productivity
  Stem Diameter
  Survival

Properties of Vegetation Communities
 Basal Cover
 Biomass
 Cover
 Cover Type
 Density

  Evenness
  Richness
  Survival
  Stratification

Landform Properties
 Accessibility
 Dispersion
 Heterogeneity
 Isolation

  Interaction
  Shape
  Size

Properties of Soil
 Classification
 Moisture

  Organic Content
  Texture

Hydrologic and Hydraulic Properties
 Flood Storage Volume
 Frequency of Flooding
 Groundwater Depth
 Groundwater Recharge Volume

  Surface Water Depth
  Surface Water Area
  Surface Water Velocity
  Surface Water Width

Aquatic Physical/Chemical Properties
 Biological Oxygen Demand
 Chlorophyll
 Turbidity
 Dissolved Solids
 Nutrients

  pH
  Salinity
  Temperature
  Toxicants
  Turbidity

Organismal Properties
 Behavior
 Bioaccumulation
 Growth and Development

  Metabolism
  Reproduction
  Tissue Health

Properties of Individual Wildlife and Fish Species
 Abundance
 Association
 Age Structure

  Density
  Mortality
  Presence/Absence

Properties of Wildlife Communities
 Abundance
 Biomass
 Density

  Evenness
  Niche Overlap
  Richness

Properties of Vegetation Communities

Just as factors which affect plant growth and maintenance are reflected in measurements of the properties of individual plants, factors which affect more than

Figure 1. Monitoring of individual plants during the appropriate season will indicate if reproduction is occurring. Productivity can be estimated by sampling leaves, flowers or seeds.

one individual plant will be reflected in measurements of the properties of vegetation communities. Therefore, measures of the properties of vegetation communities are of use in assessing the condition of natural and mitigated vegetation communities.

Measures of the properties of vegetation communities include extensions of the measures applied to individual plants, as well as measures which are unique to the characterization of communities. Measures of community survival, basal cover, cover and biomass require the accumulation of measures of individual plants. The cumulative expression of these measures, relative to the number of individuals assessed in the case of survival, or relative to the size of the area assessed in the case of basal cover and cover, provide for the description of the vegetation community.

Properties unique to vegetation communities include cover type, which is the assignment of the community, or parts of the community, to predetermined categories (Figure 2). Classification of Wetlands and Deepwater Habitats of the

Figure 2. Wetlands can be monitored as to their cover type, which is the assignment of the plant community, in this case emergents, to predetermined categories.

United States (Cowardin et al. 1979) is the most commonly used system for describing cover type, and its widespread use provides for comparison among disparate monitoring efforts. Nevertheless, the development of other descriptive systems is sometimes required in order to maximize information return. Other measures unique to the community level are density, which describes the number of individuals per unit of area, and richness, which is the list of plant species identified in the community of interest. If each individual plant within the sampling area is identified, then evenness can be determined. Evenness describes how the species abundances are distributed among the species. Another widely used measure of community structure, diversity, combines richness and evenness. Because diversity measures combine richness and evenness, they confound the number of species, the relative abundances of the species, and the homogeneity and size of the area sampled, and are therefore less useful than measures of richness and evenness. Finally, measures of stratification, a diversity index reflecting the amount of foliage at various levels above the ground, describe the vertical structure of the vegetation community.

Landform Properties

Measures of landform properties are used by landscape ecologists to identify and describe individual communities, and the relationships among communities. The measures can be of value to wetland scientists interested in local and regional planning issues, particularly as these issues relate to wildlife and trend analysis. However, the measures have been infrequently used, and therefore require precise definition, and identification of limitations, when applied.

Some measures of landform properties, such as shape and size, can be applied to studies of single wetlands. Shape is typically described as a ratio of wetland perimeter to wetland area (Bowen and Burgess 1981). Size is described as the area of the wetland or by some linear dimension such as length, width, or the ratio of length to width.

Other measures of landform properties require consideration of more than a single wetland. Accessibility describes the distance along a corridor of suitable habitat from one wetland to another, and reflects the perceived ease of species movement (Bowen and Burgess 1981). Dispersion describes the pattern (e.g. clumped, uniform, random) of spatial arrangement among wetlands (Pielou 1977). Isolation describes the distance of a wetland from other wetlands (Bowen and Burgess 1981), and interaction describes the perceived influence of a wetland on another wetland through consideration of the distance between wetlands (MacClintock et al. 1977).

Properties of Soil

Measures of the properties of soils are useful in describing wetland structure, and provide clues to wetland function. As part of mapping and trend analysis monitoring efforts, measures of soil properties help to distinguish between wetland and nonwetland areas, and provide information as to changes to these areas. If monitored as part of a wetland enhancement, restoration or creation effort, including efforts associated with the establishment of treatment wetlands, measures of the properties of soil indicate the development of hydric conditions.

Soil is typically classified according to such characters as color, texture, and size and shape of aggregates. The U.S. Department of Agriculture Soil Conservation Service system is the commonly used taxonomic classification system in the United States. Based upon the kind and character of soil properties, and the arrangement of horizons within the profile, the system also provides information about the use and management of the soil. Soil texture is based on the relative proportions of the various soil separates in a sample, and is estimated from its plasticity when extruded and by feeling its grittiness (Hays et al. 1981). Soil moisture is the percent of a given amount of soil consisting of water, and is estimated by the loss of weight on drying. Soil organic content is the percent of a given amount of soil consisting of organic matter, and is estimated by loss of weight on ignition.

Hydrologic and Hydraulic Properties

Simply stated, a wetland is a wetland because it is wet.   Therefore, determination of wetland hydrology and hydraulics is essential to understanding wetland structure and function. The measures provide useful descriptors of wetland structure, and also provide valuable information as to wetland function. Measures of hydrologic and hydraulic properties provide information about the extent of wetlands, as well as the effect of intrinsic and extrinsic changes to wetlands. Treatment monitoring benefits from measures of hydrologic and hydraulic properties in determining maximum treatment capacity.   And measures of hydrologic and hydraulic properties as part of enhanced, restored and created wetland efforts are integral to an assessment of project success.

Velocity describes the speed at which water travels, and reflects not only the depth and width of the water body, but also the topographical gradient and the extent and type of vegetation. Water depth, width and area are descriptors of wetland structure. Monitored over time, and in relation to extreme events, these measures provide an empirical estimate of the frequency of flooding and of flood storage volume.   Flood storage volume can also be estimated using one of a number of computerized hydrological models (U.S. Army Corps of Engineers 1981, Soil Conservation Service Hydrology Unit 1982 and 1986, Huber and Dickinson 1988). Model inputs include wetland and watershed slope, vegetative cover, soil type and surface type (pervious vs impervious). Groundwater depth, the distance below the ground surface at which water occurs, can be determined empirically through the installation and monitoring of wells (Figure 3). Groundwater recharge volume, the volume of surface water moving down through the soil to an underlying groundwater system or aquifer, can be estimated using the aforementioned hydrological models.

Aquatic Physical and Chemical Properties

The quality of water affects the growth, maintenance and reproduction of wetland flora and fauna. Wetland water quality is revealed by measures of aquatic physical and chemical properties.   Water quality reflects the condition of the surrounding environment, and is affected by human activities such as watershed erosion, and point and nonpoint source discharges. Wetland water quality also reflects the condition of the wetland itself.   Measures of aquatic physical and chemical properties are particularly applicable to monitoring of the effects of impacts to wetlands, and monitoring the effectiveness of treatment wetlands.

Water temperature influences the rate of metabolic reactions and the reactivity of enzymes, and the amount of oxygen which can be dissolved in water.   The pH of water affects organismal physiological reactions and membrane characteristics. Dissolved oxygen concentrations must be sufficient to enable diffusion from the water into an animal's blood.   Salinity affects water quality through its effect on the ability of species to maintain osmotic balance.   Turbidity restricts the depth to

Figure 3. Groundwater depth, the distance below the ground surface at which water occurs, can be monitored empirically through the use of wells.

which solar radiation can penetrate the water column. Dissolved solids, such as carbonates, bicarbonates, chlorides, phosphates, nitrates, and salts of calcium, magnesium, sodium and potassium, affect organismal ionic balance and other physiological processes. Biological oxygen demand, the amount of oxygen required by bacteria while stabilizing decomposable organic matter under aerobic conditions, reflects the trophic status of the aquatic body, and possibly the extent and type of inputs to the aquatic body. Trophic status is also revealed by measures of chlorophyll, and of nutrients such as nitrates, nitrites and phosphates. Measures of toxicants such as heavy metals, volatile organics and petroleum hydrocarbons provide a direct measure of contaminants.

Organismal Properties

As with plants, factors which alter the growth, reproduction and maintenance of an individual organism will affect the properties of that organism, and therefore those properties will be of use in assessing the condition of that organism. Factors which affect organismal growth, reproduction and maintenance include water quality, including the presence or absence of environmental toxins, and the availability of food and cover. Measures of the properties of individual organisms are of particular use in monitoring impacted wetlands, or in assessing the affects to a natural wetland used for treatment of a discharge.

Organismal behavior, such as predator avoidance, foraging effectiveness and intraspecific social interactions, are modified by factors which affect the wetland. So too, the rate or age of onset of reproduction, and the rate of growth and development are similarly affected. Factors may also affect organismal metabolism, such as oxygen consumption, photosynthesis, nutrient uptake or enzymatic reactivity. In a more direct sense, organisms express a response to unfavorable environmental factors by the bioaccumulation of chemical constituents. In some cases, bioaccumulated chemical constituents are evidenced by changes in tissue health, such as lesions and tumors.

Properties of Individual Wildlife and Fish Species

Factors which grossly affect organisms, particularly those factors which affect reproduction and growth and development, will be reflected in properties of individual wildlife and fish populations. Therefore, measures of the properties of individual wildlife and fish species are of use in monitoring impacted wetlands. These measures are also useful for trend analysis monitoring efforts in that they reflect the condition of the wetland relative to the focal species. Finally, measures of the properties of individual wildlife and fish species provide important information as to the value of enhanced, restored and created wetlands as wildlife or fisheries habitat.

The simplest measures of individual wildlife and fish species are presence/absence and abundance (Figure 4). Requiring relatively more effort are measures of population density, the number of individuals per unit of area. Other measures are useful in assessing the potential persistence of the species. Mortality can be expressed as either the probability of dying or as the death rate. The complement of mortality is survival, the probability of living. Natality is the production of new individuals in the population, and can be described as the maximum or physiological natality, or as the realized mortality. Changes in mortality, survival and natality are reflected in the age structure of the wildlife or fish population. Declining or stabilized populations are characterized by relatively fewer young in the reproductive age classes and a relatively larger proportion of individuals in older age classes. Conversely, growing populations are characterized by a relatively larger proportion of the younger age classes. Monitoring efforts interested in determining how a wildlife or fish species is distributed throughout the wetland will use measures of association.

Properties of Wildlife and Fish Communities

Analogous to the situation with vegetation communities, factors which affect the reproduction and growth and development of more than one wildlife or fish species will be reflected in measurements of the properties of wildlife and fish communities. As with measures of the properties of individual wildlife and fish species, measures of the properties of wildlife and fish communities are applicable

Figure 4. Wildlife presence/absence and abundance can reasonably be monitored with a minimum of effort.

to impact monitoring, trend analysis monitoring, and enhanced, restored and created wetlands monitoring efforts.

Measures of wildlife or fish community abundance and density provide course estimates of wetland condition and suitability (Figure 5). The number of species occurring in the community is the richness. Evenness refers to how the species abundances are distributed among the species. The richness and evenness measures are frequently combined to form a single measure of diversity. However, the major criticism of diversity measures is that they confound a number of variables that characterize community structure (Ludwig and Reynolds 1988). Alternatively, biomass can be used to quantitatively describe wildlife and fish communities. More commonly applied to measures of vegetation communities, biomass is of use in describing community structure, particularly energy flow. Monitoring efforts interested in determining how coexisting species use common wetland resources will use measures of niche overlap.

Figure 5. Fish community species richness and abundance can be monitored using fish traps such as this fyke net.

APPROACHES TO MONITORING

Two broad approaches are available for monitoring wetlands: remote and contact. Remote monitoring is the acquisition of information about a wetland from a distance, without physical contact. Conversely, contact monitoring is the acquisition of information about a wetland from near at hand, with physical contact.

Remote monitoring of wetlands provides a level of spatial and temporal sampling which is impractical with contact techniques. Because data are available at large and synoptic scales, large scale patterns can be discerned and large scale processes can be measured.

Space based remote sensing instruments measure electromagnetic radiation reflected and emitted by the earth's surface. Visible and thermal satellite data provide locational information about broad vegetation cover types and extent of inundation (Carter et al. 1976, Roughgarden et al. 1991, Wickland 1991). Comparison of images along a temporal gradient provides information about land use and vegetation successional changes (Mackey and Jensen 1989, Nayak et al. 1989, Byrne and Dabrowska-Zielinska 1981). Information is provided at mesoscale, macroscale and megascale levels (Delcourt and Delcourt 1988, Table

1A). Aerial photography provides similar information at the microscale and mesoscale levels (Cartmill 1973, Haddad 1990, Jean and Bouchard 1991).

Contact monitoring cannot reasonably provide information about large scale processes, but it does provide for the acquisition of more detailed structural and functional information on a site-specific basis. Contact monitoring includes measures of habitat structure (Eger and Lapakko 1988, Pritchett 1988, Conner and Toliver 1990), measures of water quality and hydrology (Cramer and Hopkins 1981, Clausen and Johnson 1990, Oberts and Osgood 1991, Walker 1991), measures of animal populations (Bosserman and Hill 1985, Hardy 1988, Shortelle et al. 1989), and measures of contaminant levels in wildlife and fish (Watson et al. 1985, Hebert et al. 1990). Specific measurement parameters used in contact monitoring are discussed in Hays et al. (1981), Cooperrider et al. (1986), Graves and Dittberner (1986), and Adamus and Brandt (1990).

Historically, many contact monitoring programs have been based upon structural parameters related to vegetation (Larson 1987, Carothers et al. 1989, Landin et al. 1989), although the recent trend is toward the measurement of appropriate indicators which reflect wetland functional condition (Brooks et al. 1989, Paul et al. 1990). Kent et al. (1992) have recommended a model index for a cost effective, scientifically responsive approach to monitoring of wetland function.

SELECTING A MONITORING APPROACH

Ideally, all monitoring programs would have adequate numbers of trained personnel using replicated quantitative techniques and large sample sizes. Monitoring would be conducted over a sufficient number of years to provide for identification of stochastic variation. Finally, the data would be subjected to parametric statistical analysis, and the results would have global application.

Realistically, the vast majority of monitoring programs, if not all monitoring programs, are constrained by the availability of resources. Government monitoring programs are understaffed, and the programs are subject to periodic reassessment of priorities. Monitoring programs conducted by academics are sensitive to available funding. Mitigation monitoring is conducted for the minimum time necessary, using minimum funding, because of the understandable disinterest of developers in long term commitment of resources to what they perceive to be an ancillary activity.

Investment and Return

Given the enormous amount of resources required to effectively and accurately monitor a single wetland, notwithstanding the resources required to monitor wetlands at the megascale, it is clear that each monitoring opportunity requires consideration of investment and return. That is, what information about the wetland is required and what resources can be applied in pursuit of this information? This consideration must bear in mind that a commitment of resources to one monitoring opportunity necessarily diminishes the resources available for subsequent monitoring opportunities.

Furthermore, the relationship between investment and return can be expected to be logarithmic; as investment increases, the rate of increase in return decreases (Figure 6). To illustrate this point, consider the monitoring of an 80 acre created wetland for wildlife species richness (unpublished data). For a single sampling technique, a pedestrian transect, one hour of sampling resulted in the identification of 8 species, 4 hours of sampling identified 15 total species, 7 hours of sampling identified 18 total species, and 10 hours of sampling identified 22 total species. A decline in return relative to the level of investment is also evident when multiple sampling techniques are used. Eighty four hours of pedestrian transect sampling, fyke net sampling, scan sampling and auditory sampling identified 32 species. An additional 84 hours (for a total of 168 h) of sampling identified 12 more species, and a third 84 hour sampling event (total of 252 h) yielded 7 more species.

Clearly, an investment of one hour is inadequate to reasonably describe species richness on the site. However, if the question is "Are wildlife species present on the site?" then one hour is adequate. Conversely, if it is important to identify nearly every species using the site then an investment of 250 or more hours is more realistic. In most cases, information needs, whether for determination of wildlife species richness, or any other wetland variable, fall somewhere in between these two extremes. Continuing with the example, an investment of more than 250 hours only results in the identification of 7 more species than an investment of 168 hours. This is a 16 percent increase in return for a 50 percent increase in investment. Again, in the real world of budget constraints and additional commitments, selecting a level of investment commensurate with the level of information required is essential. Nevertheless, the effect of a less than comprehensive monitoring effort on project output must be recognized prior to initiation of the monitoring effort.

As discussed above, information returned from a given monitoring effort is a function of the amount of resources invested in that effort. The rate of increase in return decreases as investment increases. Selection of an appropriate level of investment, and consequently, development of a pragmatic monitoring program, is expedited by considerations of space and time.

Investment, Return and Area

For a given level of investment, return decreases with increasing area to be monitored. Conversely, as the area to be monitored increases, the level of investment must be correspondingly increased to maintain a given level of information return. The relationship between area and return for a given level of effort, as it was between investment and return, is logarithmic because sampling parameters are typically area specific (Figure 7). Returning to the example of monitoring wildlife species richness, a monitoring technique which comprehensively samples a one acre site will only sample half of a 2 acre site, one fourth of a 4 acre site,

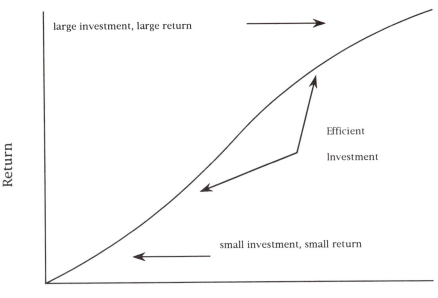

Figure 6. The relationship between investment and return can be expected to be logarithmic; that is, as investment increases, the rate of increase in return decreases.

and so on. As such, in order to monitor larger areas, either the investment must be increased accordingly, or a lesser return accepted.

The scale at which an investment is made, and at which a return is realized, is a function of the reason for monitoring. Habitat mapping and trend analysis monitoring occurs on the microscale to the megascale, with monitoring for wildlife and fisheries management occurring primarily at the microscale and mesoscale. Enhanced, restored and created wetlands monitoring occurs at the microscale. Impact analysis monitoring occurs at the microscale and mesoscale, and treatment monitoring occurs at the microscale.

Investment, Return and Time

The effect of time on monitoring investment and return has historically been restricted to consideration of the length of the monitoring program. Habitat mapping and trend analysis monitoring occurs for undefined but generally prolonged periods. Monitoring of fisheries and wildlife management occurs for extended periods corresponding with management goals, although, short term efforts do occur (Shortelle et al. 1989). Monitoring of enhanced, restored and

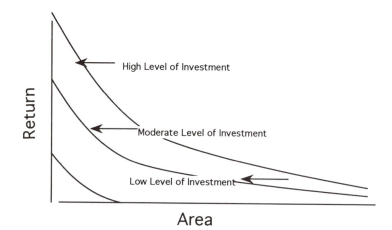

Figure 7. When sampling parameters are area specific, the relationship between area and return is inversely logarithmic. A relatively greater level of investment is required to maintain a given level of return as the size of the area to be monitored increases.

created wetlands generally occurs for finite periods of time (typically 2 to 7 years), and frequently corresponds with permit conditions. Impact analysis monitoring continues until such time as the impact source is remedied or the nature and extent of the impact is understood. Treatment monitoring occurs for extended periods corresponding with treatment needs.

Characteristic of most programs, regardless of the reason for monitoring, is establishment of a monitoring scheme which proceeds without variation until program completion. This type of effort, which can be described as continuous investment monitoring, is represented by a straight line when time is plotted against monitoring investment.

In some instances, continuous investment monitoring can reasonably be modified over time to achieve an efficient balance between investment and return. For example, monitoring of enhanced, restored and created wetlands includes tracking of structural aspects related to construction (e.g. propagule survival, surface or groundwater elevation), as well as determination of wetland function (e.g. species richness, sediment and toxicant retention). As an alternative to the maintenance of this relatively large investment throughout the monitoring period, structural and functional monitoring could be confined to an initial critical period, perhaps as little as 2 to 3 years, and subsequent years committed to monitoring of functional components. The level of investment could be reduced to levels commensurate with habitat mapping and trend analysis after a suitable period (Figure 8a). Similarly, impact analysis monitoring could be reduced to levels commensurate with habitat mapping and trend analysis after an initial investment intensive period (Figure 8b). In each case, gaining an understanding of the relationship between site specific structure and function is essential.

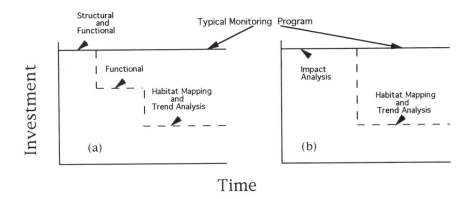

Figure 8. Most monitoring programs use a continuous level of investment throughout. Reasonably, monitoring investment for enhanced, restored and created wetlands (a) and impact analysis (b) could consist of an initial investment intensive period followed by a less intensive period(s) designed to balance investment and return.

Measures and Monitoring Approaches

A number of the measures listed in Table 1 cannot be accomplished through remote sensing, and therefore only lend themselves to contact monitoring. Properties of soil, individual organisms, and of aquatic physical and chemical properties, cannot reasonably nor effectively be measured without observations at the site. Measures of the properties of individual plants and vegetation communities also largely lend themselves to contact monitoring. However, measurement of vegetation community cover type and canopy cover can reasonably be accomplished through the use of aerial photography or satellite imagery, and individual tree canopy cover can be accomplished through the use of aerial photography when vegetation is sparse and relatively isolated. Similarly, properties of individual wildlife species and wildlife communities can in certain situations be remotely monitored. For example, presence and absence, abundance and density can be accomplished through aerial photographs or aerial flyovers of large species living in open habitats. Measures of landform properties only lend themselves to remote monitoring, either through the use of aerial photography or satellite imagery.

Measures of hydrologic and hydraulic properties are accomplished by contact monitoring, remote monitoring, or a combination of the two. Measures of groundwater and surface water depth can only be accomplished through contact monitoring. Measures of surface water area are best accomplished through remote monitoring, whereas monitoring of the frequency of flooding can reasonably be accomplished through either contact or remote monitoring. Flood storage volume and groundwater recharge volume, which require knowledge of wetland and watershed slope, vegetative cover, soil type and surface type (impervious or pervious) require both contact and remote monitoring.

Investment, Measures and Area

Any measure can be applied to any size monitoring area given a sufficient investment. Nevertheless, for many measures there is a point where either the investment required is too large to be practical, or the return is diminished to the point where the information in nonrepresentative of actual conditions. Measures of the properties of individual plants, vegetation communities, soil, organisms, and aquatic physical and chemical properties are generally most effective at the microscale level. Although, vegetation community cover type is effectively applied to the mesoscale and macroscale, and in rare cases to the megascale (Haddad 1990). Measures of landform properties are most effectively applied to the mesoscale and macroscale, although these measures are applicable to the microscale. Hydrologic and hydraulic measures are most effectively applied at the microscale and mesoscale, but measures of flood frequency and surface water area are effective at the macroscale. Properties of individual wildlife species and wildlife communities are also most effectively measured at the microscale. Although, measures of abundance, density and richness are sometimes applicable at the mesoscale.

Investment, Measures and Time

Analogous to the relationship between measure and area, any measure can be applied over any length of time given a sufficient investment. Generally, contact monitoring measures require a greater investment in time, money and skill than remote monitoring measures, and is therefore more difficult to apply over long periods of time because of inevitable staff turnover and shifting priorities. Properties of individual plants, soil, organisms and aquatic physical and chemical properties, some hydrologic and hydraulic measures, and individual and community wildlife properties, are more effectively measured over relatively short periods of time. Monitoring programs to be conducted for relatively long periods of time are more effective if they use remote measures such as properties of landforms, vegetation community cover type, or in some cases wildlife abundance, density or richness measures.

Contact monitoring measures are applicable over long periods, even given a reasonable level of investment, if sample size and frequency are reduced. However, a reduction in investment inevitably reduces the monitoring return through reduction in result representativeness. No attempt should be made to minimize sample size and frequency without first assessing the impact to the return, and then determining if the reduction in return is acceptable. For example, a reduction in sample size or sample frequency will increase the variance in the data. This variance must be evaluated to determine if it obscures the information sought, i.e., the data is too imprecise to be of practical use. Frequently, this reduction in monitoring investment results in the production of an index rather than a direct measure. An alternative approach to reducing sample size and sample frequency of contact measures is to establish a relationship between the contact measure of interest and a remote measure. In this way, the remote measure becomes an indicator of a wetland condition generally determined through a contact measure.

MONITORING DESIGN AND ANALYSIS

Design

Monitoring data is the product of either an experimental or an observational approach (Ludwig and Reynolds 1988). An experimental approach, analogous to a true experiment (Hicks 1982), presumes that the wetland is subject to experimental manipulation. That is, one or more independent variables are manipulated and their effect on one or more dependent variables is determined. Any differences among the dependent variables are attributed to the independent variables. By contrast, an observational approach uses measurements on the wetland over a range of natural conditions. The observer can either study and compare separate wetlands, or different parts of the same wetland, subject to differing conditions at the same point in time. Alternatively, the same wetland or part of a wetland can be studied or compared at two separate points in time.

Regardless of the approach, effective monitoring of wetlands requires an initial understanding of monitoring goals. This statement of the problem should include one or more criteria for assessing the results of the monitoring effort. The criteria should be measurable, and the achievable accuracy of the measure understood. If an experimental approach is used, then the independent variables must be defined, and levels established.

Consideration must be given as to how the data are to be collected. The extent of the difference to be detected and the degree of variability in the data will determine the number of observations required. In the absence of this information, the sample size should be as large as resource constraints will allow. As there are always a number of variables that cannot be controlled, the order of data collection should be randomized. Randomization averages out the effects of time, and, because randomization allows the monitoring effort to proceed as if the errors of measurement were independent, a key assumption of most statistical tests is satisfied (Hicks 1982).

Analysis

Information obtained from a monitoring effort must be examined and interpreted. Statistics facilitate this examination and interpretation, and are generally descriptive or inferential in nature. Descriptive statistics simply describe a sample, whereas inferential statistics allow inference, or the development of conclusions, based on a sample.

Descriptive statistics include measures of central tendency such as mean, mode and median. Measures of dispersion, such as standard deviation and variance, indicate how the data are distributed around the central tendency.

Spatial pattern can be an important descriptor of wetland plant and animal communities. Three basic spatial patterns, random, clumped and uniform, are recognized. Random patterns suggest environmental homogeneity or nonselective behavioral patterns, clumping suggests that individuals are aggregated in more favorable parts of the habitat, and uniform patterns suggest negative interactions between individuals (Pemberton and Frey 1984). Analysis of spatial patterns is useful in that it suggests hypotheses that might be tested to explain underlying causal factors.

Species-abundance relationships reveal information about the structure of the wetland community, and lead to theories about such issues as community stability, resource partitioning and species-area relationships (Hutchinson 1959, McGuinness 1984). Species abundances can be based upon number of individuals per species, or alternatively on other variables such as biomass or percent cover. The simplest analysis of species-abundance relationships is a frequency distribution. Other characterizations of species-abundance relationships are indicated by indices of species richness, species evenness and diversity.

Patterns in species interactions are revealed by analysis of species affinity. Niche overlap indices indicate how coexisting species use a common resource. Measures of interspecific association indicate whether or not two species select or avoid the same habitat. When a sample contains quantitative measures of species abundances, the covariation in abundance between species can be assessed.

Inferential statistics allow for testing of hypotheses about the data. If the population from which the sample was drawn can be assumed to have a known distribution, and the sample items are independent of each other and normally distributed, then parametric statistics can be used. The *t*-test is used to compare the means of two samples (or the mean of a sample to a standard), and an analysis of variance (ANOVA) test is used to compare more than two means. In experimental situations, where independent and dependent variables are defined, regression analysis is used to predict the effect of changing the independent variable on the dependent variable. Correlation analysis determines the degree of association between two factors of unknown relationship. When no assumption about the distribution of the population or the sample can be made, nonparametric statistics are appropriate. The Mann-Whitney U test and the Kruskal-Wallis *k*-Sample test are analogous to the *t*-test and ANOVA respectively, and a contingency table can be used to test for associations between variables.

Community classification is the grouping or clustering of objects based upon their resemblance (Ludwig and Reynolds 1988). Its use in monitoring of wetlands is in comparing a wetland with a reference wetland, or in comparing the same wetland at two different points in time. Community classification is accomplished through the use of resemblance functions, association analysis and cluster analysis. Community ordination is a term used to describe a set of techniques in which samples are arranged in relation to one or more coordinate axes, thereby facilitating identification of similar and dissimilar samples (Ludwig and Reynolds 1988). Ordination techniques are intended to simplify and condense large data sets, and to identify key environmental factors. Techniques include polar ordination, principle component analysis, correspondence analysis and various nonlinear ordinations. Community interpretation techniques provide for the evaluation of the effects of environmental factors on the patterns revealed by community classification and ordination. Interpretation of community classification data is accomplished with discriminate analysis, whereas interpretation of ordination data is accomplished through the use of regression and correlation statistics (Ludwig and Reynolds 1988).

REFERENCES

Adamus, P. and K. Brandt. 1990. Impacts on Quality of Inland Wetlands of the United States: A Survey of Indicators, Techniques, and Applications of Community Level Biomonitoring Data. U.S. Environmental Protection Agency Report EPA/600/3-90/073. 396 pp.

Bosserman, R.W. and P.L. Hill. 1988. Community ecology of three wetland Proceedings of the Pennsylvania State University Wetlands Water Management Mined Lands Conference.

Bowen, G.W. and R.L. Burgess. 1981. A Quantitative Analysis of Forest Island Pattern in Selected Ohio Landscapes. Oak Ridge National Laboratory Publication ORNL-TM-7759. 111 pp.

Brooks, R.P., D.E. Arnold, E.D. Bellis, C.S. Keener and M.J. Croonquist. 1989. A methodology for biological monitoring of cumulative impacts on wetland, stream and riparian components of watersheds. Draft presentation at International Symposium: Wetlands and River Corridor Management.

Byrne, G.F. and K. Dabrowska-Zielinska. 1981. Use of visible and thermal satellite data to monitor an intermittently flooding marshland. Remote Sensing of Environment 11:393-399.

Carothers, S.W., G.S. Mills and R.R. Johnson. 1989. The creation and restoration of riparian habitat in southeastern arid and semiarid regions. pp. 359-376 In Wetland Creation and Restoration: The Status of the Science. Eds. J.A. Kusler and M.E. Kentula. U.S. Environmental Protection Agency, Corvallis, Oregon, Report No. 600/3-89/038.

Carter, V., L. Alsid and R.R. Anderson. 1976. Man's impact upon wetlands. pp. 293-302 In ERTS-1: A New Window On Our Planet. Geological Survey Professional Paper 929, Eds. R.S. Williams, Jr. and W.D. Carter. 362 pp.

Cartmill, R.H. 1973. Evaluation of remote sensing and automatic data techniques for characterization of wetlands. pp. 1257-1277 In Proceedings of the ERTS-1 Symposium, Washington, D.C.

Choate, K.D., J.T. Watson and G.R. Steiner. 1990. Demonstration of constructed wetlands for treatment of municipal wastewaters: monitoring report for the period March 1988-October 1989. Tennessee Valley Authority, Knoxville, Tennessee. Report No. TVA/WR/WQ-90/11. 128 pp.

Clausen, J.C. and G.D. Johnson. 1990. Lake level influences on sediment and nutrient retention in a lakeside wetland. Journal of Environmental Quality 19(1):83-88.

Conner, W.H. and J.R. Toliver. 1990. Observations on the regeneration of bald cypress (*Taxodium distichum*) in Louisiana swamps. Southern Journal of Applied Forestry 14(3):115-118.

Cooperrider, A.Y., R.J. Boyd and H.R. Stuart. 1986. Inventory and Monitoring of Wildlife Habitat. U.S. Department of the Interior, Bureau of Land Management. Service Center. Denver, Colorado. 858 pp.

Costello, C.J. 1991. Wetlands treatment of dairy animal wastes in Irish drumlin landscape. pp. 702-709 In Constructed Wetlands for Wastewater Treatment. Ed. D.A. Hammer. Lewis Publishers, Inc., Chelsea, Michigan.

Cowardin, L.M., V. Carter, F.C. Golet and E.T. LaRoe. 1979. Classification of Wetlands and Deepwater Habitats of the United States. Fish Wildlife Service Biological Report FWS/OBS-79/31. 103 pp.

Cramer, G.H. and W.C. Hopkins. 1981. The effects of elevated highway construction on water quality in Louisiana wetlands. Federal Highway Administration Report No.: LA-75-4G-F.

Dahl, T.E. and H.R. Pywell. 1989. National status and trends study: Estimating wetland resources in the 1980's. pp. 25-31 In AWRA Wetlands: Concerns and successes symposium.

Delcourt, H.R. and P.A. Delcourt. 1988. Quaternary landscape ecology: Relevant scales in space and time. Landscape Ecology 2:23-44.

Eger, P. and K. Kapakko. 1988. Use of wetlands to remove nickel and copper from mine drainage. pp. 780-787 In Proceedings of the Tennessee Valley Authority First Annual Conference on Constructed Wetlands for Wastewater Treatment.

Erwin, K.L. 1991. An evaluation of wetland mitigation in the South Florida Water Management District. Report to the South Florida Water Management District, Contract No. C89-0082-A1.

Forman, R.T.T. and M. Godron. 1986. Landscape Ecology. John Wiley and Sons, New York. 619 pp.

Graves, B.M. and P.L. Dittberner. 1986. Variables for Monitoring Aquatic and Terrestrial Environments. U.S. Fish and Wildlife Service Biological Report 86(5).

Guntenspergen, G., J. Keough, F. Stearns and D. Wikum. 1989. ELF communications system ecological monitoring program: Wetland studies-final report. Prepared for Submarine Communications Project Office, Washington, D.C. Technical Report E06620-2, Contract No. N00039-88-C-0065. 250 pp.

Haddad, K. 1990. Marine wetland mapping and monitoring in Florida. Fish Wildlife Service Biological Report 90(18), pp. 145-150.

Hardy, J.W. 1988. Land Treatment of municipal wastewater on Mississippi Sandhill Crane National Refuge for wetlands/crane habitat management: A status report. pp. 186-190 In Proceedings of the Tennessee Valley Authority First Annual Conference on Constructed Wetlands for Wastewater Treatment.

Hawkins, A.S. and J.L. Stewart. 1990. Environmental management program: Long term resource monitoring program for the upper Mississippi River system. Prepared for the Winona County Soil and Water Conservation District, Lewiston, Minnesota and the Fish and Wildlife Service, Onalaska, Wisconsin. Report No. EMTC-90/05. 49 pp.

Hays, R.L., C. Summers and W. Seitz. 1981. Estimating Wildlife Habitat Variables. Fish Wildlife Service Biological Report FWS/OBS-81/47. 111 pp.

Hebert, C.E., G.D. Haffner, I.M. Weis, R. Lazar and L. Montour. 1990. Organochlorine contaminants in duck populations of Walpole Island. Journal of Great Lakes Research 16(1):21-26.

Henny, C.J., D.R. Anderson and R.S. Pospahala. 1972. Aerial surveys of waterfowl production in North America, 1955-71. U.S. Department of the Interior, Fish Wildlife Service Special Scientific Report-Wildlife 160. 48 pp.

Hicks, C.R. 1982. Fundamental Concepts in the Design of Experiments. Holt, Rinehart and Winston, New York. 425 pp.

Hink, V. and R.D. Ohmart. 1984. Middle Rio Grande biological survey final report. U.S. Army Corps of Engineers, Albuquerque, New Mexico.

Huber, W.C. and R.E. Dickinson. 1988. Stormwater Management Model, Version 4: User's Manual. Cooperative Agreement CR-811607. U.S. Environmental Protection Agency, Athens, Georgia. 569 pp.

Hutchinson, G.E. 1959. Homage to Santa Roasalia, or why are there some many kinds of animals? American Naturalist 93:145-159.

Jean, M. and A. Bouchard. 1991. Temporal changes in wetland landscapes of a section of the St. Lawrence River, Canada. Environmental Management 15:241-250.

Johnston, J.B. and L.R. Handley. 1990. Coastal mapping programs at the U.S. Fish and Wildlife Service's National Wetlands Research Center. Fish Wildlife Service Biological Report 90(18): Federal coastal wetland mapping programs. pp. 105-109.

Kent, D.M., R.J. Reimold, J. Kelly and C. Tammi. 1992. Coupling wetlands structure and function: Developing a condition index for wetlands monitoring. pp. 559-570 In Ecological Indicators Volume 1. McKenzie, D.H., D.E. Hyatt and V. Janet McDonald (Eds.) Elsevier Science Publishers Ltd., Essex, England.

Koeln, G.T., J.E. Jacobson, D.E. Wesley and R.S. Rempel. 1988. Wetland inventories derived from Landsat data for waterfowl management planning. pp. 303-310 in Proceedings of the 53rd Wildlife Management Institute of North American Wildlife and Natural Resources Conference.

Kusler, J.A. and M.E. Kentula. 1990. Wetland Creation and Restoration: The Status of the Science. Washington, D.C. 594 pp.

Landin, M.C., E.J. Clairain and C.J. Newling. 1989. Wetland habitat development and longterm monitoring at Windmill Point, Virginia. Wetlands 9:13-25.

Larson, J.S. 1987. Wetland mitigation in the glaciated northeast: risks and uncertainties. In Mitigating Freshwater Wetland Alterations in the Glaciated Northeastern United States: An Assessment of the Science Base. Eds. J.S. Larson and C.S. Neill. University of Massachusetts Environmental Institute Publication No. 87-1.

Leibowitz, N.C., L. Squires and J.P. Baker. 1991. Environmental Monitoring and Assessment Program: Research plan for monitoring wetland ecosystems. EPA Report No. EPA/600/3-91/010. 191 pp.

Ludwig, J.A. and J.F. Reynolds. 1988. Statistical Ecology. John Wiley & Sons, Inc., New York. 337 pp.

MacClintock, L., R.F. Whitcomb and B.L. Whitcomb. 1977. Evidence for the value of corridors and minimization of isolation in preservation of biotic diversity. American Birds 31:6-16.

Mackey, H.E. and J.R. Jensen. 1989. Wetlands mapping with spot multispectral scanner data. In Proceedings of the 9th Annual Convention on Surveying and Mapping Auto-Cartography, Baltimore, Maryland. 21 pp.

McFarlane, C. and R.D. Watson. 1977. The detection and mapping of oil on a marshy area by a remote luminescent sensor. pp. 197-201 In Proceedings of the American Petroleum Institute Oil Spill Conference. New Orleans, Louisiana.

McGuinness, K.A. 1984. Equations and explanations in the study of species-area curves. Biological Reviews 59:423-440.

Meiorin, E.C. 1991. Urban runoff treatment in a fresh/brackish water marsh in Fremont, California. pp. 677-685 In Constructed Wetlands for Wastewater Treatment. Ed. D.A. Hammer. Lewis Publishers, Inc. Chelsea, Michigan.

Molini, W.A. 1989. Pacific flyway perspectives and expectations. pp. 529-536 In Proceedings of the 54th Conference of the Wildlife Management Institute for North American Wildlife and Natural Resources.

Nayak, S., A. Pandeya, M.C. Gupta, C.R. Trivedi, K.N. Prasad and S.A. Kadri. 1989. Application of satellite data for monitoring degradation of tidal wetlands of the Gulf of Kachchh Western India. Acta Astronautica 20:171-178.

Neilson, J.D. and G.H. Green. 1981. Enumeration of spawning salmon from spawner residence time and aerial counts. Transaction of the American Fisheries Society 110(4):554-556.

New Hampshire Water Supply Pollution Control Commission. 1989. Assessment of wetlands management and sediment phosphorous inactivation, Kezar Lake, New Hampshire: Phase 2 implementation and monitoring. Concord, New Hampshire. 109 pp.

Newling, C.J. and M.C. Landin. 1985. Long-term monitoring of habitat development at upland and wetland dredged material disposal sites 1974-1982. Dredging Operations Technical Support Program Report No.: WES/TR/D-85-5. 228 pp.

Oberts, G.L. and R.A. Osgood. 1991. Water quality effectiveness of a detention/wetland treatment system and its effect on an urban lake. Environmental Management 15(1):131-137.

Orth, R.J., K.A. Moore and J.F. Nowak. 1990. Monitoring seagrass distribution and abundance patterns: A case study from the Chesapeake Bay. pp. 111-123 In Fish Wildlife Service Biological Report 90(18): Federal coastal wetland mapping programs.

Pacific Estuarine Research Laboratory. 1990. A manual for assessing restored and natural coastal wetlands with examples from southern California. California Sea Grant Report No. T-CSGCP-021. La Jolla, California.

Paul, J.F., A.F. Holland, S.C. Schimmel, J.K. Summers and K.J. Scott. 1990. USA EPA Environmental Monitoring and Assessment Program: An ecological status and trends program. U.S. Fish and Wildlife Service Biological Report 90(18):71-78.

Pemberton, S.G. and R.W. Frey. 1984. Quantitative methods in ichnology: spatial distribution among populations. Lethaia 17:33-49.

Peters, D.D. 1989. Status and trends of wetlands in the California central valley. pp. 33-45 In AWRA Wetlands: Concerns and Successes.

Pielou, E.C. 1977. Mathematical Ecology. Wiley & Sons, New York.

Pritchett, D.A. 1988. Creation and monitoring of vernal pools at Santa Barbara, California. pp. 282-292 In Proceedings of Environmental Restoration: Science and Strategies for Restoring the Earth, Berkeley, California.

Roughgarden, J., S.W. Running and P.A. Matson. 1991. What does remote sensing do for ecology? Ecology 72(6):1918-1922.

Rump, P.C. 1987. The state of Canada's wetlands. pp. 259-266 In Proceedings of the Peatlands Symposium, Edmonton, Canada.

Shortelle, A.B., J.L. Dudley, B. Prynoski and M. Boyajian. 1989. Vernal pool wetlands: Wildlife values, acidification and a need for management. pp. 463-471 In AWRA Wetlands: Concerns and Successes Symposium. Tampa, Florida.

Soil Conservation Service Hydrology Unit. 1982. Technical Release Number 20 (TR-20). National Technical Information Service.

Soil Conservation Service Hydrology Unit. 1986. Technical Release Number 55 TR-55). National Technical Information Service.

Stark, L.R., R.L. Kolbash, H.I. Webster, S.E. Stevens, K.A. Dionis, and E.R. Murphy. 1988. The Simco #4 Wetland: Biological patterns and performance of a wetland receiving mine drainage. pp. 332-334 In Mine Drainage and Surface Mine Reclamation, Vol. I. Information Circular No. 9183. U.S. Bureau of Mines.

Stillings, L.L., J.J. Gryta and T.A. Ronning. 1988. Iron and manganese removal in a *Typha* dominated wetland during ten months following its construction. pp. 317-324 In Mine Drainage and Surface Mine Reclamation, Vol. I. Information Circular No. 9183. U.S. Bureau of Mines.

Tennessee Valley Authority. 1990. Design and performance of the constructed wetland wastewater treatment system at Phillips High School, Bear Creek, Alabama. Tennessee Valley Authority, Chattanooga, Tennessee. Report No. TVA/WR/WQ-90/5. 73 pp.

U.S. Army Corps of Engineers. 1981. HEC-1 Flood Hydorgraph Package: User's Manual. Water Resources Support Center. The Hydrologic Engineering Center, Davis, California. 78 pp.

U.S. Army Corps of Engineers. 1989. Evaluation of freshwater wetland replacement projects in Massachusetts. New England Division, Waltham, Massachusetts. 23 pp.

U.S. Environmental Protection Agency. 1990. Environmental monitoring and assessment program. EPA/600/3-90/060.

Walker, W.W. 1991. Water quality trends at inflow to Everglades National Park. Water Resources Bulletin 27:59-72.

Watson, M.R., W.B. Stone, J.C. Okoniewski and L.M. Smith. 1985. Wildlife as monitors of the movement of polychlorinated biphenyls and other organochlorine compounds from a hazardous waste site. pp. 91-104 In Proceedings of the Northeast Fish and Wildlife Conference.

Weller, M.W. 1979. Birds of some Iowa wetlands in relation to concepts of faunal preservation. Proceedings of the Iowa Academy of Science 86:81-88.

Weller, M.W. and L.H. Fredrickson. 1974. Avian ecology of a managed glacial marsh. The Living Bird 12:269-291.

Weller, M.W. and D.K. Voigts. 1983. Changes in the vegetation and wildlife use of a small prairie wetland following a drought. Proceedings of the Iowa Academy of Science 90:50-54.

Wickland, D.E. 1991. Mission to planet earth: The ecological perspective. Ecology 72(6):1923-1933.

Woodward, D.F., E. Snyder-Conn, R.G. Riley and T.R. Garland. 1988. Drilling fluids and the Arctic tundra of Alaska, U.S.A.: Assessing Contamination of wetlands habitat and the toxicity to aquatic invertebrates and fish. Archives of Environmental Contamination and Toxicology 17(5):683-697.

Wynn, S.L. and R.W. Kiefer. 1977. Monitoring vegetation changes in a large impacted wetland using quantitative field data and quantitative remote sensing data. pp. 178-180 In Proceedings of the Sensing of Environmental Pollutants 4th Joint Conference, New Orleans, Louisiana.

Young, D.A. 1987. Petroleum extraction and waterfowl utilization within a major wetland complex: Are they compatible? pp. 165-171 In Proceedings of the Peatlands Symposium, Edmonton, Canada.

Appendix 1

Spatial hierarchy of Delcourt and Delcourt (1988).

Hierarchical Domains	Sublevels	Area (m^2)	Map Scale
		1.5×10^{14}	Smaller Scale
	Global		
Megascale		10^{14}	1:20,000,000
	Continent		
		10^{12}	1:2,000,000
Macroscale	Macroregion		
		10^{10}	1:200,000
	Mesoregion		
Mesoscale		10^{8}	1:20,000
	Microregion		
		10^{6}	1:2,000
	Macrosite		
		10^{4}	1:200
Microscale	Mesosite		
		10^{2}	1:20
	Microsite		
		10^{0}	1:2

CHAPTER 10

CONSTRUCTED WETLANDS FOR WASTEWATER TREATMENT

Robert A. Corbitt and Paul T. Bowen

Throughout history, wetlands have provided wastewater treatment for pollutants from sources ranging from rainfall runoff in forested areas to intentional discharges of community sewage. During the 1970s, focused wetland studies began to provide an understanding of the phenomena of wetlands (wastewater) treatment processes. Both European and United States investigators provided research into wetlands treatment in northern climates. The Europeans seemed to concentrate on man-made or constructed wetland systems, while the United States' investigators focused more on natural systems.

Natural and constructed wetland treatment systems have distinctive characteristics that strengthen their application on a site specific basis. They are relatively easy systems to operate and maintain, they are extremely energy efficient when compared to mechanical systems, and they provide beneficial habitat for a variety of wildlife. Wetland systems provide effective wastewater treatment and are especially efficient in the removal, and beneficial reuse, of nutrients. Constructed wetland systems, including restoration of degraded wetlands, enhance the aesthetic value of an area, and may be used by the public as a nature study area.

Nonetheless, in the United States, the use of natural wetlands for wastewater treatment ran head-on into the swell of public interest and attention to environmental protection which had begun in the 1960s. The water quality problems from nutrients and oxygen-demanding pollutants were well publicized. This alarm was easily transferred to any consideration of natural wetlands for wastewater treatment.

In the meantime, wastewaters have been treated and reused successfully in agriculture, silviculture, aquaculture, and golf course and landscape irrigation. Also, more attention has been given to development of an understanding of constructed wetlands. As a result, wetlands treatment systems, especially those man-made, are no longer a startling environmental liability. Site specific determinations should be sought early on from the regulatory agencies for consideration of any allowance for treatment in an existing natural system. The best opportunity would be afforded when the overall function and value of an existing, natural wetland is enhanced.

BASIS FOR TREATMENT OF WASTEWATER BY WETLANDS

Wetland systems have distinctive characteristics. The U.S. Fish and Wildlife Service has developed the most widely accepted classifications of wetlands (Cowardin et al. 1979). The classification hierarchy extends to system, subsystem, and class levels. The four principal systems are the estuarine, riverine, lacustrine, and palustrine systems. Plant communities, hydroperiod, soils, and other factors interact to establish the uniqueness of individual systems.

The study of wetland functional values is the essential first step in understanding how wetlands perform as a wastewater treatment facility (U.S. Environmental Protection Agency 1983, Tiner 1984, Hammer 1989). Wetland functions are the physical, chemical, and biological characteristics of a wetland. Wetland values are those characteristics that are beneficial to society. Principal functions and values of wetlands are presented in Table 1.

Table 1.   Important Functions and Values of Natural Wetlands.

Flood Water Storage	Nutrient Removal/Retention
Flooding Reduction	Chemical and Nutrient Sorption
Erosion Control	Food Chain Support
Sediment Stabilization	Fish and Wildlife Habitat
Sediment/Toxicant Retention	Migratory Waterfowl Usage
Groundwater Recharge/Discharge	Recreation
Natural Water Purification	Uniqueness/Heritage Values

These functions and values have obvious natural benefits but also identify how wetlands may be used as an effective mechanism for wastewater treatment, either in their natural state or as a constructed system. Furthermore, much of the wastewater treatment in a constructed wetland occurs by means of bacterial metabolism and physical sedimentation, as it does in an activated sludge or trickling filter treatment facility. Wetland plants serve an important role by providing structure to support the algae and bacterial population that provide wastewater treatment capability and reliability in a created wetland community environment.

Constructed wetlands treatment systems range from the creation of a marsh in a natural setting to extensive construction involving considerable earth moving and use of impermeable barriers. Vegetation that is introduced or naturally established in the constructed system will be influenced by many factors, including the characteristics of the wastewater passing through the system.

The ability to provide internal controls is the most significant difference in natural and constructed wetland treatment systems. For example, design features for flexible water management normally would disturb plant communities in a natural system, and would necessitate increased land area, to achieve treatment goals comparable to constructed systems. Design of the constructed wetland system can ensure specific hydrologic and hydraulics conditions and provide for targeted treatment of specific parameters.

Constructed wetlands for wastewater treatment are classified by hydrologic regime (Table 2) (U.S. Environmental Protection Agency 1988, Hammer 1989, Water Pollution Control Federation 1990). Free water surface systems (FWS) consist of basins or channels and barriers to seepage, and have relatively shallow water depths and low flow velocities. Subsurface flow systems (SFS), sometimes called vegetated submerged bed systems (VSB), consist entirely of subsurface flow conveyed to the system via a trench or bed underdrain network. As with SFS systems, they have barriers to seepage and very low flow velocities. Both FWS and SFS systems are characterized by emergent vegetation.

TREATMENT PROCESSES

Natural and constructed wetland systems provide wastewater treatment by significantly reducing oxygen-demanding substances such as BOD and ammonia, suspended solids, nutrients such as nitrogen and phosphorus, and other pollutants such as metals. Primary, secondary, and incremental removal mechanisms include, physical, chemical and biological operations as well as plant uptake (Sherwood et al. 1979, Reed et al. 1988, U.S. Environmental Protection Agency 1988, Hammer 1989, Water Pollution Control Federation 1990).

BOD

Biological metabolism is the primary process for removal of BOD in a wetland system. (BOD refers to biochemical oxygen demand, which is a standard test used to assess organic matter/oxygen demand of wastewater.) The diverse microbial populations in the water column consume the soluble organic waste products. In the soil and sediment (organic) layer, the organics are sorbed in the soil, and later, biologically oxidized into stable end products. Also, sedimentation in FWS systems, and filtration in SFS systems, are processes that remove BOD associated with suspended organic materials. Typically, the wastewater will receive some level of preapplication treatment which would have removed essentially all BOD associated with suspended matter.

The principal biological activity in FWS wetlands is from algae and bacteria where microbial growths are attached to structures such as plant stems and litter, in the wetland. The phenomena is akin to BOD removal in oxidation ponds and by attached growth (trickling filter) treatment facilities. Loading rates are lower on SFS wetland treatment systems due to the limited supply of oxygen which is characteristic of these systems. The plant root systems are key to treatment performance making the selected plant species especially important.

Table 2. Typical Classifications of Constructed Wetlands for Wastewater Treatment.

Constructed Wetland Types	Diagram	Description
Free Water Surface	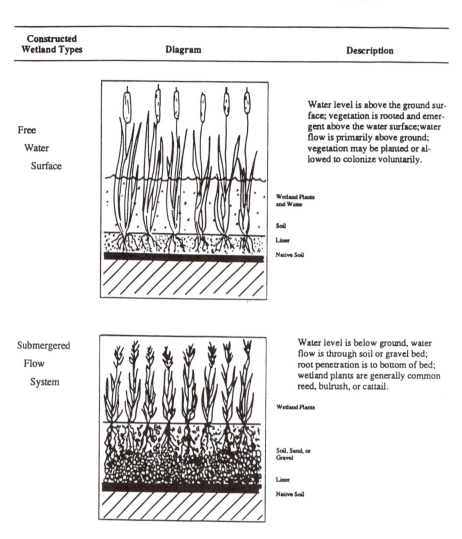	Water level is above the ground surface; vegetation is rooted and emergent above the water surface; water flow is primarily above ground; vegetation may be planted or allowed to colonize voluntarily.
Submergered Flow System		Water level is below ground, water flow is through soil or gravel bed; root penetration is to bottom of bed; wetland plants are generally common reed, bulrush, or cattail.

Suspended Solids

Settleable suspended solids are readily removed in the early stages of a FWS by gravity settling in quiescent, shallow areas downstream of the system inlet. In the case of wastewaters receiving little pretreatment, the distributed inflow design will safeguard against development of adverse and detrimental conditions from sludge buildups. In the SFS wetland, settleable solids are removed by filtration as the wastewater moves through the underdrain system. For colloidal suspended solids, the primary removal mechanism is bacterial metabolism in both types of wetland treatment systems. Some filtering and sorption will occur in an SFS system.

In the FWS wetland suspended solids may be produced in the form of algae, which also will produce diurnal swings in dissolved oxygen concentrations. In the initial system cells, algae performs important treatment functions, as in an oxidation pond, but the system design should limit the algae population in the later stages of the wetland system so as to avoid permit violations. Nutrient reductions through the system will help effect this requirement.

Nutrients

Constructed wetlands are an effective means to control the discharge of nutrients such as nitrogen and phosphorus. The mechanisms for nutrient removal in a wetland system are:

- Direct plant uptake
- Chemical precipitation
- Uptake by algae and bacteria
- Soil absorption
- Denitrification
- Loss by insect and fish uptake

The density and types of plants in the system will determine the direct and indirect performance rates of these mechanisms. Thus, it is extremely important to maintain a healthy plant community throughout the year to ensure effective performance of the nutrient removal mechanisms. Plant selection should be based on the species ability to establish a stable community that will survive over the spectrum of the system's hydroperiod and other environmental conditions.

Both nitrogen and phosphorus must be considered simultaneously. The wetlands community is composed of many different, interacting populations of organisms functioning in a way unique to the site's environmental conditions. Often, either nitrogen or phosphorus is the limiting factor in an aquatic system, and the addition of excess nutrients causes eutrophication or other undesirable changes in the ecological community.

Nitrogen

A primary mechanism for removal of nitrogen is through nitrification and subsequent denitrification processes performed by the system's microorganisms. Other mechanisms include nitrogen fixation, mineralization, and chemical transformation.

Under natural conditions, living emergent plants exhibit an average nitrogen to phosphorous (N:P) ratio of approximately 7:1. It follows that the nutrient uptake of these plants is limited to an assimilation ratio. When high concentrations of nitrogen and phosphorus are present, the plants may increase their nutrient uptake, by a specific ratio, in excess of current needs. This phenomena is commonly referred to as luxury uptake.

Macrophytes compete among themselves, and with the system's microorganisms, for excess nitrogen. Nitrogen removal objectives favor the microorganisms, because they provide the ultimate nitrogen sink through release of nitrogen gas in the denitrification process. Key factors in the wetland system that affect the denitrification rate are pH, temperature, organic carbon, nitrate levels, and the ecological interactions and exposure times of the denitrifying bacteria within the system.

In a FWS system, nitrification occurs under aerobic conditions in the water column. The nitrates then diffuse into the organically rich sediments and soil where anaerobic conditions necessary for denitrification exist. Nitrification and denitrification processes in SFS systems can occur, but the availability of oxygen for nitrification may be limiting. Denitrification requires anaerobic conditions and the presence of a readily available carbon source. In wetland systems for nutrient removal after secondary or advanced treatment, nitrogen removals may be limited by reduced levels of carbon sources.

Phosphorus

In a natural and constructed wetland system, the principal phosphorus removal mechanisms are precipitation of phosphates under elevated pH conditions, sorption within the bottom soils, uptake by the macrophytic plants and fixation by algae and bacteria. Unlike denitrification for nitrogen removal, there is no single long-term sink for phosphorus removal or retention. This deficiency requires that all the phosphorus removal mechanisms be considered in the system design process.

In the wetland system, both organic and inorganic forms of phosphorus may occur in a dissolved or particulate state in the water column and soil. The interchange of phosphorus between the water and soil phases depends on several factors, such as concentration gradients, pH, dissolved oxygen, and system hydraulics. The latter includes velocity suspension of particulates and bottom scouring to re-suspend settled particulates.

Soil and bottom sediments can be a significant phosphorus sink in mineral soils. In this instance, dissolved inorganic phosphorus is removed through adsorption to clays and precipitation as insoluble complexes of aluminum, calcium, and iron. The aluminum and iron reactions prefer acid to neutral soils, whereas the calcium reactions require alkaline conditions. Plants may then uptake the phosphorus, especially that adsorbed to the soil.

To take advantage of mineral soils as a sink, the water in the system must be moving in the soil column to afford reaction opportunities. The SFS wetlands reduce the influence of water-soil interchanges of phosphorus and force movement through the soil column. However, this approach limits the system's hydraulic capacity without a significant increase in wetland treatment area. If the area soils are mostly organic, special provisions would be required to seek deeper, more mineral soils in order to use the soil as a significant phosphorus removal mechanism.

Plant uptake influences the removal of nutrients and provides a significant short-term phosphorus sink during establishment of the wetland system, but it is not the principal phosphorus removal mechanism. Longer-term storage occurs when the plants die and form a litter zone which forms peat (organic soils). Likewise, fixation of phosphorus in algae and bacteria cells results in storage of phosphorus in the bottom sediments. The stability of these phosphorus sinks depends on several factors including the plant tissue decay rate, peat formation rate, and occurrence of scouring.

Metals

The mechanisms for metal removal in wetland systems are chemical oxidation or reduction, resulting in precipitation, sorption by plants or soil, and simple filtration or sedimentation. The presence of clay soils in an SFS enhances the opportunities of removal by adsorption. Filtration also will remove suspended metals in an SFS wetland system. In an FWS system, sedimentation will occur, and the development of neutral to alkaline conditions will favor precipitation of metals. In both systems, metal uptake by plants may occur.

SITING CONSIDERATIONS

Existing Wetlands

Whether it be a commercial development or a wastewater treatment facility, wetlands have become crucial in the planning process. Depending on the type of natural wetland involved, environmental permitting can be very costly and time consuming. The first step is to identify candidate wetlands for renovation of wastewater. To facilitate permitability and public support, identification of highly disturbed or otherwise impacted wetlands is important. These sites have relatively less functional value than undisturbed or unimpacted sites, and are prime candidates for restoration credits.

Natural wetlands are considered as "waters of the United States" and are subject to federal water quality jurisdiction, including water quality standards and permitted discharges. Prior to approval of a treated wastewater discharge, determinations must be made that the designated, and existing, wetlands and downstream waters will be maintained and protected. Ambient monitoring is necessary to ensure that permitted discharges to natural wetlands preserve and protect wetland functions and values (U.S. Environmental Protection Agency 1990).

Constructed Wetlands

Due to the relatively large area required for a constructed wetland wastewater treatment system, siting analyses are especially important. Prior to initiating field studies, potential sites may be identified and screened using published data. Typical information gathered includes aerial photographs, Soil Conservation Service soils maps, Federal Emergency Management Administration flood studies, U.S.

Geological Survey topographic maps, and local land use plans. (U.S. Environmental Protection Agency 1988, Water Pollution Control Federation 1990). Essential considerations for siting potential wetland treatment systems include identification of existing wetlands, habitat for listed species, existing hydrology, soils, and topography. The type of wetland treatment many be pre-planned or may be a part of the alternative site selection process.

The selected site must be suitable for development of the ecological community required to provide desired treatment. It must be within an affordable distance from existing sewerage facilities. Moreover, the site must be located so as to gain public and regulatory acceptance. In most cases, the planning process should anticipate 1-2 years, and in some cases longer, for completion of the site selection process.

Constructed wetland systems offer the potential for a multiple purpose project which will not only serve as a wastewater treatment facility but also provide other valuable functions, such as a protected natural area and wildlife habitat. From a public acceptance perspective, the latter functions are uniquely important. This "designed" habitat can be developed to include public access as an integral part of the project.

In contrast to existing wetlands, constructed wetlands for wastewater treatment are generally not subject to water quality standards, but are subject to discharge permitting requirements. However, a multiple use project may be considered as "waters of the United States" and therefore subject to water quality jurisdiction. Based on such factors as the size and degree of created wetland isolation, a case-by-case jurisdiction determination should be made as early as possible in a proposed project.

Listed Species

The potential presence of habitat for state and federally listed species should be investigated. The presence of habit for listed species generally complicates the permitting process, and critical habitat should be avoided. Nevertheless, a constructed wetland system could potentially be designed to support listed species. Federal regulations may prohibit continuation of the project or may require mitigative actions for the loss of a specific habitat during development of the project.

Hydrology

The review of conditions at wetlands which may receive treated wastewaters is necessary to evaluate the impacts of changing the hydroperiod, which is the relationship of water depth and period of inundation over an annual cycle. Due to their composition and structure, wetland systems are very sensitive to hydrologic modifications. The existing plant community is acclimated to aerobic, anaerobic, or both conditions, and to seasonal fluctuations of water levels including dry periods.

The successful performance of the wetland system for wastewater treatment is dependent on the ability to retain water and control hydraulics through the system. In the site selection phase, attention to hydrology should focus on the potential site's runon, runoff, and infiltration characteristics, and its relation to both surface water and groundwater resources. Local climate factors (precipitation, evaporation, and temperature) and the influence of these factors on hydrology, are considered in more detail in the final wetland system design.

Soils

Soil is one of the most important physical components of both natural and constructed wetlands. Mineral composition, organic matter, moisture regime, temperature regime, chemistry, and depth interact to greatly influence the types of plants that will succeed on the surface, and the microorganisms that will live in the soil phase.

For the wetland treatment system, soil types and characteristics such as permeability and drainage are influential in development of the site (Table 3). Permeability is especially important in that it reflects the ease with which water and air move through the soil. Infiltration into, and drainage within the soil, are or can be affected by subsurface impervious layers. Of course, the soils must support wetlands vegetation.

Soil type directly influences wastewater treatment performance. For example, the sorption capability of clays can be important in the removal of metals. Also, mineral soils allow chemical precipitation of phosphorus. On the other hand, organic soils offer little or no potential for removal of metals or phosphorus.

As water level control is imperative, FWS wetland systems require soils with very low or no permeability unless there is a natural impervious strata to restrict drainage. When siting a constructed wetland system, considerations should be given to compaction of medium texture soils, such as a loam or silt loam, or installation of a clay or synthetic impermeable barrier below coarse textured (sandy) soils to maintain the system's water controls. For a SFS wetland system, coarse textured (sandy), highly permeable soils are desirable. However, there must be a natural or installed impervious subsurface barrier to restrict vertical wastewater migration. The barrier serves the primary purpose of maintaining hydrology for the emergent wetland vegetation. In areas with shallow groundwater, the impervious barrier also protects the groundwater from contamination.

Topography

In general, topography is not a siting constraint for constructed FWS wetlands, as soil and drainage factors almost always favor farm fields, pastures, or bottom lands with negligible vertical relief relative to alternate sites. However, such sites may sustain dike, berm and other physical damages from uncontrolled external flooding. For SFS wetland systems, topography is important. A slope of one percent or more is needed to function as a gravity system. Site preparation often represents the major cost element for either type of constructed wetland system.

Table 3. Comparison of Mineral and Organic Soil Characteristics (Cowardin et al. 1979, U.S. Environmental Protection Agency 1988, Hammer 1989).

Attribute	Mineral Soil[1]	Organic Soil[2]
Density	High	Low
Drainability	Poor	Moderate
Organic carbon	Less than 12-20%	Greater than 12-20%
Permeability	Low	Moderate
pH	Slightly acid to neutral	Acid, pH less than 6
Sorptive capacity	High	Low

[1] For mineral soils with high clay content. Attributes do not apply to those with high sand or gravel content.
[2] For most organic soils. Attributes do not apply to well decomposed organic soils, such as peat.

Still other factors, such as historic land uses, current land use, and zoning and other local ordinances are important on a site specific basis and in some way, influence the site selection process. For example, if a records search shows that a candidate site was once a wetland, it quickly becomes an alternative deserving serious consideration, for the lost resource may now be reclaimed.

BASIS OF DESIGN

With the definition of wastewater characteristics and effluent limits and the selection of a site, design of the constructed wetlands treatment system may begin. Design development includes the determination of treatment process criteria, definition of physical features of the wetland, and selection of the wetland plant species. The period for the wetland communities to become established also is considered.

The guidelines that follow are intended to put design data into perspective rather than to establish universal criteria. Designers should conduct additional literature reviews and consider pilot testing before actual site specific design development.

Process Criteria

Design of a constructed wetland treatment system uses criteria with names analogous to those of conventional (mechanical) facilities, such as BOD and nutrient loading rates, hydraulic loading rates, and detention time. For FWS systems, water depth is an added criteria. An important, early determination is the degree of pre-application treatment before the wetlands system (U.S. Environmental Protection Agency 1983, Reed et al. 1988, Hammer 1989, Metcalf & Eddy 1991).

Pretreatment

The degree of wastewater treatment before application to the wetland system is a major project decision. Pretreatment not only affects the process criteria for the constructed wetland system, but also determines structural and other physical feature needs for the project. Typically, the greater the level of pretreatment the lower the land requirements for the wetland system. Conversely, the capital and operating cost of the total treatment works often increases with expanding pre-application treatment facilities.

At a minimum, preliminary treatment should remove grit and large solids. Treatment to remove settleable solids and some BOD is also very desirable, though not essential. Pretreatment to secondary or even advanced levels is most desirable. These lower strength wastewaters more closely approach the dilute wastes historically assimilated by natural wetlands as one of their fundamental functional values.

BOD Loading Rate

Selection of the BOD loading rate requires consideration of the organic load/oxygen demand to be applied on an area basis, with respect to the oxygen required for microbial metabolism and provided by the atmosphere (FWS systems) and the vegetation. Also, the BOD load must provide the carbon source required by denitrifying bacteria.

In an FWS system, the treatment process is best characterized as an attached growth biological reactor under plug flow conditions. This characterization is due to the fact that the primary treatment mechanism, especially for BOD, is provided by microorganisms attached to the wetland vegetation exposed to a single pass of wastewater flow. Moreover, algae also contribute to the treatment performance as they do in a facultative/oxidation pond system. As in conventional treatment facilities, an overload of BOD with the depletion of oxygen creates an odor problem and provides little or ineffective treatment. In a wetland system, these stress conditions may alter, or even kill, the emergent vegetation and the microbial population. Likewise, settleable solids may produce pockets of high load or oxygen demand. Design criteria, operational practices, and attention to wetland system performance should minimize such problems.

By current practice, BOD loading rates to constructed wetland treatment systems approach 45.4 kilograms per day per acre for both FWS or SFS systems. As a point of comparison, this loading rate is twice the maximum rate typically applied to facultative ponds. The attached microbial population and the oxygen transfer capability of the wetland plants contribute to the success of higher loadings to wetland systems.

In practice, with secondary and higher levels of pretreatment, the actual BOD loading rate may be very much lower. Other more limiting factors, such as nutrient loadings, typically dictate land area requirements. With advance pretreatment and stringent nutrient effluent limits, the BOD loading rate may only be a few kilograms per day per acre.

Nutrient Loading Rate

The nutrient loading rate is particularly important for wetland systems discharging to sensitive watersheds where nitrogen or phosphorus permit limits are established. The type of permit limit is significant in that controls of ammonia-nitrogen, total nitrogen, and phosphorus vary not only with the constituent but also with the magnitude of the limit.

Reduction of ammonia-nitrogen by nitrification is expected to occur in the presence of an adequate supply of oxygen, and an important consideration in wetland nitrogen removal is plant uptake. Denitrification for nitrogen removal requires a carbon source to complete the conversion. Phosphorus reductions are more complicated, and include plant uptake. However, plant uptake does not function as the control mechanism.

Even though nitrogen-to-phosphorus ratios are important to the functions of a wetland, both may not be listed in permit limits. Furthermore, reported design loading rates vary widely. Nutrient loading rates of 1 to 2 and 45 to 150 kilograms per day per acre for nitrogen and phosphorus, respectively, may be used. Wetlands tend to become nitrogen-limited quickly with the application of wastewaters.

Hydraulic Loading Rate

The hydraulic loading rate is closely tied to site specific conditions, such as climate, soil characteristics such as permeability, and selected plant species. Also, the BOD loading rate and evaporation rate inversely influence the hydraulic loading rate. As an order of magnitude guide, a hydraulic loading rate of 75,800 liters per day per acre may be used. This rate is about half that typically used for rotating biological contactor design, and about two-thirds the hydraulic loading rate to a facultative pond.

Detention Time

To achieve the desired treatment performance, the wastewater must be in the wetland system a sufficient time for the primary biological, and any secondary chemical and physical (sorption), actions to take place. Other factors, such as water depth and plant selection, also influence determination of detention times. Biological activities typically determine detention time, and detention times on the order of 10 to 20 days should be expected. This is about half that of a facultative pond. The physical design, via internal flow depth and routing controls, can provide flexibility to satisfy varying detention time requirements.

Water Depth

In an FWS treatment system, the selection of water depth depends on the relationship between system detention time, vegetation type, and the pollutant character. Typically, depths of 0.6 to 0.9 meters may be used in a deep marsh community. A shallow marsh community will have water depths of 0.5 meters or less. Most emergent plant species perform best in depths less than 0.3 meters. On the whole, an average system water depth may range from 0.5 to 0.6 meters.

In an SFS wetland treatment system, the water flow is maintained below the ground surface, and the bed depth is controlled by the depth of penetration of the plant root systems. It is the root system that provides the oxygen for the SFS system.

Conceptual Design

In conjunction with the selection of process design criteria, the conceptual design begins to define and develop the process flow diagram and physical features of the constructed wetlands treatment system. The site selection process, and follow-up investigations of the selected site, identify some of the conceptual design needs. Examples include needs to provide impermeable (soil or liner) bottoms to control vertical migration of flows requiring treatment, and elevated dikes or berms to prevent excess inflow of surface runoff.

The selected site also influences the layout of the treatment system and the development of other physical features. Inlet, outlet, and internal control of flows is especially important. Also, emergency overflows or diversions are needed to manage excess water and protect dikes and berms. Additional considerations are necessary for system operation and maintenance. For example, access must be provided for water quality and vegetation monitoring and special attention is needed for potential or planned harvesting techniques.

Layout

The layout of the constructed wetland treatment system varies by site. In an FWS system, multiple cells are developed in both parallel and series configurations. Long length-to-width ratios may be used to enhance plug flow characteristic and/or prevailing topographic features may be modified to have a more natural effect. For the later, channeling may be included to promote plug flow.

The configuration of an SFS wetland system includes multiple beds, arranged in parallel. Since very low velocities are maintained in these systems, short length-to-width ratios are necessary.

Flow Controls

Influent flow distribution should be considered, and its importance is inversely proportional to the degree of pretreatment (Reed et al. 1988, U.S. Environmental Protection Agency 1988, Hammer 1989, Water Pollution Control Federation 1990). A point discharge of only screened wastewater will quickly establish odorous and objectionable sludge deposits in an FWS system, and possibly bind the media in an SFS system. Conversely, an influent for polishing, after advanced treatment works, is more amenable to point discharge. In constructed FWS wetlands, swales and/or perforated pipe may be used to disperse the influent. The inlet zone of an SFS wetland system should include large gravel that will distribute but not "filter" the influent. Table 4 lists typical characteristics of an SFS system media. Often a lake will be considered as a means to collect flows from the system, but also to add polishing to the treatment process and to improve the aesthetic value of the project. Implementation may be easy if a contiguous borrow area is needed during construction for dike and berm materials.

Internal flow controls are especially critical to successful operation and treatment performance of the wetlands system. Weirs, gates, or other regulating devices are used to manage water depths, control velocities, divert flows during abnormal flow periods, or isolate cells for maintenance. For example, water levels must be fluctuated to optimize the performance of individual cells. After steady state conditions are achieved, significant water level changes are stopped, and weir settlings are changed infrequently. This eliminates the potential of flushing fines from the sediments and litter zones, which would impact performance of the downstream cell. Most importantly, the steady state hydroperiod is established for development of the long term vegetative communities.

The discharge from the wetlands treatment system will be subject to effluent permit requirements. Typically, this means a point discharge with flow monitoring and effluent sampling. Receiving water sampling also may be required by the permit. When discharges from cells in parallel are planned, as in an SFS system, a common effluent collection must be included in the design. If the final discharge is unique, such as to a contiguous natural wetland, effluent distribution from the final cell(s) should be considered, in which case, an alternate point or means of verifying permit compliance must be negotiated.

Plant Selection

Many plant species are available, and have been used, for wetland wastewater treatment systems (Reed and Bastian 1980, Reed et al. 1988, Hammer 1989, Water Pollution Control Federation 1990). The two principal groups are free floating plants such as hyacinths and duckweed, and rooted macrophytes such as cattails and reeds. Operationally, the main difference between the two groups is that the floating plants require regular harvesting to maintain effectiveness. The rooted macrophytes may be used to create shallow or deep marsh communities of cattails or bulrushes, or alternatively shallow marshes of grass/herbaceous species. Less commonly, forested (hardwood) swamps are used in a constructed wetland system.

Structuring mitigation wetlands into communities similar to endemic wetlands is desirable in that nature has completed a successful plant selection process. Also, the presence of the endemic plant communities may well expedite the establishment of the mitigation wetland. For a wetland constructed for wastewater treatment, competing design criteria usually favors decisions based on requirements for land area and operating depth rather than endemic plant selection.

Fundamental to the emergent plant selection process is the identification of plant species that exhibit sustained (and preferably rapid) growth in wastewater-rich flows, are disease resistant, and are not vulnerable to minor changes in their environment. Also, it is often desirable that they provide high nutrient and mineral sorption capability. Compatibility with hydrological design factors is required. Table 5 includes a variety of emergent plants that may be considered for use in cells of differing hydroperiods within a constructed wetland system.

Table 4. Typical Media Characteristics for Constructed SFS Wetland Treatment Systems (U.S. Environmental Protection Agency 1988).

Media type	Maximum 10% grain size, mm	Porosity	Hydraulic conductivity k_s, ft^3/ft^{2}d	K_{20}
Medium sand	1	0.42	1380	1.84
Coarse sand	2	0.39	1575	1.35
Gravelly sand	8	0.35	1640	0.86

Table 5. Candidate Emergent Plants for Constructed Wetland Treatment Systems.

Common Name	Scientific Name
Carex spp.	Sedges
Cladium jamaicense	Sawgrass
Eleocharis spp.	Spikerushes
Elodea spp.	Waterweeds
Glyceria spp.	Manna Grass
Juncus spp.	Rush
Nyssa spp.	Tupelo
Panicum spp.	Maidencane
Phragmites australis	Reed
Pontederia cordata	Pickerelweed
Potamogeton spp.	Pondweeds
Sagittari spp.	Arrowheads
Scirpus spp.	Bulrush
Sparganium spp.	Burreeds
Sphagnum spp.	Mosses
Taxodium distichum	Bald Cypress
Typha spp.	Cattails
Utricularia spp.	Bladderworts

As there are relatively few emergent plant species that are adapted to deeper waters, manipulation of operating water depths can create a limiting condition affecting plant diversity. With multiple cells and water level controls, the designer or operator may selectively influence the distribution of plant species. Maintaining a relatively stable water level for extended periods will further limit diversity by preventing the germination of seeds from annuals and perennials.

Cattails and bulrushes are among the most researched plant species for wetlands treatment systems. Both are capable of growing in shallow and relatively deep waters, achieve similar heights, and form very dense stands and well developed litter zones. Furthermore, they comprise the most common plant communities utilized in the deep marsh portion of a constructed wetlands treatment system. As cattails respond favorably after a disturbance, such as construction activities and altered hydrology, and spread rapidly by vegetative growth, they provide the benefit of a lessened establishment period. Cattails may be an unwelcome invader to portions of the planned wetland system. Conversely, cattails do not compete well with established plant communities where conditions tend to inhibit cattail seed germination.

Bulrush have the ability to compete with cattails in that bulrush will grow and persist even when flooded in deep waters. An important advantage of bulrush species is they are less of a threat to contiguous natural wetland communities than the more aggressive, invader cattail species. Bulrush communities also tend to be more aesthetically pleasing than cattails.

With shallower operating water levels, a grass/herbaceous marsh (also referred to as mixed or shallow marsh) community may be developed. The shallow marsh provides increased surface area for attached microorganisms and material for the litter zone. These communities have an increased plant species diversity, the capability to reduce and stabilize nutrient concentrations, and provide expanded potential for wildlife habitat. Some of the shallow marsh plant species are adapted to continuous submergence. Shallow depths reduce detention time. Therefore, a grass/herbaceous marsh will have greater land requirements than deeper marshes.

Forested swamps have the greatest land requirements, as water depths are lowest and seasonal drawdowns are needed. Few tree species, such as cypress, gum, and tupelo, can survive prolonged periods of inundation. Most other wetland tree species require periodic, complete water drawdown to survive. Forested swamps have the capability to serve as a relatively long-term sink for nutrients, and provide additional, and different, wildlife habitat than marsh systems.

Establishment

An establishment period is required to achieve full operability and treatment performance of a constructed wetland system (U.S. Environmental Protection Agency 1988, Hammer 1989). Animal foraging may impact plantings during this time. Intrusion by undesirable plant species may account for significant plant loss, and may be minimized by controlling water levels. Flooding favors the wetland plant species and often is an especially effective control technique. Also, during this time, plant diversity will likely increase with the emergence of desirable, volunteer plant species.

The establishment period may vary from a few months to years. The principal influences on the time period are the type and size of the wetland system, and the prevailing environmental conditions. For example, in favorable weather, a hyacinth system can become operational in only a few months. Rooted plants may require eight to sixteen months to provide full area coverage, and depending on the size originally purchased, tree species are likely to take a year or more to become well established.

OPERATION AND MAINTENANCE

Regardless of the selected vegetation, wetland wastewater treatment systems require operation and maintenance of physical features and monitoring of wetland vegetation and water quality (U.S. Environmental Protection Agency 1988, Hammer 1989). Routine operation and maintenance procedures are necessary for the system as a whole. Individual pieces of equipment, such as pumps and water quality samplers, will have prescribed maintenance procedures. Gates and valves should be exercised periodically, and weirs should be kept clean to remain effective. Other requirements include simple visual inspections of the structural condition of dikes and berms. Special attention should be given to the detection and removal of burrowing animals.

Vegetation Management

The common wetland system categories, based on vegetation and in descending order of operating and maintenance costs, are floating plants (hyacinth), cattail or bulrush monocultures, grass/herbaceous marshes, and forested swamps. In floating plant systems, there is no doubt that harvesting is a critical procedure for continued nutrient removal, as plant tissue is the primary nutrient sink. For emergent vegetation, needs for harvesting are not well supported.

Several techniques that may be used to harvest wetland plants include thinning or weeding vegetation using modified farm equipment or floating harvesters, controlled burning on an area basis to maintain the system's hydraulic profile, and complete removal of the excessive accumulations of litter and the upper sediment or soil layer using construction equipment. The process requires that wetland cells be taken out of service, and will cause moderate to severe disturbance to the wetland treatment performance. Furthermore, the wetlands must go through a re-establishment period before effective wastewater treatment can be expected. As the plant tissue of emergent plants is not a primary pollutant sink, there is no apparent value in harvesting in a functioning FWS or SFS system.

Regulation of water levels is indispensable for operation of emergent vegetation systems. Grass/herbaceous marshes and forested wetland communities need seasonal drawdowns to preserve their community structure. Moreover, controls are needed to maintain operable, surface or subsurface water levels during uncommonly dry and wet periods. In FWS systems, fluctuating water levels may used as an effective means to control pest insects, especially mosquitoes.

System Monitoring

Monitoring of the biological communities and water quality must be conducted to ensure effective wastewater treatment performance in the constructed wetland system. During the first year of operation, a baseline condition will be established from seasonal growth and successional patterns for the specific wetland system. The data are then used to establish expected seasonal trends in productivi-ty and density for assessment of the status of the plants and their impact on water quality.

At least once a year, vegetation maps should be prepared to record the extent of plant communities. Normally, fixed transects of the wetland system are used. For very large systems, aerial photography with ground verification may be implemented. Ongoing water quality monitoring is then correlated to the vegetative mapping to accept or reject shifts in characteristics of the plant communities.

Water quality monitoring will include influent and effluent sufficient to satisfy permitting requirements. At least some internal sampling is needed to establish performance trends within the wetland system. Typical water quality parameters include BOD, solids, nitrogen, phosphorus, dissolved oxygen, pH, and other pollutants of local concern such as specific heavy metals.

CASE STUDIES

The following case studies are used to illustrate the application of wetlands systems for wastewater treatment.

Gustine, California

A small agricultural town, Gustine treats approximately 1.2 mgd of wastewater (U.S. Environmental Protection Agency 1988). One third of the wastewater originates from domestic and commercial sources, and the remainder from three dairy products industries. The wastewater is of high strength, averaging over 1200 milligrams per liter of BOD_5. Until recently, the City's wastewater treatment facilities consisted of 14 oxidation ponds operated in series. The ponds covered approximately 54 acres and provided some 70 days of detention time. The effluent was then discharged without disinfection to a small stream leading to the San Joaquin River.

As a result of mandated secondary treatment levels, alternatives were evaluated. It was determined that oxidation ponds combined with a constructed marsh was the most cost-effective system. The advantages were that suitable land was available, the treatment method was compatible with the surrounding area, and there was consumption of very little energy. Following completion of the oxidation ponds, construction began in 1985 for 24 marsh cells, each approximately one acre in size, to operate in series. Flexibility was added to the system by incorporating the capability to withdraw from any of one of the last seven oxidation ponds directly into the marsh system. In total, the marsh system required some 36 acres, including all interior cell dividers, levees, and an outer flood protection levee.

Columbia, Missouri

In early 1989, the City of Columbia began the development of a constructed wetland (120 acres ultimate) as part of its 18 mgd wastewater treatment program (Brunner et al. 1992). The treatment goal was to produce a secondary quality effluent from the wetlands system. A key aspect of the project was strong public support for the wetlands as a means to eliminate an outfall pipeline to the Missouri River. Also, the reclaimed wastewater would be used along with river water to supply 2700 acres of newly constructed conservation wetlands. The project helped accelerate the Missouri Department of Conservation's plan to acquire property and construct this wildlife habitat.

Orlando, Florida

The Orlando Easterly Wetlands Reclamation Project is a 1200 acre constructed wetlands treatment system that began operation in 1987 (Jackson 1989). Faced with a no discharge order from the regulatory agencies and a compounded problem due to sensitive soil and groundwater conditions that made land disposal unfeasible, the City of Orlando initiated this 20 mgd constructed wetlands project. Effluent from the Iron Bridge water pollution control facility is pumped to the site via 16 miles of 48-inch-diameter force main. The constructed wetland is segmented into 18 cells. Control structures between cells ensure proper operating depths and even flow distribution and allow individual cells to be isolated and drained for maintenance.

There are three major areas within the wetland, each containing a different emergent vegetative community. The first area is a 420 acre shallow marsh/wet prairie, followed by a 380 acre mixed marsh and a 400 acre hardwood swamp. Approximately two million rooted wetland plants and 160,000 trees were planted throughout the site.

The wetlands solution provides environmental protection by reducing nitrogen (0.70 mg/L) and phosphorus (0.08 mg/L) concentrations in the effluent to natural background levels. The project reclaimed a wet prairie drained in the 1850s for pasture land. The project now provides a reliable water source for adjacent wetlands, located between the site and the St. Johns River, that were deteriorating due to man-made drainage of the area.

Savannah, Georgia

The undeveloped portions of western Chatham County will not develop without additional capacity at the Travis Field Wastewater Treatment Facility, and a no discharge requirement is in effect (EMC Engineering Services). However, expansion of the Savannah International Airport complex, as well as a proposed industrial park and cross-county access highway create a need for additional wastewater treatment capacity. In 1990, Savannah conducted a feasibility study of some 45,000 acres for a site for a 10 mgd constructed wetlands treatment system. Three candidate sites passed a fatal flaw analysis and were identified for further study. The Georgia Environmental Protection Division accepted the City's concept to provide a discharge meeting background water quality as a no discharge system.

Lakeland, Florida

In the early 1900s, a wetland site near Lakeland was altered by phosphate mining operations (Jackson 1989). For a period, phosphate processing wastewater containing mostly clay and sand with some gypsum was treated in a series of settling ponds. Subsequently, the mining operation was discontinued. In the mid-1980s, the City of Lakeland began a process to acquire some 1400 acres, including the abandoned clay settling ponds, and to construct a wetlands treatment system for 14 mgd. The system used two overland flow cells ahead of four mixed marsh cells. The water then flowed into a shallow lake and then into deeper lake prior to discharge to the Alafia River. In addition to a federal grant for the wastewater treatment aspects of the project, the City also obtained a grant from the Florida Department of Natural Resources Old Mines Reclamation Fund. The wetland system began operating in March, 1987.

REFERENCES

Brunner, C.W., J.W Smith, N.P. Stucky, D.W. Kempf, R.K. Baskett, R.H. Kadlec, L.B. Patterson, J.E. White and J.A. Zdeb. 1992. Design of constructed wetlands with effluent reuse wetlands in Columbia, Missouri, Proceedings 65th Annual Conference & Exposition, Water Environment Federation, Alexandria, Virginia.

Cowardin, L.M., V. Carter, F.C. Golet, and E.T. LaRoe. 1979. Classification of Wetlands and Deepwater Habitats of the United States, FWS/OBS-79/31, Fish and Wildlife Service, U.S. Department of Interior, Washington, DC.

EMC Engineering Services, Inc. 1990. Travis field wetlands treatment system feasibility study, Savannah, Georgia.

Hammer, D.A. (ed.). 1989. Constructed Wetlands for Wastewater Treatment: Municipal, Industrial and Agricultural. Lewis Publishers, Inc., Chelsea, Michigan.

Jackson, J. 1989. Man made wetlands for wastewater treatment: Two case studies. Pp. 574-580 In Hammer, D. A. (ed.) Constructed Wetlands for Wastewater Treatment: Municipal, Industrial and Agricultural. Lewis Publishers, Inc., Chelsea, Michigan.

Metcalf & Eddy, Inc. 1991. Wastewater Engineering - Treatment, Disposal, and Reuse. Tchobanoglous, G. and F.L. Burton (eds.). McGraw-Hill, Inc., New York, New York.

Reed, S.C. and R.K. Bastian (eds.) 1980. Aquaculture systems for Wastewater treatment: an engineering assessment. EPA 430/9-80-007, U.S. Environmental Protection Agency, Washington, DC.

Reed, S.C., E.J. Middlebrooks and R.W. Crites. 1988. Natural Systems for Waste Management and Treatment, McGraw-Hill, Inc., New York, New York.

Sherwood, R., G. Tchobanoglous, J. Colt and A. Knight. 1979. The use of aquatic plants and animals for the treatment of wastewater. Departments of Civil Engineering and Land, Air, and Water Resources, University of California, Davis, California.

Tiner, R.W., Jr. 1984. Wetlands of the United States: Current Status and Recent Trends, Fish and Wildlife Service, U.S. Department of Interior, Washington, DC.

U.S. Environmental Protection Agency. 1983. Freshwater wetlands for wastewater management, environmental impact statement, phase 1 report. EPA 904/9-83-107, Region 4, Atlanta, Georgia.

U.S. Environmental Protection Agency. 1988. Constructed wetlands and aquatic plant systems for municipal wastewater treatment. EPA/625/1-88/022, Cincinnati, Ohio.

U.S. Environmental Protection Agency.   1990. Water Quality Standards for Wetlands - National Guidance, Washington, DC.

Water Pollution Control Federation.   1990.   Natural Systems for Wastewater Treatment, Manual of Practice FD-16. Alexandria, Virginia.

CHAPTER 11

WETLANDS FOR STORMWATER TREATMENT

David R. Bingham

Stormwater runoff pollution has been found to prevent attainment of the desired uses of water resources in many locations. Pollution in stormwater is derived from widespread sources which come in contact with rainfall and snow melt runoff. The runoff, with its associated pollutants, is transported through natural or man-made systems to the water resource. Constructed wetlands can be used to control some of the pollutants contained in stormwater runoff. Another benefit of using wetlands for treatment of stormwater runoff is the mitigation of the hydrologic effects of urbanization.

Point source pollution is regulated by the Clean Water Act (CWA) through the National Pollutant Discharge Elimination System (NPDES) program. The effectiveness of point source effluent limitations imposed by discharge permits on restoring water quality has been well documented (U.S. Environmental Protection Agency 1990a). As the contribution of pollutants from point sources diminishes because of implementation of the controls mandated by the CWA, however, the relative contribution from nonpoint source pollution increases in significance. Nonpoint source pollutant loadings are considered a major contributor to degradation of surface waters and have been cited as responsible for nonattainment of water quality goals (U.S. Environmental Protection Agency 1990b). Stormwater is the major transport mechanism which conveys nonpoint source pollution to receiving waters. To fully attain the beneficial uses identified for the nation's water resources, implementation of stormwater pollution controls is thus necessary.

In the 1987 amendments to the CWA, Congress introduced new provisions to address diffuse pollution sources such as stormwater runoff. The development of a workable program to regulate stormwater discharges has been difficult because of the large numbers of individual discharges, the diffuse nature of the discharges and related water quality effects, and limited resources of the regulatory agencies and regulated entities. After extended development and review, however, the provisions led to the development of the stormwater regulations promulgated by U.S. EPA in November 1990. These regulations require that NPDES permits be obtained for discharges of stormwater runoff. The regulations affect municipalities with populations served by separate storm drainage systems greater than 100,000 (medium and large municipalities), and industrial sites falling within one of eleven industrial categories (U.S. Environmental Protection Agency 1990a). In addition, they require the development of stormwater runoff control programs for certain construction sites. U.S. EPA is also developing regulations that address stormwater runoff pollution in communities with populations less than 100,000 (CWA Article 402).

Due to the recent advent and continued evolution of stormwater pollution control regulations, extensive infrastructure to control this category of pollution does not exist. Historically, stormwater was considered to be a flood control issue. There has been limited implementation of pollution control measures, and there is limited funding to address such issues. However, a growing emphasis on the need for stormwater pollution control is expected to continue in the coming years.

There are numerous options for treatment of pollution from stormwater (Table 1). Options include source controls which prevent or reduce availability of pollutants which come in contact with stormwater, to regulatory controls which control land use and development in a manner designed to minimize stormwater pollution, and structural controls. Constructed wetlands are one of many possible structural controls.

The stormwater pollution controls shown in Table 1 are also frequently referred to as BMPs (best management practices). Structural controls are generally more expensive than source and regulatory controls, but are far less expensive than most point source control facilities (such as wastewater treatment plants). While they involve site construction work (drainage, grading, planting, etc.) and some structural facilities (outlet structures, small dams), they do not generally involve the use of buildings and facilities with mechanical and electrical equipment. This is in part because comparatively low-cost control facilities such as constructed wetlands can achieve reasonable pollutant removal rates. It is also in part because of the aforementioned limited focus on, and funding for, stormwater pollution control projects.

Structural stormwater pollution controls such as constructed wetlands may be implemented as part of a new development or redevelopment, or may be *retrofit* into an area to provide control for an existing development. The process of retrofitting structural stormwater controls is difficult and expensive due to site constraints, lack of available land and the need to modify existing drainage patterns.

Naturally occurring wetlands are not generally recommended as a means of stormwater pollution control. While they can provide a pollution control benefit, the need to protect existing wetlands resources and the associated regulatory restrictions frequently make this an undesirable option.

The remainder of this chapter addresses the hydrologic and chemical characteristics of stormwater, the function of constructed wetlands and their effectiveness in treating stormwater, the design of constructed wetlands for stormwater treatment, and costs of such systems.

Table 1. Options for treatment of pollution from stormwater.

Source Controls	
Animal Waste Removal	Reduced Roadway Sanding
Catch Basin Cleaning	and Salting
Cross Connection Identification	Solid Waste Management
and Removal	Street Sweeping
Proper Construction Activities	Toxic and Hazardous
Reduced Fertilizer, Pesticide	Pollution Prevention
and Herbicide Use	
Regulatory Controls	
Land Acquisition	Protection of Natural Resources
Land Use Regulations	
Structural Controls	
Constructed Wetlands	*Infiltration Facilities*
Detention Facilities	Dry Wells
Extended Detention Dry Ponds	Infiltration Basins
Wet Ponds	Infiltration Trenches
Filtration Practices	Porous Pavement
Filtration Basins	*Vegetative Practices*
Sand Filters	Grassed Swales
Water Quality Inlets	Filter Strips

STORMWATER CHARACTERISTICS

The characteristics of stormwater described in this section are its hydrology and its pollutant content. These characteristics are important in determining the applicability of and the design of constructed wetlands for stormwater treatment.

Hydrologic Characteristics

When precipitation contacts the ground surface, it can take several paths. These include returning to the atmosphere by evapotranspiration (which includes direct evaporation and transpiration from plant surfaces), infiltration into the ground, storage in depressions on the ground surface, and traveling over the ground surface (runoff). Altering the surface that precipitation contacts alters the fate and transport of the runoff. Urbanization replaces permeable surfaces with impervious surfaces (roof tops, roads, sidewalks, parking lots, etc.), which are often designed to remove rainfall as quickly as possible from an area. Increasing the proportion of paved areas decreases infiltration, storage, and evapotranspiration, and increases the volume and rate of precipitation leaving the area as runoff (Figure 1).

In addition to increased runoff volumes, the travel time of the runoff decreases. Mechanisms which delay flow of runoff to receiving waters, such as vegetation, are replaced with systems designed to remove and convey stormwater from the surface. The travel time for the stormwater to reach the receiving waters is thus greatly reduced. Figure 2 illustrates typical pre-development and post-development discharge rates over time for an urban area. The hydrologic effects of urbanization can result in increased frequency and severity of flooding, reduced streamflow during periods of prolonged dry weather (loss of base flow), and greater runoff and stream velocity during storm events.

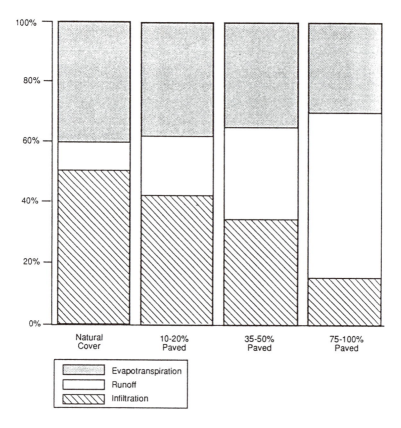

Figure 1. Typical changes in runoff flow resulting from paved surfaces.

Many of these hydrologic changes lead to increased pollutant transport and loading to receiving waters. Increases in runoff volume result in greater discharges of pollutants. As peak discharge rates increase, erosion and channel scouring become greater problems. The eroded sediments also carry nutrients, metals, and other pollutants associated with them. These and other effects result in increasing stormwater-related pollution problems as increased urbanization occurs.

Pollutant Characteristics

In addition to the effects of hydrologic changes on stormwater pollution loads, land development leads to an increase in pollutant loadings in runoff. The increased pollutant loadings result from a combination of higher total and peak runoff volumes caused by the hydrologic changes discussed above, and an increase in the concentrations and amounts of pollution associated with developed land. Table 2 lists the primary categories of pollutants that result from urban runoff,

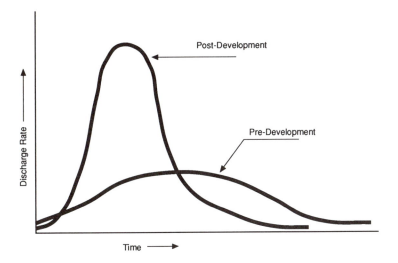

Figure 2. Pre- and post-development hydrographs.

parameters typically monitored for each category, potential pollutant sources, and associated potential effects (U.S. Environmental Protection Agency 1993).

Suspended Solids

Suspended solids are made up of particulate matter that settles and fills in the bottoms of ditches, streams, lakes, rivers, and wetlands. Sediment loading occurs primarily from soil erosion and runoff from construction sites, urban land, agricultural areas, and stream banks. Although some sedimentation is natural, construction, farming, and urbanization can accelerate the process by increasing the rates of stormwater runoff, by removing cover vegetation, and by changing slopes and affecting soil stability. Increased runoff from developed areas transports solids from various sources, including soil erosion, deposition from wind erosion, litter (both manmade and naturally produced), and road sanding. These solids also carry nutrients, metals, and other substances that can impact the water resources and aquatic life. Sedimentation can have substantial biological, chemical, and physical effects in receiving waters. Suspended solids can make the water look cloudy or turbid, diminishing its aesthetic quality and recreational uses. Once the solids settle out and become deposited as sediment, they can smother bottom dwelling species and act as a sink for pollutants which can later be resuspended or leached into the overlying waters.

Table 2. Pollutants associated with urban runoff, monitoring parameters, pollutant sources and potential effects.

Category	Parameters	Possible Sources	Effects
Suspended Solids	**Organic and Inorganic** TSS Turbidity Dissolved Solids	Urban/Agricultural Runoff Combined Sewer Overflows (CSO's)	Turbidity Habitat Alteration Recreational & Aesthetic Loss Contaminant Transport
Nutrients	Nitrate Nitrite Ammonia Organic Nitrogen Phosphorus	Urban/Agricultural Runoff Landfills Atmospheric Degradation Erosion Septic Systems	**Surface Waters** Algal Blooms Ammonia Toxicity **Groundwater** Nitrate Toxicity
Pathogens	Indicator Bacteria (Fecal Coliforms and Streptococci, Enterococci) Viruses	Urban/Agricultural Runoff Septic Systems CSOs Boat Discharges Domestic/Wild Animals	Ear/Intestinal Infections Shellfish Bed Closure Recreational/Aesthetic Loss
Organic Enrichment	BOD COD TOC	Urban/Agricultural Runoff CSOs Landfills, Septic Systems	DO Depletion Odors Fishkills
Toxic Pollutants	Trace Metals Organics	Pesticides/Herbicides Underground Tanks Hazardous Waste Sites Landfills Illegal Oil Disposal Industrial Discharges Urban Runoff, CSO's	Lethal and Sublethal to humans and other organisms
Salt	Sodium Chloride	Road Salting	Drinking Water Contamination Impact on non-salt tolerant species

Nutrients

The nutrients that are of primary concern to water quality and that are present in high concentrations in stormwater are nitrogen and phosphorus. Nutrients are associated with agricultural and urban runoff, atmospheric deposition, leachate from landfills and septic systems, and erosion. Nutrient additions can result in eutrophication of receiving waters, which can cause excessive algal growth. Some nutrients contributed from urban runoff originate from chemical fertilizers and thus are in a dissolved form readily utilized by algae in the receiving waters. Phosphorus is generally the most important nutrient in freshwater because it is growth-limiting, while nitrogen tends to limit growth in marine waters.

Nutrient enrichment can result in severe algal blooms, either in the water column or in stream and lake beds, which can cause unpleasant odors and otherwise detract from aesthetic value. High densities of certain algal species can result in taste and odor problems in drinking water from reservoirs. More seriously, some marine algal species contain toxins that can be harmful to humans eating affected fish or shellfish. Also, die-off and decay of excessive amounts of algae can result in depletion of dissolved oxygen in the water column.

Pathogens

Pathogens are bacteria, protozoa, and viruses that can cause disease in humans. Although not pathogenic themselves, the presence of indicator bacteria, such as fecal coliform or enterococci, are used as indicators of pathogens and of potential risk to human health. Potential health risks are associated with primary and secondary contact recreation, such as swimming, and with consumption of contaminated fish and shellfish in areas affected by urban runoff. Bacterial and viral pathogens are for the most part attributed to livestock in agricultural areas and runoff in urban areas. Other sources of these organisms can include failed septic systems, landfills, or combined sewer overflows and unauthorized sanitary sewer cross connections in storm drains. Pathogens generally cause water quality degradation in resources used by humans for primary and secondary contact recreation or shellfishing.

Oxygen-Demanding Matter

Microorganisms consume organic matter that is deposited in water bodies by stormwater runoff. In the process, oxygen is removed from the water. Organic enrichment can be caused by agricultural and urban runoff, combined sewer overflows, and leachate from septic systems and landfills. A sudden release of oxygen-demanding substances into a water body during a storm can result in the depletion of oxygen and in fish kills. Organic enrichment can also have long-term effects on sediment quality, increasing organic content and the tendency of sediments to deplete surface waters of oxygen (referred to as sediment oxygen demand [SOD]). The solid and dissolved organic content of water and its potential to deplete oxygen are measured by its biochemical oxygen demand (BOD).

Toxic Pollutants

Toxic pollutants include trace metals and organic chemicals. Heavy metals in urban runoff can result from the breakdown of products such as trash cans and car bumpers, fallout from automobile emissions, and other metal products. Potential sources of toxic pollutants include vehicular residues, industrial areas, landfills, hazardous waste sites, leaking underground and aboveground fuel storage tanks, and agricultural areas which use chemicals. Other potentially toxic compounds in stormwater include oil and grease from vehicles and construction equipment. These compounds are generally in runoff from roads, parking lots, service areas, and construction sites, and can be a constituent of landfill leachate. Oil and grease products can be adsorbed to sediment particles and be deposited in the bottom sediments. These compounds can be toxic to aquatic organisms and can bioaccumulate in fish and shellfish, potentially resulting in toxic effects to humans eating this tainted food.

Sodium and Chloride

Discharges of sodium and chloride to surface waters result primarily from road salting operations during winter months and snowmelt during spring thaws. These discharges can affect the taste of drinking water supplies and be harmful to people on low sodium diets. Of secondary importance is its potential impact on salt-intolerant plant species. However, the concentrations of sodium and chloride in runoff are typically insignificant, especially in resources with continuous flushing (rivers and streams). Sodium and chloride discharges are, therefore, of greater concern in drinking water supplies, groundwater and lakes.

Summary of Quality Characteristics

The U.S. EPA's Nationwide Urban Runoff Program (NURP), conducted from 1978 to 1983, provided a comprehensive study of stormwater runoff from residential, commercial, and light industrial areas throughout the United States. A large database of pollutant concentrations and loads measured during various storm events was developed. These data can be applied using pollution load estimation techniques. Table 3 presents median and mean values of event mean concentrations derived from urban runoff data collected in U.S. EPA's NURP study (U.S. Environmental Protection Agency 1993). The values in this table can be used as a rough estimate of stormwater pollutant concentrations. However, site-specific data are required because of the high variability of concentrations common in stormwater.

Ranges of concentrations of various pollutants found in rainfall, stormwater, combined wastewater, and wastewater effluent are presented in Table 4. Stormwater contains some pollutants at levels comparable to or higher than that of treated wastewater effluent. In addition, it frequently contains toxic pollutants as shown in Table 5.

FUNCTION OF CONSTRUCTED WETLANDS FOR STORMWATER TREATMENT

A constructed wetland for purposes of stormwater treatment generally consists of several components:

- Stormwater inflow control
- Pretreatment area

Table 3. Water quality characteristics of urban runoff for the median Nationwide Urban Runoff Program (U.S. Environmental Protection Agency 1983).

Constituents	Site Median EMC	Site Mean EMC
Total Suspended Solids (mg/l)	100	141 to 224
Biochemical Oxygen Demand (mg/l)	9	10 to 13
Chemical Oxygen Demand (mg/l)	65	73 to 92
Total Phosphorus (mg/l)	0.33	0.37 to 0.47
Soluble Phosphorus (mg/l)	0.12	0.13 to 0.17
Total Kjeldahl Nitrogen (mg/l)	1.5	1.68 to 2.12
Nitrate and Nitrite Nitrogen (mg/l)	0.68	0.76 to 0.96
Total Copper (µg/l)	34	38 to 48
Total Lead (µg/l)	144	161 to 204
Total Zinc (µg/l)	160	179 to 226

EMC is the Event Mean Concentration.

- Wetland area including plants and substrate
- Ground and surface water
- Wetland outflow control

These components are combined to provide a mechanism for the removal of pollutants from urban runoff. Removal is affected in part through the hydrologic function of the constructed wetland, but primarily through other removal functions.

Hydrologic Functions

Stormwater flows are intermittent and highly variable depending on watershed characteristics and the depth, duration and intensity of the rainfall event. As discussed earlier, increased flow rates and total volumes of runoff occur with development. Wetland systems provide storage volume, resulting in hydraulic detention which decreases the rate of flow. As the flow rate decreases, the velocity of the water is also reduced. Decreases in volume, although small, may result from evapotranspiration. Limited infiltration may also occur, although generally the water level in the wetland must be maintained at or near the ground surface.

The ability of the constructed wetland to provide detention storage, and to a lesser extent volume retention, enhances its ability to control pollutants. In particular, decreased flow rates and velocities promote sedimentation. The erosive force of the water is also reduced so that possible downstream effects are mitigated. In addition, the decreased flow rates tend to both reduce and elongate the pollutant loadings to the downstream resources, thus reducing the initial pulse of pollutant load, which tends to result from intermittent stormwater flows.

Table 4. Characteristics of rainfall, stormwater, combined wastewater and treated effluent (Metcalf & Eddy, Inc. 1991). Single values represent averages.

Parameter	Rainfall	Stormwater	Combined Wastewater	Primary Effluent	Secondary Effluent
Suspended Solids (mg/l)	--	141 to 224	270 to 550	40 to 120	10 to 30
5-Day Biochemical Oxygen Demand (mg/l)	1 to 13	10 to 13	60 to 220	70 to 200	15 to 45
Chemical Oxygen Demand (mg/l)	9 to 16	73 to 92	260 to 480	165 to 600	25 to 80
Fecal Coliform Bacteria (MPN)/100ml)	--	1,000 to 21,000	200,000 to 1,100,000	--	--
Total Phosphorus (mg/l)	0.02 to 0.15	0.37 to 0.47	1.2 to 2.8	7.5	6
Total Nitrogen (mg/l)	--	3 to 24	4 to 17	35	30
Total Kjeldahl Nitrogen	--	1.68 to 2.12	--	--	--
Nitrate Nitrogen (mg/l)	0.05 to 1.0	0 to 4.2	--	--	--
Total Lead (µg/l)	30 to 70	161 to 204	140 to 600	--	--

Table 5. Priority pollutants in at least ten percent of Nationwide Urban Runoff Program samples (U.S. Environmental Protection Agency 1983).

Metals and Inorganics

Antimony	Cyanides
Arsenic	Lead
Beryllium	Nickel
Cadmium	Selenium
Chromium	Zinc
Copper	

Pesticides

Alpha-hexachlorocyclohexane	Chlordane
Alpha-endosulfane	Lindane

Halogenated Aliphatics

Methane, dichloro

Phenols and Cresols

Phenol	Phenol, 4-nitro
Phenol, pentachloro	

Polycyclic Aromatic Hydrocarbons

Chrysene	Phenanthrene
Flouranthene	Pyrene

Pollutant Removal Functions

There are several major mechanisms by which wetlands can remove stormwater pollutants.

Sedimentation

Sedimentation is the most important process by which particulate pollutants are removed from the stormwater. Many of the pollutants in stormwater discussed earlier, such as nutrients, metals and toxics, are associated with particulate matter. As the rate of sedimentation is related to the size of the particles, slower velocities and flow rates will result in more sedimentation and greater pollutant removal.

Filtration

Particulates in the stormwater are physically obstructed by wetland vegetation and thus can be removed by filtration. This is promoted by the reduced velocity of flow and the dense vegetation in the wetland.

Adsorption

Dissolved elements, including nutrients and metals, can be adsorbed onto particulate matter by various chemical and physical reactions. Conditions which exist in wetlands such as longer hydraulic residence times and shallow water depths increase the opportunity for dissolved elements to come in contact with particulates, thus enhancing adsorption. Once adsorbed, the formerly dissolved constituent is then subject to sedimentation and filtration.

Precipitation

Certain dissolved elements such as metals can form precipitates and settle, depending on a number of factors including pH, oxygen content and temperature. For example, metals can form oxide and hydroxide precipitates under aerobic conditions, and sulfide precipitates under anaerobic or reduced conditions, thus allowing them to be removed from the stormwater (Livingston et al. 1992).

Volatilization

Pollutants may enter the atmosphere by evapotranspiration as well as by aerosol formation during windy conditions (Livingston et al. 1992). This process can remove oils, chlorinated hydrocarbons and mercury, although for larger weight, petroleum hydrocarbons it is not significant.

Microbial Decomposition

Microorganisms can utilize the soluble organic load carried in stormwater. As they use oxygen, the soils tend to become anaerobic. This facilitates precipitation and may also result in loss of nitrogen by denitrification (conversion of nitrate nitrogen to nitrogen gas which escapes to the atmosphere).

Vegetative Uptake

Wetland plants are capable of the uptake of nutrients, heavy metals and organics. Uptake can occur from the soil through roots and rhizomes, and from the water column directly in dissolved form. Plant uptake varies seasonally, being highest during the growing season. Chemicals can be released back into the water column if the plants are allowed to die and decay.

POLLUTANT REMOVAL EFFECTIVENESS

Given the various physical, chemical and biological mechanisms which can contribute to pollutant removal in constructed wetlands, it is of interest to assess their pollutant removal effectiveness. Although use of constructed wetlands to treat stormwater is increasing, there are few systems where effectiveness monitoring data have been collected and clearly documented. Therefore, generalizations on pollutant removal efficiencies must be made with caution. However, there is reason to believe that systems can be designed and operated to provide good removals for particular constituents.

The data in Table 6 were compiled from a variety of geographic locations, climates, and watershed types and sizes. There are also numerous variations on the type and size of wetland system used. The length of the sampling period and the number of sampling events are provided to give an indication of the extent of each project. Data on several natural wetland systems are also included for purposes of comparison. Despite the variety of projects presented, there is reasonable consistency in the type of pollutant removals obtained.

Removals of total suspended solids (TSS) are variable, but generally are not below 63 percent. Removals in the eighty to ninety percent range are common. Removals of phosphorus range from 37 to 90 percent, and most results are centered in the 40 to 60 percent range. Maximum total nitrogen removals are less than for phosphorus (30 to 76 percent). Ammonia removals are generally lower, and nitrate removals are similar to those of total nitrogen. Metals removals are variable but consistently high for lead, which tends to be associated with particulate matter. Mechanisms such as sedimentation, filtration and adsorption all assist in promoting good removal rates for suspended solids, and the nutrients and metals which tend to be associated with solids. Dissolved constituents such as ammonia tend to have lower removal rates.

Table 6. Effectiveness data for constructed and natural wetlands (Maryland Department of Natural Resources 1987, Meiorin 1989, Hochheimer et al. 1991).

Project Site/Location	Monitoring Program	% Removal			
		solids TSS	phosphorus TP OP	nitrogen TN NH3 NO3 TKN	metals Cu Pb Zn
CONSTRUCTED WETLANDS					
DUST Marsh Fremont, CA	11 storms 17 months	64	48 56	- 10 15 28	31 88 33
Tampa Office Pond Tampa, FL	20 storms 2 years	63	54 63	- 34 75 -	- - 33
McCarrons MWTA Roseville, MN	25 storms 21 months	90	55 33	50 - 56 48	- 75 -
Orange County Orlando, FL	11 storms 2 years	83	37 21	30 32 - -	- 81 62
Lake Jackson Tallahassee, FL	1 storm 13 days	94	90 78	76 37 70 -	- - -
NATURAL WETLANDS					
Island Lake Orlando, FL	5 storms 2 years	-	87 89	- 96 - 41	87 83 67
Swift Run Wetland Ann Arbor, MI	6 events 6 months	64	42 43	- - 56 9	- 50 -
Wayzata Wetland Wayzata, MN	multiple 1 year	97	78 -	- 44 - -	73-83 90-97 78-86
Clear Lake Marsh	- 5 years	79	55 29	- large 53 9 in-crease	- - -

The data in Table 6 indicate that pollutant removals are generally at least as good for constructed systems as for natural wetlands. This suggests that wetlands can be constructed to treat stormwater runoff at least as well as a natural system might be expected to function, and probably better if good design practices are used.

A recent review of pollutant removal effectiveness data involving the use of wetlands for stormwater treatment included 15 constructed wetlands projects and 11 natural wetlands projects (Strecker et al. 1992). This review included most of the projects listed in Table 6 as well as others. Constructed wetlands were found to have higher average removal rates than natural systems, as well as less variability. The site median removal rates and coefficients of variation are presented in Table 7.

The coefficient of variation is a measure of the statistical variability in the data. The higher removal rates and reduced variations for constructed systems were attributed to the fact that these systems are designed for better removals and more consistent performance (Strecker et al. 1992). It was also noted, however, that differences in project data collection and reporting procedures from the various studies make comparisons difficult.

Table 7. Summary of effectiveness data for constructed and natural wetlands (Strecker et al. 1992).

	Total Suspended Solids (TSS)	Ammonia (NH₃)	Total Phosphorus (TP)	Lead (Pb)	Zinc (Zn)
Constructed Wetlands Site Median Removal Rate (%)	80	44	58	83	42
Coefficient of Variation (%)	28	49	48	56	39
Natural Wetlands Site Median Removal Rate (%)	76	25	5	69	62
Coefficient of Variation (%)	62	168	1,900	67	47

As evidenced by the aforementioned information, wetlands, including constructed wetlands, can provide good removals of certain types of pollutants. These removal rates can improve water quality and may be sufficient to protect the desired uses of water resources.

DESIGN OF CONSTRUCTED WETLANDS FOR STORMWATER TREATMENT

As constructing wetlands for stormwater treatment is still an emerging technology, there are no widely accepted design criteria, methods or rules of thumb. The systems currently in place have been adapted to site-specific conditions and constraints. Many of these systems are combined with wet detention ponds, or are part of a treatment train (a group of treatment processes) used to treat stormwater. A summary of the characteristics of those constructed wetlands projects for which effectiveness data were presented (Table 6) is provided in Table 8. A wide variation of watershed sizes, configurations, inlets, treatment unit sizes and outlets is evident. Two typical constructed wetland systems are illustrated in Figure 3.

Although there is no general agreement on specific design criteria or factors, there is some general agreement on what should be considered or evaluated during the design. This can be presented by comparing factors and design criteria from some of the design guidelines which currently exist (Table 9). The Florida design criteria apply to combination wet detention ponds and constructed wetlands which are common in that state (Livingston et al. 1992). The Washington guidelines are based on work conducted on the east coast as modified for local conditions (Washington State Department of Ecology 1991).

Table 8. Characteristics of constructed wetlands projects (Meiorin 1989, Hockheimer et al 1991).

Project Site/ Location	Watershed Area (ac)	Configuration	Inlet Description	Unit	Treatment System Area (ac)	Volume (ac-ft)	Vegetation Type	Outlet Description
Dust Marsh Fremont, CA	2,965	lagoon and pond in parallel followed by wetland channel system	0.37 ac debris basin	lagoon overland flow pond braided channel	4.94 3.95 1.68 25.4	2.91 not reported 5.5 76.6	not reported not reported cattail cattail, bulrush	not reported
Office Pond Tampa, FL	6.3	combination pond and wetland	grassed swale	pond, wetland	0.35	0.32	not reported	not reported
McCarrons Roseville, MN	543	detention pond followed by 6 wetland cells in series	not reported	pond wetland	2.39 6.17	2.9 3.1	not reported	not reported
Orange County Orlando, FL	41.6	detention pond followed by wetland	1.5 m diam. storm inlet	pond wetland	0.2 0.73	1.6 2.19	not reported hyacinth, duckweed, cattail	earth spillway, compound weir
Lake Jackson Tallahassee, FL	2,230	detention pond with sand filter followed by wetland	not reported	pond/filter wetland	20 6.2	1.5 9.9	not reported cattail, rush, pickerelweed	filter under-drain to box culvert concrete spillway with stop log weir

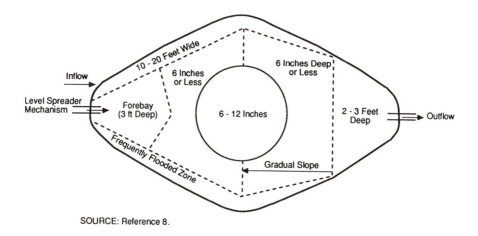

SOURCE: Reference 8.

A. Example Shallow Constructed Wetland System Design for Stormwater Treatment

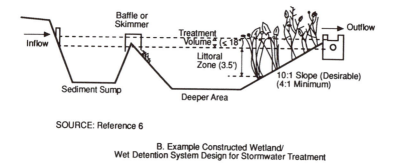

SOURCE: Reference 6

B. Example Constructed Wetland/
Wet Detention System Design for Stormwater Treatment

Figure 3. Example of a constructed Wetland Systems for stormwater treatment.

Many of the hydraulic or hydrologic criteria in Table 9 are aimed at promoting pollutant removal as well as mitigating flow increases from development and urbanization. Other criteria such as depth of water and soils are designed to allow appropriate plants to grow to enhance the pollutant removal functions described earlier in this paper. There are also numerous general inlet, pond and outlet design criteria which result in a well functioning system. Finally, the operation and maintenance of the system, once constructed, is critical for continued performance of its intended functions.

Table 9. Design criteria comparison for selected constructed wetlands for stormwater treatment.

Design Criteria or Factor	Florida[1]	Maryland[2]	Washington[3]
Runoff volume storage	1 inch of runoff above pool elevation	1 year-24 hour storm, or surface area	6 month-24 hour storm below pool elevation; 100-year storm
Hydraulic residence time	>14 days; all volume released >120 hours; half volume released >60 hours	--	--
Depth	Shallow marsh 1/2-2 ft. Deep pond <6-8 ft.	<6 inches (50% area) 6 inches (50% area) 6 in.-1 ft. (25% area) about 3 ft. (25% area)	6 inches (50% area) 0.5-1 ft. (15% area) 2-3 ft. (15% area) >3 ft. (20% area)
Surface Area	<70% open water	>3% watershed area	>1.5% watershed area
Side Slope	<6:1 to 2-3 ft. below permanent pool	--	<3:1
Short circuiting	>3:1 length to width ratio	>2:1 length to width ratio if possible; use baffles, islands or peninsulas if inlet is near outlet	>3:1 length to width ratio; 5:1 preferred; maximize inlet to outlet distance; use baffles, islands and peninsulas
Soils	>6 inch depth wetlands soils	>4 inch depth; use hydric soils when available	Use soil appropriate for species planted
Planting	Use native aquatic species (species list provided)	Use 2 aggressive species, 3 additional species (species list provided); plant 30% of shallow (<1 ft. deep) area w/3 ft. spacing and 40 clumps/acre wetland of each aggressive species; plant 10 clumps of 5 individuals per acre wetland near edge of pond for each additional species	Submit vegetation plan; use acceptable species which vary with water depth or at at shoreline (species list provided)
Inlet Considerations	Use landscape retention areas and swales to promote infiltration and aesthetics; dissipate energy of entering water	Dissipate energy of incoming water; use forebay to trap large sediments (depth >3 ft., volume about 10% of total pool volume)	Use oil spill control device or oil/water separator prior to forebay; use forebay (depth >3 ft., area about 25% of total pool area)

Table 9 (cont.). Design criteria comparison for selected constructed wetlands for stormwater treatment.

Design Criteria or Factor	Florida[1]	Maryland[2]	Washington[3]
Pond/Wetland General Considerations	Provide capability to be completely drained; sod areas above normal water level	Use liner if needed to maintain pool level; include a frequently flooded area 10-20 ft. from edge of normal water level; avoid standing water for mosquito and algae control	Use liner if needed to maintain pool level; provide island for nesting birds if possible; use dense vegetation in forebay and near outlet to prevent erosion
Outlet Considerations	Use single or multi-stage control device; include skimmer or other means to prevent oil and grease from exiting; provide scour protection and emergency spillway for 100-year storm when applicable	Site structure in deepest part of wetland; protect against outlet pipe blockage	--
Operation & Maintenance	Responsible party must submit a plan; document plant species survival (normally 85% survival or replant); remove nuisance species; remove sediment and trash; inspect inlet, outlet and vegetation; vary water level and control shoreline grass for insert protection, maintain erosion and sediment controls	Remove accumulated solids in forebay and near outlet	Clean pretreatment devices; clean forebay Once every 5 years, or when 6 inches of sediment accumulates; remove floatables annually; maintain shoreline to prevent erosion

[1] Livingston et al. 1992.
[2] Maryland Department of Natural Resources 1987.
[3] Washington State Department of Ecology 1991.

COST OF CONSTRUCTED WETLAND SYSTEMS

System costs will vary widely according to site-specific issues. The major elements to be considered include:

- Planning, engineering and legal
- Land acquisition and easements
- Construction
  - Site preparation

- Excavation
- Trenching
- Placement of soil (including hydric)
- Planting
- Drainage and piping
- Inlet and outlet structures
- Pretreatment devices
- Placement of fill
- Riprap and crushed stone
- Erosion control measures
- Mulching and seeding
- General landscaping
- Construction supervision and inspection
- Operations and Maintenance
  - Monitoring
  - Inspection
  - Cleaning
  - Disposal of waste materials

Due to these numerous factors and the many different ways of conducting or contracting the work, it is not possible to give a rule of thumb for estimating costs of constructed wetlands for stormwater treatment. Variation in construction cost (1990 dollars) may range from $5,560 to over $80,000 per acre (Hochheimer et al. 1991).

REFERENCES

Hochheimer, J., Yates, C. and Cavacas, A. 1991. Constructed Wetlands as a Nonpoint Source Control Practice. Draft Report prepared for U.S. EPA.

Livingston, E., E. McCarron, J. Cox, and P. Sanzone. 1992. The Florida Development Manual: A Guide to Sound Land and Water Management. Florida Department of Environmental Regulation. Stormwater/Nonpoint Source Management Section.

Maryland Dept. of Natural Resources, Sediment and Stormwater Division, Water Resources Administration. 1987. Guidelines for Constructing Wetland Stormwater Basins.

Meiorin, E. 1989. Urban Runoff Treatment in a Fresh/Brackish Water Marsh in Fremont, California. Pp. 677-685 In Hammer, D. (ed.) Constructed Wetlands for Wastewater Treatment. Lewis Publishers, Chelsea, Michigan.

Metcalf & Eddy, Inc. 1991. Wastewater Engineering: Treatment, Disposal, and Reuse. McGraw-Hill, New York.

Strecker, E., Kersnar, J., Driscoll, E. and Horner, R. 1992. The Use of Wetlands for Controlling Stormwater Pollution.

U.S. Environmental Protection Agency. 1983. Nationwide Urban Runoff Program, Water Planning Division. Washington, D.C.

U.S. Environmental Protection Agency. 1990a. National Water Quality Inventory - 1988 Report to Congress. EPA 440-4-90-003. Washington, D.C.

U.S. Environmental Protection Agency. 1990b. Nonpoint Source Pollution in the U.S. Report to Congress. Washington, D.C.

U.S. Environmental Protection Agency. 1993. Process for Urban Runoff Pollution Prevention and Control Planning. Center for Environmental Research Information, Washington, D.C.

Washington State Dept. of Ecology. 1991. Stormwater Management Manual for the Puget Sound Basin (draft report).

CHAPTER 12

ACID MINE DRAINAGE TREATMENT WITH WETLANDS AND ANOXIC LIMESTONE DRAINS

Jeffrey Skousen, Alan Sexstone, Keith Garbutt and John Sencindiver

Coal and metal mining disturb large volumes of geologic material and expose them to the environment. Through this exposure to air and water, sulfide minerals commonly associated with coal and metal deposits are oxidized and hydrolyzed resulting in acid mine drainage (AMD). AMD is a low pH, sulfate-rich water with high amounts of total acidity. Total acidity is comprised of mineral acidity, predominately iron, but also aluminum, manganese, zinc, and other metals varying with the geologic deposit, and hydrogen ion acidity. Other metals and trace elements in the geologic material may also be solubilized due to the acid leaching conditions. As contaminated streams flow into larger streams or lakes, dilution occurs making the water less toxic. Also, chemical and biological reactions cause some neutralization of acidity and precipitation of metals. Oxidized iron precipitates as ferric hydroxide (yellow boy) as the pH increases above 3.5, while aluminum and manganese hydroxides require a pH of at least 5 and 7, respectively, to precipitate.

Approximately 20,000 km of streams and rivers in the United States are impacted by AMD, and the majority of these streams receive AMD from old, abandoned surface and deep mines (Girts and Kleinmann 1986). Since no company or individual claims responsibility for reclaiming abandoned mine lands (AML), no treatment of the AMD occurs and continual contamination of surface and groundwater resources results. Besides degrading water quality, AMD impacts fish populations and other aquatic life, wildlife species which use the stream for habitat and water, stream aesthetics and land values, and the quality of life for humans who reside near the polluted stream. Recently, legislation has allowed funds generated from taxes on coal production, called the AML Reclamation Fund, to be used for addressing water problems on AMLs through federal and state AML programs.

Since 1977, the Surface Mining Control and Reclamation Act (SMCRA) has required coal mining operations to collect all water on the site and treat contaminated drainage before discharge into receiving streams. Where AMD occurs, treatment with alkaline reagents to neutralize acidity, precipitate metals, and raise pH is the conventional approach (Skousen 1988). This treatment method, often called an active system, is costly in terms of equipment, chemicals, and manpower (Skousen et al. 1990). In addition, the treatment system must be continued for an indefinite time period. Estimates of this cost to the coal industry exceed $1 million per day (Kleinmann 1990).

Wetlands and anoxic limestone drains (ALDs) are passive treatment systems in contrast to active chemical treatment systems. Wetlands have been used for decades in the treatment of municipal wastewater (Water Pollution Control Federation 1990) but only within the last 10 years have they received serious attention in the treatment of AMD. Researchers at Wright State University and West Virginia University independently noted that AMD from AML was improved after passing through natural *Sphagnum* wetlands in Ohio and West Virginia (Huntsman et al. 1978, Wieder and Lang 1982, Figure 1). Since then, investigators have documented many other sites where the same phenomenon was observed (Brooks et al. 1985, Burris 1984, Samuel et al. 1988). The next step in using passive treatment processes for reclamation was to construct artificial wetlands on mine sites and determine whether the same treatment benefits were realized. Criteria for building wetlands were developed and eventually refined by trial and error by numerous wetland builders. Research has also helped in the development of construction specifications (Figure 2).

TREATMENT OF AMD BY NATURAL WETLANDS

Many studies of AMD treatment by natural and volunteer wetlands show a substantial drop in sulfate and iron concentrations, and a concomitant pH rise in the effluent. For example, Tub Run Bog in northern West Virginia receives mine drainage from several abandoned deep mines in the area (Wieder and Lang 1982). After flowing through the wetland, pH rises from 3.0 to 5.5, sulfate concentrations are lowered from 250 mg/l to 10 mg/l, while iron is decreased from about 50 mg/l to less than 2 mg/l. Another natural wetland in Pennsylvania receives AMD with a pH of 5.5. Although pH is not altered, the wetland reduces iron from 25 mg/l to less than 1 mg/l, and manganese from 35 to 2 mg/l (Kleinmann 1985).

Not all natural wetlands are capable of tolerating AMD. In a study of *volunteer* natural wetlands in West Virginia, remnants of wetlands were found which had been destroyed by AMD (Garbutt, unpublished data). In the same study, an old healthy wetland which had been receiving AMD for approximately 20 years was found to have more iron in the outflow than in the inflow. This suggests that after a certain level of metal accumulation, wetlands may "unload" the metal accumulated and may decrease rather than increase the water quality over a long time period.

Although natural wetlands have been used for AMD treatment, limitations exist for their continued use as AMD treatment systems (Wildeman et al. 1991).

Figure 1. Acid mine drainage treatment by natural wetlands was observed in the late 1970s and has led to the development of constructed wetland technology. (Photo courtesy of Robert Kleinmann).

Figure 2. Wetlands have been constructed in greenhouse troughs where temperature and water flows can be controlled. Some of the troughs receive acid mine drainage while others receive rain water at the same rate. With special equipment and procedures, chemical and physical reactions, microbial processes, and plant productivity can be monitored in these wetlands. An understanding of these processes may help in the building of more efficient wetland systems.

Transmission of water through natural wetland substrates is often restricted, and the water flows primarily across the surface decreasing the potential for anaerobic processes. Also, natural wetlands may be rich in humic acids that lower the wetland's capability to neutralize acid drainage. On the other hand, alkaline mine drainage, which may contain significant amounts of some metals, readily kills natural *Sphagnum* wetlands. Finally, intentional contamination of a natural water ecosystem by AMD will likely require approval of state and federal wetland regulatory agencies and, in fact, will probably not be approved.

TREATMENT OF AMD BY CONSTRUCTED WETLANDS

Construction of wetlands on mined lands is beneficial to reclamation for several reasons (Figure 3). First, wetlands have been recognized as a precious ecological system and, as such, have become a carefully regulated resource. Laws and regulations regarding wetland preservation and use have been passed and promulgated over the past several decades, and several federal agencies have responsibility to carry out legislative action. If wetlands are present on an area to be mined, a special permit is required in order to disturb the site (Skousen 1989). Mitigation after disturbance is required, so wetland construction criteria may be important for reestablishment of a functioning wetland ecosystem. Therefore, creation of wetlands on mined lands during reclamation, whether or not a wetland was originally present on the site, is a pro-active environmental response to wetland disturbance. Second, wetlands provide important habitat for a number of wildlife species and sometimes enhance the aesthetic appeal of an area. Third, wetlands can be used as a relatively inexpensive tool for treating acid discharges or drainage with dissolved or suspended solids. However, after AMD passes through a wetland, the water may still require additional chemical treatment, but will probably require lower amounts of chemicals, thereby saving money. The U.S. Office of Surface Mining (OSM) currently requires a conventional AMD treatment system to be on site where wetland systems have been built to treat AMD discharges. OSM's policy reflects reservations concerning a wetland's seasonal and long-term capacity to treat AMD (OSM 1988).

The first attempts at AMD treatment with wetlands consisted of simply planting a few cattails in old sediment ponds on mined sites (Figure 4). The hope was that wetland vegetation would spread and AMD would be "magically" treated. In many of these early attempts, the vegetation either died or did not proliferate, and the wetland was ineffective in treating AMD (Wheeler et al. 1991). Since then, wetland builders have tried to simulate more accurately the substrates and vegetation in effective natural wetlands.

The U.S. Bureau of Mines estimates that over 400 wetlands have been constructed for the purpose of AMD treatment (Kleinmann 1991). Water treatment by constructed wetlands in the U.S. showed iron decreases of 28 to 99% (Girts and Kleinmann 1986, Brodie et al. 1988, Stark et al. 1988, Stillings et al. 1988, Brodie 1991, Hellier 1991, Duddleston et al. 1992, Hendricks 1991, Wieder

Figure 3. Constructed wetlands are especially practical when dealing with acid mine drainage from abandoned mine lands. The wetland provides some treatment with little or no maintenance costs.

Figure 4. Early attempts at wetland treatment of acid mine drainage involved planting cattails into old sediment ponds on surface mines and allowing the cattails to spread.

1992). During initial stages after construction and while vegetation was becoming established, iron removal was generally from 30 to 50%. Following vegetation development, iron removal increased to 50 to 90%. Manganese removal by wetlands was erratic and ranged from 8 to 98%, but was generally less than 30%. Strong evidence suggests a sequential removal of metals: iron must be precipitated first, then manganese may be decreased (Sexstone, unpublished data). The same reports found water pH to remain approximately the same after flowing through most wetland sites. The water on a few sites showed lower pH due to iron oxidation and acidity generation. Other sites which showed increased water pH generally had a layer of limestone in the wetland substrate. On sites where the water pH was increased, the system was better at decreasing manganese and aluminum concentrations.

Sorption onto organic materials such as peat and sawdust was responsible for removing 50 to 80% of the metals in AMD. Brodie et al. (1988) reported that five different substrates (clay, topsoil, mine spoil, acid wetland soil, and non-acid wetland soil) with associated vegetation were no different in iron removal after one growing season. This finding suggests that a particular substrate material, as long as it has metal retention capabilities, may not be as important as the conditions created by wetland plants and the presence of an anaerobic zone. On the other hand, Wieder (1992) reported that straw/manure and mushroom compost wetlands were superior in AMD treatment to *Sphagnum* peat or sawdust wetlands. Substrate recipes are as numerous as wetland builders, however a mixture of 50% peat/manure, 30% hay, and 20% soil will generally provide sufficient permeability, exchange capacity, and organic matter to produce a functional system. In addition, a sediment slurry from a functioning wetland is beneficial to inoculate the newly constructed wetland with appropriate microorganisms.

Several studies report on the effects of different plant species in wetlands. *Sphagnum* was the first wetland species examined because it was the most common species found, and *Sphagnum* has a well-documented capacity to accumulate iron (Gerber et al. 1985, Wenerick et al. 1989). However, Spratt and Wieder (1988) found that saturation of *Sphagnum* moss with iron could occur within one growing season, and Wieder (1988) suggested that metal retention capacity in a wetland is finite. Some have suggested that metal retention is limited in wetlands because organic matter inputs by wetland plants are limited (Kleinmann 1990). Many of the original constructed wetlands were vegetated with *Sphagnum* but few remained effective. Cattails (*Typha*) have been found to have a greater environmental tolerance than *Sphagnum* (Samuel et al. 1988). Sencindiver and Bhumbla (1988) found that cattails growing in AMD-fed wetlands did not have elevated concentrations of metals in plant tissues, indicating little accumulation of metals in tissues. Algae and a few other wetland species have also received attention due to the observation that enhanced metal removal was associated with algal blooms (Pesavento and Stark 1986, Kepler 1988, Hedin 1989). In Colorado, algal mixtures were found to aerobically remove manganese from mine drainage (Duggan et al. 1992), presumably due to elevated pH resulting from algal growth. Probably the most important role that wetland plants serve in AMD treatment systems may be their ability to stimulate microbial processes. Kleinmann (1991) explains that plants provide sites for microbial attachment, release oxygen from their roots, and supply organic matter for heterotrophic microorganisms.

FACTORS IMPORTANT TO TREATMENT OF AMD IN WETLANDS

Several mechanisms of metal retention, listed in their relative order of increasing importance, operate in wetlands including: 1) direct uptake by living plants, 2) exchange plus organic complexation reactions with the wetland matrix, and 3) chemical and microbiological oxidation/reduction reactions which lead to precipitation reactions. Retention of iron by these mechanisms has received the most detailed study.

Wetland plant species vary in their ability to accumulate metals (Fernandez and Henriques 1990). Some reports document elevated tissue concentrations (Spratt and Wieder 1988), while others show little metal accumulation (Folsom et al. 1981). On an annual basis, uptake by *Typha* accounted for less than 1% of the iron removal by a volunteer wetland treating AMD (Sencindiver and Bhumbla 1988).

While it may be true that metal concentration and accumulation in plant tissues may be small in any one year, plant tissue is a renewable resource. Old tissue is senesced yearly and new tissue, with new sites for metal accumulation, is produced. Thus, over the entire life of the wetland, plants may accumulate a portion of metals received in a wetland. Metal release through decomposition of wetland plant tissue is undetermined.

Exchange reactions and organic complexation can account for over 50% of metal retention in the wetland (Kleinmann 1991, Bhumbla et al. 1990). These chemical and physical processes are abiotic and do not require microorganisms. Kleinmann (1991) states that adsorption and complexation by organic substrates help compensate for limited initial biological activity during the first few months of operation in a new wetland system. Although some natural inputs of organic matter occur annually at plant senescence, the physical filtering capacity of a wetland will ultimately be finite as all exchange and complexation sites become metal saturated. Substantial artificial inputs of organic matter have been used as a successful strategy to temporarily renew this filtering capacity, following an observed decline in wetland performance (Eger and Melchert 1992, Haffner 1992, Stark et al. 1991).

Insoluble precipitates represent a major sink for metal retention in wetlands. Approximately 40 to 70% of the total iron removed from AMD by wetlands is found as ferric hydroxides from hydrolysis of ferric iron or from microbial oxidation of ferrous iron (Henrot and Wieder 1990, Calabrese et al. 1991, Wieder 1992). Unlike exchange and complexation reactions, iron precipitate formation has no theoretical maximum, but is limited in reality by the physical density and volume of the material produced.

Up to 30% of the iron may be found as reduced iron and may be combined with sulfides (Calabrese et al. 1991, McIntyre and Edenborn 1990, Wieder 1992). Mono and disulfides form as a result of microbial sulfate reduction in the presence of an oxidizable carbon source. In addition to its metal removal potential, sulfate reduction consumes acidity and raises water pH (Hedin and Nairn 1992a, Rabenhorst et al. 1992).

The long-term stability of metal precipitates is unclear. Ferric hydroxides can be reduced with time to ferrous iron by anaerobic iron reducing bacteria in the wetland. Similarly, the use of ferric iron as a pyrite oxidant under anaerobic conditions would result in ferrous iron in wetland effluents.

Wetlands may also be used as post treatment "polishing" systems to remove trace metals or nutrients from previously treated AMD. Currently under investigation is the addition of AMD, pretreated with anhydrous ammonia, to wetlands. Complete precipitation of metals occurs at pH > 9 with ammonia addition, but the resulting effluent is high in pH, ammonia, and sulfate. Theoretically, wetlands should be able to handle high nitrogen loads as has been previously demonstrated in wastewater treatment systems (Hammer 1989). Combined nitrification and denitrification coupled with plant uptake should remove significant amounts of excess N and result in a more acceptable effluent following ammonia treatment. The primary limiting factor is establishing conditions which optimize the microbial conversion of ammonia to nitrate.

ALKALINITY GENERATION AND ANOXIC LIMESTONE DRAINS

Alkalinity generation in a wetland is important for precipitating metals and raising pH. One method of generating alkalinity is through sulfate reduction. This process can generate up to 200 mg/l of sulfides which are available for metal complexation (Rabenhorst et al. 1992). This process requires a source of sulfate (common in AMD), organic matter, and a reducing environment. Factors such as climate, season, and organic matter inputs can dramatically affect the amount of sulfate reduction.

Anoxic limestone drains (ALDs) have received much attention recently because they have the potential to add alkalinity to mine water without biological limitations (Brodie et al. 1990, Nairn et al. 1990, Skousen 1991, and Turner and McCoy 1990). An ALD consists of high quality limestone buried in a trench, underdrain, or cell (Figure 5). Acid water is intercepted while it is anoxic (underground) and directed into the buried cell of limestone. By keeping the limestone and mine water anoxic, limestone can continue to dissolve without becoming armored. ALDs precondition water by the addition of alkalinity which raises pH and buffers the water. The water can then be diverted into a wetland or ditch where oxidation, hydrolysis, and precipitation reactions can occur. By adding alkalinity to AMD, wetland systems can be more efficient in their removal activities.

Figure 5. Anoxic limestone drains are buried cells or trenches of limestone. When properly constructed, and with the right type of water, alkalinity can be introduced into the water to help passively treat acid mine drainage.

ALDs require certain water conditions for long-term effectiveness. For example, dissolved oxygen (DO) concentrations should be below 2 mg/l, and ferric iron and aluminum concentrations should be below 25 mg/l. Under these conditions, the limestone can continue to dissolve, and armoring and clogging with metal hydroxides will be minimized. Selection of the appropriate water and environmental conditions is critical for long-term alkalinity generation in an ALD.

Preliminary results showed that eight out of eleven ALDs were functioning properly in West Virginia (Skousen and Faulkner 1992). In all cases, water pH was raised after ALD treatment but three of the sites had pH values less than 5.0, indicating that the ALD was not fully functioning because the target pH was 6.2. Water acidity was decreased 50 to 80% while iron concentrations were also decreased, presumably due to some iron precipitation in the drain. Aluminum in inflow water was also not found in outflow water, again indicating aluminum precipitation in the drain. With iron and aluminum decreases in outflow water, it is probable that some coating or clogging of limestone is occurring inside the ALD. It is also possible that inflow DO levels are above 2 mg/l, suggesting improper measurement, or inadequate design and construction to restrict oxygen invasion.

Utilization of an ALD is complicated by mixed ferric/ferrous iron concentrations in AMD and/or when DO is found in the water. Current research involves pretreatment of AMD with an anaerobic wetland to strip DO concentrations, and to either aerobically precipitate ferric iron or convert it to ferrous iron through iron reduction. The water, after passing through an anaerobic wetland, can then be introduced into an ALD. There are still many questions relative to the longevity of ALDs and the factors involved in limestone dissolution and metal precipitation in ALD environments. Like wetlands, ALDs may be a solution for AMD treatment for a finite period after which the system must be replenished or replaced.

SIZING AND DESIGN PARAMETERS

The most comprehensive approach for design of passive treatment systems for treating AMD has been prepared by Hedin and Nairn (1992b). They developed a decision chart based on AMD flow, water chemistry, and calculated contaminant loadings (Figure 6). Flows should be measured, not visually estimated, and variation in flows should be accounted for (Figure 7). AMD should be collected as close to the mine opening or seep outlet as possible and analyzed for pH, acidity, alkalinity, iron, and manganese. If an ALD is being considered, water analysis should also determine DO, ferrous iron, and aluminum concentrations. Once the contaminant loadings are calculated based on flow and water analyses, the water can be classified and an appropriate passive treatment system designed.

Alkaline Water

If the water is classified as alkaline (the water has enough alkalinity to buffer the acidity produced by metal hydrolysis reactions), the metals contained in the AMD will precipitate when given sufficient residence time. In this case, only an aerobic system or settling pond is necessary. No organic substrate is required in the pond but planting cattails is helpful in physical filtering processes and aesthetics. Aeration, which is often limiting in such systems, should be maximized by constructing waterfalls or rip rap ditches, and also by maintaining shallow ponds. The key is to retain the water long enough for oxidation and precipitation reactions to occur. Wetland size should be based on 20 grams of iron removal per square meter per day of wetland (20 gmd). For manganese removal, 0.5 gmd is the estimated value for wetland sizing.

Acid Water

If the water is classified as acid, then a system for generating alkalinity must be provided.

Option 1

ALDs require specific water conditions for long-term effectiveness. Factors that may limit the long-term effectiveness of ALDs are the presence of ferric iron, aluminum, and oxygen in the AMD. When water containing any ferric iron or aluminum contacts limestone, both metals hydrolyze and precipitate. Oxidation is not necessary. These metal hydroxides can coat the limestone and cause the drain

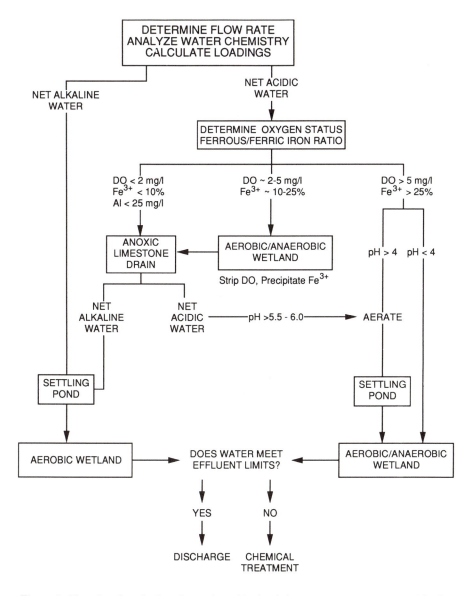

Figure 6. Flow chart for selection of a passive acid mine drainage treatment system or combination of systems (adapted from Hedin and Nairn 1992a). See text for explanation and sizing parameters.

Figure 7. Residence time of water in wetlands is important to maximize water contact with the wetland substrate where microbial transformations and oxidation/reduction reactions occur. Sometimes baffles are necessary to facilitate water movement through the wetland.

to plug. Small amounts of DO will allow ferrous iron to oxidize to ferric iron causing the above problems. Therefore, determination of these parameters is critical before installing an ALD. Even when these criteria are met and accurate measurements were presumably taken, some ALDs have failed (Skousen and Faulkner 1992). Guidelines for ALD construction are given in Skousen (1991) and also Hedin and Nairn (1992a) and are abbreviated here.

A high grade of limestone should be used because it will dissolve quickly and contain less impurities. A calcium carbonate equivalent (CCE) of 80% or greater is recommended. Particle size should range from 4 to 10 cm, thus maintaining water transmission through the material and allowing adequate surface area for reaction. Limestone may be placed in trenches or cells depending on the slope of the ground surface. Water should cover most, if not all, of the limestone in the trench or cell year round. Ideally, the limestone should lay in a pool of water, and as water flows in, water flows out on the other end through an airtrap outlet. Trenches can be 0.6 to 6.0 m wide, 0.6 to 1.5 m deep, and vary in length depending on site constraints. Plastic is placed on top of the limestone to prevent oxygen movement into the limestone. Plastics from 10 to 20 mil thickness have been used with good success. Some ALDs have plastic completely surrounding the limestone. A minimum of 0.6 m of soil should cover the ALD, and the soil should be composed of fine earth materials and carefully compacted. Revegetation with shallow-rooted plant species, preferably grasses, controls erosion and protects the drain. Control of water during construction is critical because limestone can be partially armored as installation progresses. Diversion of water, control of sediment, and careful reintroduction of water after the ALD is completed will help in its longevity.

Calculating the amount of limestone and the size of the drain is accomplished as follows:

Flow (gpm) $\times$ acidity (mg/l) $\times$ 0.0022 = A tons/year of acid
A tons/year $\times$ years of life = B tons of limestone
B tons/CCE (%) = C tons
C tons/expected dissolution (%) = D total tons

A ton of limestone has a volume of about 0.4 m^3. So if 250 total tons of limestone is calculated to treat an AMD source for 20 years, a volume of area about 100 m^3 would be necessary. This equates to a limestone cell 6 m wide, 1.5 m deep, and 11 m long. Similarly, a trench 2 m wide, 1 deep, and about 50 m long could also be used if a trench is more suitable for the site. The goal, given the right conditions, is to use limestone to generate excess alkalinity in the water, and direct the ALD-treated water into an aerobic system where metals can oxidize, hydrolyze, and precipitate.

Option 2

If large amounts of dissolved oxygen (greater than 5 mg/l), ferric iron (> 25% ferric vs ferrous), and aluminum are found in acid water, then construction of an anaerobic/compost wetland is recommended, as use of an ALD will not be helpful. Anaerobic wetlands generate alkalinity through a combination of bacterial sulfate reduction reactions and limestone dissolution. With a suitable organic substrate, and limestone placed in the anaerobic zone, alkalinity can be generated. Various substrate materials have been used but, as mentioned, those with manure or composted materials from 15 to 30 cm thick have shown suitable microbial activities. Limestone can also be placed in the substrate if anaerobic conditions prevail. Sizing is based on 5 gmd. If the pH of the water is less than 3.5, then the area calculated above should be multiplied by 4 to account for the much greater acid loadings.

Option 3

If small amounts of oxygen (2 to 5 mg/l) and ferric iron (10 to 25%) are found in the water, a specialized aerobic/anaerobic wetland and ALD combination may be used. Conveyance of the slightly oxidized water to the wetland should be accomplished in pipes or other enclosed system to restrict additional oxygen from invading the water. In the oxidized surface portion of the wetland, the small percentage of ferric iron may be hydrolyzed and precipitated if water pH is above 3.5. As the water moves downward into the anaerobic zone, oxygen will be used by microbes for organic matter decomposition, thereby scavenging oxygen from the water. After passing through the wetland, water from the anaerobic zone can be directed through closed pipes to an ALD where it may pick up additional alkalinity. In this type of wetland, given proper functioning, ferric iron will precipitate or convert to ferrous iron and DO will be consumed, thereby making the water suitable for ALD treatment.

The passive treatment systems as described here (aerobic wetland, anaerobic wetland, and ALD) provide for matching a technology, or combination of technologies, to particular types of AMD. However, not all AMD is suitable for passive treatment systems.

SELECTION AND PLANTING OF MACROFLORA IN CONSTRUCTED WETLANDS

Little work exists on the appropriate selection of plant species for wetlands, yet this can have important implications for the long-term success of a project. In constructed wetlands, higher plants serve several purposes.

- Substrate consolidation - the roots of higher plants hold the substrate together, prevent channels from forming, and increase residence time of water in the wetland.
- Stimulation of microbial processes - plants supply sites of attachment for microbes, release oxygen from their roots, and provide a source of organic matter for heterotrophic microorganisms.
- Wildlife habitat - wetland plants supply food and shelter for a wide range of game and nongame animals (especially birds).
- Aesthetics - a wetland filled with plants is considerably more pleasing to the eye than a treatment pond.
- Metal accumulation - this may play a minor role in the total metal removal of the wetland, and will vary with the species and different water and seasonal circumstances.

The choice of wetland species should be based on a knowledge of local conditions, the relative ability of species to tolerate the conditions found in a particular watershed, and their ability to provide the functions listed above. For example, *Typha angustifolia* is extremely rare in West Virginia, although it is common in surrounding states. Thus a wetland planted with *T. angustifolia* in West Virginia is unlikely to flourish, while *Typha latifolia*, a close cousin, does well. Similarly *T. angustifolia* and *T. latifolia* have different niches with respect to water depth. They form concentric circles where they grow together as a function of water depth. In general, the water depths usually employed in wetland systems (shallow) are more suitable for *T. latifolia* than *T. angustifolia*. It should also be noted that since AMD is extremely variable, a plant community that thrives in one wetland location may be completely inappropriate in a second nearby location if the composition of the AMD is different. The importation of exotics, even when they have proven effective elsewhere, should be avoided. The choice of local species is also likely to maximize the wildlife habitat component of the wetland.

Polycultures may be preferable to monocultures. Investigations of natural wetlands which are receiving AMD show a tendency for treatment to be higher in wetlands with a higher diversity of species (Garbutt, unpublished data). This is probably due to the increased number of niches found in these wetlands as a function of the metal loading from inflow to outflow, but it also provides a higher diversity of microhabitats for the microbial community.

Where information on metal accumulation is available, it should be used in the selection of appropriate species. However, if a plant species is accumulating heavy metals in plant parts which form an important food resource for wildlife, there is a real possibility of transmission and concentration of these toxic metals through the food chain. For example, an AMD source with high lead content may be an important problem, while bioaccumulation of iron or manganese may be of little concern.

Current practice is to plant a fairly dense monoculture of *Typha*. As passive treatment systems get bigger, the number of plants needed and the manpower required to plant them can lead to prohibitive expense. So some planting designs which do not fill the wetland with plants at its creation should be considered. Slow-growing or annual species may be important components in some wetland systems but these will not form an initial matrix from which a wetland will develop in a useful time frame.

Finally, it is important to allow the plants to establish for some time with uncontaminated water before AMD is applied to the system. When plants are transplanted, they suffer "transplantation shock" which makes them particularly susceptible to any environmental shock. Immediate inundation with AMD will certainly slow growth and development and in some cases may kill the entire wetland. If a wetland is built during the early to middle of the growing season, the plants should be given one to two months establishment time prior to AMD introduction into the system. If a system is constructed late in the growing season, AMD flow should not begin until the next growing season.

SUMMARY AND CONCLUSIONS

Acid mine drainage is a common, intractable problem in many areas of the world where coal and other minerals are mined. Chemical treatment methods are effective, but passive systems offer a low-cost, more natural approach to AMD treatment. Mine water flowing through natural and constructed wetlands has generally shown a decrease in metals and total acidity. Factors important to AMD treatment by wetlands are initial flow and water chemistry, wetland substrate, wetland vegetation, and microbial composition and activity. Metal retention in wetlands may be due to direct uptake by plants, exchange reactions with the wetland matrix, and chemical and microbiological oxidation/reduction reactions which lead to precipitation reactions. Based on flows and chemistry, a passive water treatment system (which may include an aerobic wetland, an anaerobic wetland, and anoxic limestone drain or combinations) may be developed to either completely or partially treat the water thereby reducing AMD treatment costs with chemicals. Based on current knowledge, passive treatment technologies may be a treatment solution for many AMD discharges. However, AMD treatment by such systems appears to be finite after which the system must be replaced or replenished.

REFERENCES

Bhumbla, D.K., J.P. Calabrese, J.C. Sencindiver, A.J. Sexstone, and G.K. Bissonnette. 1990. Attenuation of Fe and Mn in AMD applied to constructed wetlands. Agronomy Abstracts. p. 33.

Brodie, G.A. 1991. Achieving compliance with staged, aerobic, constructed wetlands to treat acid drainage. pp. 151-174. In: Proceedings of 1991 American Society for Surface Mining and Reclamation Conference, Durango, CO.

Brodie, G.A., C.R. Britt, H. Taylor, T. Tomaszewski, and D. Turner. 1990. Passive anoxic limestone drains to increase effectiveness of wetlands acid drainage treatment systems. In: Proceedings, 12th National Association of Abandoned Mined Land Conference, Breckinridge, CO.

Brodie, G.A., D.A. Hammer, and D.A. Tomjanovich. 1988. An evaluation of substrate types in constructed wetlands acid drainage treatment systems. pp. 389-398. In: Mine Drainage and Surface Mine Reclamation. Vol. I. Info. Circular 9183, U.S. Bureau of Mines, Pittsburgh, PA.

Brooks, R., D.E. Samuel, and J.B. Hill. 1985. Wetlands and water management on mined lands. Proceedings of a Conference. The Pennsylvania State University, University Park, PA.

Burris, J.E. 1984. Treatment of mine drainage by wetlands. Contribution No. 264, Department of Biology, The Pennsylvania State University, University Park, PA.

Calabrese, J.P., A.J. Sexstone, D.K. Bhumbla, G.K. Bissonnette, and J.C. Sencindiver. 1991. Application of constructed cattail wetlands for the removal of iron from acid mine drainage. pp. 559-575. In: Proceedings, Second International Conference on the Abatement of Acidic Drainage, Vol. 3. MEND, Montreal, Canada.

Duggan, L.A., T.R. Wildeman, and D.M. Updegraff. 1992. The aerobic removal of manganese from mine drainage by an algal mixture containing *Cladophora*. pp. 241-248. In: Proceedings, 1992 American Society for Surface Mining and Reclamation Conference, Duluth, MN.

Duddleston, K.N., E. Fritz, A.C. Hendricks, and K. Roddenberry. 1992. Anoxic cattail wetland for treatment of water associated with coal mining activities. pp. 249-254. In: Proceedings, 1992 American Society for Surface Mining and Reclamation Conference, Duluth, MN.

Eger, P., and G. Melchert. 1992. The design of a wetland treatment system to remove trace metals from mine drainage. pp. 98-107. In: Proceedings, 1992 American Society for Surface Mining and Reclamation Conference, Duluth, MN.

Fernandez, J.C., and F.F. Henriques. 1990. Metal levels in soil and cattails (*Typha latifolia* L.) plants in a pyrite mine, Lousal Portugal. Int. J. of Env. Studies 36:205-210.

Folsom, B.L., C.R. Lee, and D.J. Bates. 1981. Influence of disposal environment on availability and plant uptake of heavy metals in dredged material. In: Tech. Report EL-81-12, U.S. Army Corps of Engineers, Waterway Exp. Stat., ICE, Vicksburg, MS.

Gerber, D.W., J.E. Burris, and R.W. Sone. 1985. Removal of dissolved iron and manganese ions by a *Sphagnum* moss system. In: Wetlands and Water Management on Mined Lands. The Pennsylvania State University, University Park, PA.

Girts, M.A., and R.L.P. Kleinmann. 1986. Constructed wetlands for treatment of acid mine drainage: A preliminary review. p 165-171. In: National Symposium on Mining, Hydrology, Sedimentology, and Reclamation. Univ. of Kentucky, Lexington, KY.

Haffner, W.M. 1992. Palmerton zinc superfund site constructed wetlands. pp. 260-267. In: Proceedings, 1992 American Society for Surface Mining and Reclamation, Duluth, MN.

Hammer, D.A. 1989. Constructed wetlands for wastewater treatment. Lewis Publishers, Chelsea, MI.

Hedin, R.S. 1989. Treatment of acid coal mine drainage with constructed wetlands. pp. 349-362. In: Wetlands Ecology, Productivity and Values: Emphasis on Pennsylvania. Pennsylvania Academy of Science, Easton, PA.

Hedin, R.S., and R.W. Nairn. 1992a. Passive treatment of coal mine drainage. Course Notes for Workshop. U.S. Bureau of Mines, Pittsburgh, PA.

Hedin, R.S., and R.W. Nairn. 1992b. Designing and sizing passive mine drainage treatment systems. In: Proceedings, 13th West Virginia Surface Mine Drainage Task Force Symposium, Morgantown, WV.

Hellier, W.W. 1991. Pennsylvania's constructed wetlands: an overview. In: Microbial Technologies of Mineral Pollutants Removal from Industrial Wastewater, Moscow, USSR.

Hendricks, A.C. 1991. The use of an artificial wetland to treat acid mine drainage. pp. 550-558. In: Proceedings of Second International Conference on the Abatement of Acidic Drainage. MEND, Montreal, Canada.

Henrot, J., and R.K. Wieder. 1990. Processes of iron and manganese retention in laboratory peat microcosms subjected to acid mine drainage. J. Env. Qual. 19:312-320.

Huntsman, B.E., J.B. Solch, and M.D. Porter. 1978. Utilization of a *Sphagnum* species dominated bog for coal acid mine drainage abatement. Abstracts, 91st Annual Meeting Geologic Society America, Ottawa, Ontario, Canada.

Kepler, D.A. 1988. An overview of the role of algae in the treatment of acid mine drainage. pp. 286-290. In: Mine Drainage and Surface Mine Reclamation, Vol. I. Info. Circular No. 9183. U.S. Bureau of Mines, Pittsburgh, PA.

Kleinmann, R.L.P. 1985. Treatment of acid mine water by wetlands. p 48-52. In: Control of AMD. Info. Circular No. 9027. U.S. Bureau of Mines, Pittsburgh, PA.

Kleinmann, R.L.P. 1990. Acid mine drainage in the United States. In: Proceedings of the First Midwestern Region Reclamation Conference, Southern Illinois University, Carbondale, IL.

Kleinmann, R.L.P. 1991. Biological treatment of mine water--an overview. pp. 27-42. In: Proceedings of Second International Conference on the Abatement of Acidic Drainage. MEND, Montreal, Canada.

McIntyre, P.E., and H.M. Edenborn. 1990. The use of bacterial sulfate reduction in the treatment of drainage from coal mines. pp. 409-415. In: Proceedings of the 1990 Mining and Reclamation Conference. West Virginia University, Morgantown, WV.

Nairn, R.W., R.S. Hedin, and G.R. Watzlaf. 1990. A preliminary review of the use of anoxic limestone drains in the passive treatment of acid mine drainage. In: Proceedings, 12th West Virginia Surface Mine Drainage Task Force Symposium, Morgantown, WV.

Office of Surface Mining. 1988. Use of wetland treatment systems for coal mine drainage. OSMRE Dir. TSR-10. U.S. OSMRE, Washington, DC.

Pesavento, B.G., and L.R. Stark. 1986. Utilization of wetlands and some aspects of aqueous conditions. In: Proceedings of 7th West Virginia Surface Mine Drainage Task Force Symposium, Morgantown, WV.

Rabenhorst, M.C., B.R. James, and J.N. Shaw. 1992. Evaluation of potential wetland substrates for optimizing sulfate reduction. pp. 90-97. In: Proceedings, 1992 American Society for Surface Mining and Reclamation, Duluth, MN.

Samuel, D.E., J.C. Sencindiver, and H.W. Rauch. 1988. Water and soil parameters affecting growth of cattails. pp. 367-374. In: Mine Drainage and Surface Mine Reclamation, Vol. I. Info. Circular No. 9183. U.S. Bureau of Mines, Pittsburgh, PA.

Sencindiver, J.C., and D.K. Bhumbla. 1988. Effects of cattails (*Typha*) on metal removal from mine drainage. pp. 359-366. In: Mine Drainage and Surface Mine Reclamation, Vol. I. Info. Circular No. 9183. U.S. Bureau of Mines, Pittsburgh, PA.

Skousen, J.G. 1988. Chemicals for treating acid mine drainage. Green Lands 18(3): 36-40.

Skousen, J.G. 1989. Wetlands are more than two cattails. Green Lands 19(2): 45-47.

Skousen, J.G., K. Politan, T. Hilton, and A. Meek. 1990. Acid mine drainage treatment systems: chemicals and costs. Green Lands 20(4):31-37.

Skousen, J.G. 1991. Anoxic limestone drains for acid mine drainage treatment. Green Lands 21(4): 30-35.

Skousen, J.G., and B.B. Faulkner. 1992. Preliminary results of anoxic limestone drains in West Virginia. In: Proceedings, 13th West Virginia Surface Mine Drainage Task Force Symposium, Morgantown, WV.

Spratt, A.K., and R.K. Wieder. 1988. Growth responses and iron uptake in *Sphagnum* plants and their relation to acid mine drainage treatment. pp. 279-285. In: Mine Drainage and Surface Mine Reclamation, Vol. I. Info. Circular No. 9183. U.S. Bureau of Mines, Pittsburgh, PA.

Stark, L.R., R.L. Kolbash, H.I. Webster, S.E. Stevens, K.A. Dionis, and E.R. Murphy. 1988. The Simco #4 wetland: biological patterns and performance of a wetland receiving mine drainage. pp. 332-344. In: Mine Drainage and Surface Mine Reclamation, Vol. I. Info. Circular No. 9183. U.S. Bureau of Mines, Pittsburgh, PA.

Stark, L.R., W.R. Wenerick, P.J. West, F.M. Williams, and S.E. Stevens. 1991. Adding a carbon supplement to simulated treatment wetlands improves mine water quality. pp. 465-483. In: Proceedings, Second International Conference on the Abatement of Acidic Drainage. MEND, Montreal, Canada.

Stillings, L.L., J.J. Gryta, and T.A. Ronning. 1988. Iron and manganese removal in a *Typha*-dominated wetland during ten months following its construction. pp. 317-324. In: Mine Drainage and Surface Mine Reclamation, Vol. I. Info. Circular No. 9183. U.S. Bureau of Mines, Pittsburgh, PA.

Turner, D., and D. McCoy. 1990. Anoxic alkaline drain treatment system, a low cost acid mine drainage treatment alternative. In: 1990 National Symposium on Mining. University of Kentucky, Lexington, KY.

Wenerick, W.R., S.E. Stevens, H.J. Webster, L.R. Stark, and E. DeVeau. 1989. Tolerance of three wetland plant species to acid mine drainage: a greenhouse study. In: Constructed Wetlands for Wastewater Treatment, Lewis Publishers, Chelsea, MI.

Wheeler, W.N., M. Kalin, and J.E. Cairns. 1991. The ecological response of a bog to acidic coal mine drainage - deterioration and subsequent initiation of recovery. pp. 449-464. In: Proceedings, Second International Conference on the Abatement of Acidic Drainage. MEND, Montreal, Canada.

Wieder, R.K. 1988. Determining the capacity for metal retention in man-made wetlands constructed for treatment of coal mine drainage. pp. 375-381. In: Mine Drainage and Surface Mine Reclamation, Vol. I. Info. Circular No. 9183. U.S. Bureau of Mines, Pittsburgh, PA.

Wieder, R.K. 1992. The Kentucky wetlands project: a field study to evaluate man-made wetlands for acid coal mine drainage treatment. Final Report to the U.S. Office of Surface Mining. Villanova University, Villanova, PA.

Wieder, R.K., and G.E. Lang. 1982. Modification of acid mine drainage in a fresh water wetland. pp. 43-53. In: Symposium on Wetlands of the Unglaciated Appalachian Region. West Virginia University, Morgantown, WV.

Wildeman, T., J. Gusek, and G. Brodie. 1991. Wetland design for mining operations. Draft handbook presented at the 1991 American Society for Surface Mining and Reclamation Conference, Durango, CO.

Water Pollution Control Federation. 1990. Natural systems for wastewater treatment. Water Pollution Control Federation, Washington, D.C.

CHAPTER 13

DESIGNING WETLANDS FOR WILDLIFE

Donald M. Kent

Wetlands have been noted for their high value in providing habitats for a wide variety of wildlife (Shaw and Fredine 1956), and exceed all other land types in wildlife productivity (Payne 1992). A wetland designed for wildlife is the combination of details and features which, when implemented, results in the provision of habitat for wildlife which use wetlands to satisfy all or part of their life requisites. The design should be a preliminary sketch or plan for work to be executed, conceived in the mind and fashioned skillfully. In practice, designs for wetlands can take several forms.

The simplest and earliest efforts at designing wetlands for wildlife were characterized by the preservation of existing wildlife habitat. The most prominent effort among these was the establishment of the National Wildlife Refuge system, which protects uplands as well as wetlands. Florida's Pelican Island was the first refuge, established in 1903 by President Theodore Roosevelt to protect egrets, herons and other birds that were being killed for their feathers. There are presently over 450 National Wildlife Refuges, comprising a network that encompasses over 90 million acres of lands and waters. Southern bayous, bottomland hardwood forests, swamps, prairie potholes, estuaries and coastal wetlands are represented. Preservation of wetland wildlife habitat continues, although at a slower pace, through the efforts of federal and state initiatives, and private organizations.

More complex approaches to the design of wetlands for wildlife are characterized by the restoration and enhancement of historic wetlands, and the creation of new wetlands. Illustrative of large-scale restoration efforts, the 1985 Food Security Act has provided for wetlands restoration on Farmers Home Administration and Conservation Reserve Program lands. Almost 55,000 acres of agricultural lands were restored to wetlands between 1987 and 1989, and another 90,000 acres were targeted for restoration in 1990 and 1991 (Mitchell in Kusler and Kentula 1990). The North American Wetlands Conservation Act enacted in 1989 will provide 25 million dollars annually in federal matching funds over the next 15 years for restoration of wetlands vital to waterfowl and other migratory birds. Other restoration and enhancement efforts have been effected for years by wildlife managers (Weller 1987, 1990; Weller et al. 1991), in some cases to counteract the effects of previous management efforts (Talbot et al. 1986, Newling 1990, Rey et al. 1990). At a generally smaller scale, restoration designs occur as mitigation requirements for regulated wetland fills (Kusler and Kentula 1990).

Designing wetlands for wildlife through creation of wetlands is undoubtedly a greater challenge than preservation or restoration. Whereas design through preservation is accomplished through observation of current wildlife use, and design through restoration is accomplished through historical knowledge of wildlife use, creation requires the attraction of wildlife to a new resource. Prominent among creation efforts is the Dredged Material Research Program of the U. S. Army Corps of Engineers (U.S. Army Corps of Engineers 1976). Authorized by the River and Harbor Act of 1970, the U.S. Army Corps of Engineers Waterways Experiment Station (WES) initiated research in 1973 which included testing and evaluation of concepts for wetland and upland habitat development (Garbisch 1977, Lunz et al. 1978a). WES has since designed and constructed thousands of acres of freshwater and coastal wetlands from dredged material, and demonstrated the value of these wetlands to wildlife (Cole 1978, Crawford and Edwards 1978, Lunz et al. 1978b, Webb et al. 1988, Landin et al. 1989). As is the case with restoration, the wetland regulatory process has resulted in a large number of small-scale wetlands designed at least in part for wildlife (Michael and Smith 1985, U.S. Army Corps of Engineers 1989, Kusler and Kentula 1990).

Regardless of whether a design for wetland wildlife is accomplished through preservation of existing wildlife habitat, restoration or enhancement of historic wetland habitat, or the creation of new wetland habitat, the fundamental principles for effective design are the same. These principles are related to minimum habitat area, minimum viable population and tolerance of the wildlife species for disturbance. Therefore, the objective of this chapter is to discuss the effects of wetland size, the relationship of the wetland to other wetlands, and anthropogenic disturbance on wetland effectiveness for providing wildlife habitat.

SIZE

Size is generally the first factor considered in designing a wetland for wildlife. Ideally, the objectives of the design, for example provision of all life requisites for the species of interest, determine wetland size. More often, the size of the wetland is preordained by land or financial constraints, or even mitigation requirements, and an assessment of what wildlife species can reasonably be supported is made.

Grinnell and Swarth (1913) were the first to note the relationship between the number of species and the size of the habitat in their study of montane peaks. Following attempts to quantify the relationship for terrestrial habitats (Cain 1938), it was the application of the concept to true islands which led to its widespread recognition (MacArthur and Wilson 1963, 1967). In what has become known as the theory of island biogeography, the greater the size of the island the greater the species richness. This relationship is described by $S = CA^z$, where S is the number

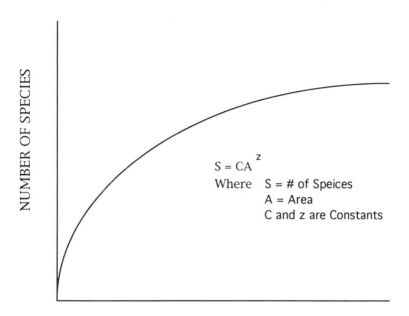

$$S = CA^z$$

Where   S = # of Speices
A = Area
C and z are Constants

Figure 1. The theory of island biogeography suggests that the greater the size of the island, the greater the species richness (MacArthur and Wilson 1967). The relationship is thought to occur because larger islands have more habitat and greater habitat diversity.

of species, A is the area, and C and z are dimensionless constants which need to be fitted for each set of species-area data (Figure 1, MacArthur and Wilson 1967). The relationship is thought to occur primarily because larger islands have more habitat and greater habitat diversity.

Although the theory has its origins in terrestrial ecology, there are reservations about the applicability of island biogeography to terrestrial reserves (Kushlan 1979, Harris 1984, Forman and Godron 1986). Certainly there are inherent differences between the two systems, as the nature of surrounding habitat is far more distinct for oceanic islands than terrestrial islands. This should result in differences in true island and terrestrial island immigration rates. Nevertheless, the relationship between terrestrial island size and species richness has been demonstrated to hold for birds (Peterken 1974, 1977; Moore and Hooper 1975, Galli et al. 1976, Whitcomb 1977, Gottfried 1979, Robbins 1979, Whitcomb et al. 1981, Ambuel and Temple 1983 and Lynch and Whigham 1984) and large mammal species (Bekele 1980), and for habitat types such as forests (Whitcomb et al. 1981), urban parkland (Gavareski 1976), caves (Culver 1970, Vuilleumier 1973) and mountains (Brown 1971, Vuilleumier 1970, Thompson 1978, Fritz 1979). Therefore, despite inherent differences between oceanic and terrestrial islands, there is evidence that the same isolating mechanisms are operating. The degree to which these mechanisms operate is of course dependent upon the degree of habitat insularity, which in turn depends on species-specific habitat specificity, tolerance and vagility. Harris (1984) has suggested that the isolating mechanisms operate most strongly on amphibians and reptiles, followed by mammals, resident birds, and then migratory birds. The degree to which the latter group is susceptible depends on breeding site fidelity, and the extent to which reproduction is restricted to the breeding site.

Between 1780 and 1980, an estimated 117 million acres of wetlands were lost in the contiguous United States (Dahl 1990). This loss has fragmented and insularized remaining wetlands, producing in many cases relatively small terrestrial islands. Wildlife populations are increasingly isolated and reduced in size, leading inevitably to extinction (Senner 1980). Extinction occurs for several reasons. First, small, closed populations are more susceptible to extrinsic factors such as predation, disease and parasitism, and to changes to the physical environment. Second, demographic stochasticity, the random variation in sex ratio and birth and death rates, contributes to fluctuations in population size (Steinhart 1986), increasing the susceptibility of small, closed populations to random extinction events. Finally, small, closed populations suffer genetic deterioration, primarily due to inbreeding, leading to a decrease in population fitness (Soulé 1980, Allendorf and Leary 1986, Ralls et al. 1986, Soulé and Simberloff 1986).

The effects of inbreeding depression can be illustrated by considering the fate of a small, closed population (Senner 1980). For an effective population size (N_e, number of breeding individuals) of 4, constrained by extrinsic factors to a maximum of 10 individuals, genetic heterozygosity declines with each successive generation (Figure 2). As heterozygosity declines, the average survival of offspring declines due to inbreeding depression. Inbreeding depression includes viability depression, which is the failure of offspring to survive to maturity, and fecundity depression, which is the tendency for inbred animals to be sterile. Mammals, in which the male X chromosome is always hemizygous, also suffer sex-ratio depression by way of a relative increase in males. As fecundity decreases, the population size can no longer be maintained at its limit. The probability of survival, while initially very high, drops very sharply after approximately 15 generations. The population approaches extinction after approximately 25 generations.

The rate of loss of heterozygosity per generation (f) for inbreeding populations is equal to $1/2N_e$, and animal breeders note an obvious effect on fecundity as f approaches 0.5 or 0.6 (Soulé 1980). As

$$\Delta f = 1 - (1 - 1/2N_e)^t,$$

substituting 0.6 for Δf and solving for t (number of generations) indicates that the number of generations to the extinction threshold is approximately 1.5 times N_e (Soulé 1980). Smaller populations and those with shorter generation times become extinct in less time than larger populations and those with longer generation times (Figure 3).

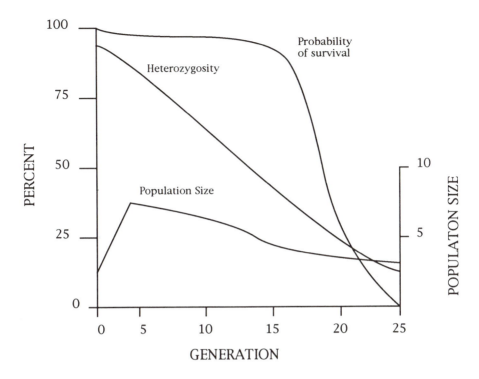

Figure 2. The fate of a small (Ne = 4), closed population constrained by extrinsic factors to few individuals (10 in this example) is a decline in genetic heterozygosity, a decline in offspring survival and a decline in population size (Senner 1980).

Domestic animal breeders have determined that an inbreeding rate of 2 or 3 percent per generation is sufficient for selection to eliminate deleterious genes (Stephenson et al. 1953, Dickerson et al. 1954). Citing differences between domestic and natural populations, Soulé (1980) has recommended that a one percent inbreeding rate be adapted as the threshold for natural populations. As

$$f = 1/2N_e,$$

the minimum effective population size is 50 if the inbreeding rate is to be maintained at one percent. However, even at this rate a population of $N_e = 50$ will lose about one fourth of its genetic variation in 20 to 30 generations (Soulé 1980). Setting the inbreeding rate at 0.1 percent, Franklin (1980) has recommended a minimum effective population size of 500 individuals for long-term survival. As only a percentage of individuals in a population contribute to the gene pool in any given generation, the actual minimum viable population size (Shaffer 1981) will be much larger.

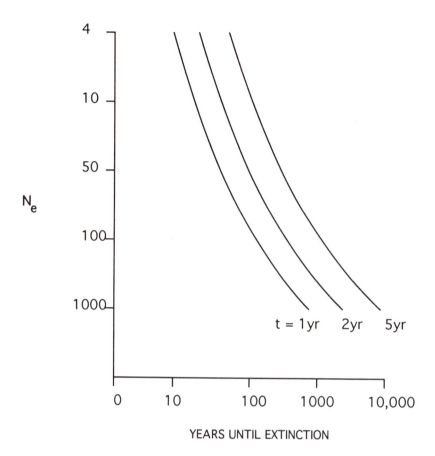

Figure 3. Based on the rate of loss of heterozygosity per generation for inbreeding populations, the number of generations to the extinction threshold is 1.5 time N_e (Soulé). Smaller populations and those with shorter generation times become extinct in less time than larger population and those with longer generation times.

Of course, genetic risks are not the only threat to long-term population survival. Demographic risks such as disease, meteorological catastrophes, and populations too dispersed to effect breeding can also be important contributors to the determination of minimum viable population size when populations are very small (Goodman 1987). However, with few exceptions, genetic deterioration should occur well in advance of demographic extinction, and demographic risks will be seen to exacerbate genetic risks. Also, many demographic risks are largely unpredictable, therefore negating the development of effective design criteria discrete from those derived from genetic considerations. Therefore, it seems reasonable to emphasize genetic risks when estimating minimum viable population size.

For reproductively isolated populations, minimum refuge size is a product of home range size and minimum effective population size. The home range size for many wetlands wildlife species is poorly understood (as is the degree of reproductive isolation). Nevertheless, for purposes of illustration, consider 3 distinct wetland species: the bullfrog (*Rana catesbeiana*), the Pacific water shrew (*Sorex bendirei*), and the mink (*Mustela vison*). Bullfrogs are territorial only during breeding, and have been observed living and breeding in permanent ponds as small as 1.5 m diameter (Graves and Anderson 1987). Pacific water shrew are territorial and have a home range of approximately 1 ha (Harris 1984). Male and female mink have generally distinct home ranges of approximately 12 ha (Allen 1986). Minimum refuge sizes for a minimum effective population size of 50 are 1.9×10^{-2} ha, 54.5 ha and 600 ha respectively. For a minimum effective population size of 500 individuals, refuge sizes are 0.19 ha, 545 ha and 6000 ha. These estimates are conservative, as they assume all individuals in the populations are contributing to the gene pool. Given that generally only a subset of any given population breeds at any one time, minimum refuge sizes are likely to be significantly larger.

Some existing preserves appear to be large enough to support minimum viable populations of at least some species. National Wildlife Refuges range in size from 0.4 ha Mille Lacs, Alaska to 8 million ha Yukon Delta, Alaska, and average approximately 80,000 ha. Although many refuges consist of uplands as well as wetlands, it is clear that at least some National Wildlife Refuges are large enough to support minimum viable populations of some species. However, few opportunities remain for the preservation of such large tracts, and wetlands outside the refuge system may not be large enough to support minimum viable populations. For example, 28 percent of wetland habitats in the east and central Florida region are less than 2 ha in size, and 40 to 60 percent are less than 20 ha in size (Gilbrook 1989). Restoration, enhancement and creation of riverine wetlands has resulted in relatively large contiguous habitats (Baskett 1987, Weller et al. 1991). More frequently however, these efforts result in wetlands of 100s of ha, 10s of ha, and even areas of less than 1 ha (Michael and Smith 1988, Ray and Woodroof 1988, Reimold and Thompson 1988, U.S. Army Corps of Engineers 1989, Landin and Webb 1989).

RELATIONSHIP TO OTHER WETLANDS

Given the paucity of large wetlands available for preservation, and the very real possibility that smaller wetlands will not support minimum viable populations of many species, it is necessary to provide mechanisms for interpopulation movement. Interpopulation movement increases effective habitat size and creates a metapopulation (Gilpin 1987). The metapopulation is composed of an interacting system of local populations that suffer extinction and are recolonized from within the region. The metapopulation will be sustained if the source(s) of colonists is proximally located, and the immigration rate is greater than the reciprocal of the time to extinction (Brown and Kodric-Brown 1977). The metapopulation has a decreased danger of accidental extinction compared to individual populations, and an ability to counter genetic drift through occasional migration. In the absence of a metapopulation, the wetland internal disturbance regime becomes the critical design feature, and the minimum dynamic area must be large enough to support internal colonization sources (Pickett and Thompson 1978). As discussed above, this minimum dynamic area is likely to be unattainable in many instances.

Metapopulations can reasonably be established and maintained if interpopulation movement can be affected at a minimum rate of every few generations (Wright 1969, Nei et al. 1975, Kiester et al. 1982, Allendorf 1986). The rate of movement is a function of the distance between populations and the quality of the intervening habitat, and will vary among species based upon dispersal ability, habitat specificity and habitat tolerance.

The rate of movement between populations varies inversely with the distance between habitats (McArthur and Wilson 1967, Diamond 1975, Gilpin 1987, Soulé 1991). Animals find it more difficult to disperse from one habitat to another as the distance between habitats increases, and habitats are more likely to be recolonized following extinction events if habitats occur in close proximity (Figure 4, Wilson and Willis 1975, Brown and Kodric-Brown 1977, Soulé 1991). For example, unoccupied spruce grouse (*Dendragapus canadensis*) habitat is significantly further from occupied habitat than other occupied habitat (Fritz 1979). Rodents, rabbits and hares appear to benefit from proximally located habitats (Soulé 1991), and mammals typically disperse less than 5 home range diameters (Chepko-Sade and Halpin 1987). Generally, small, sedentary, cursorial species with narrow habitat tolerances will be more greatly affected by the distance between habitats than will large pedestrian species, volant species, migratory species, and species with broad habitat tolerances.

Ultimately, interpopulation movement is determined by the quality of the intervening habitat, with quality a function of species-specific habitat tolerance (Harris 1984, Forman and Godron 1986, Noss 1987). At its simplest, intervening habitat can be viewed as either unsuitable or suitable, with unsuitable habitat constituting barriers to movement and suitable habitat constituting corridors. Roadways are classic barriers which inhibit the movement of mammals (Oxley et al. 1974, Madsen 1990), and are responsible for the death of an estimated 100 million amphibians, reptiles, birds and mammals per year (Arnold 1990, Anonymous 1992, Lopez 1992). Some bird species have an intrinsic aversion to abandoning cover (Diamond 1973, Soulé 1991), and rodents tend to stay within suitable habitat (Holekamp 1984, Garrett and Franklin 1988, Wiggett and Boag 1989). At the mesoscale and macroscale lakes, mountains and valleys affect mammal dispersal (Shirer and Downhower 1968, Seidensticker et al. 1973, Storm et al. 1976). For species such as these with poor dispersal abilities or narrow habitat tolerances, individual populations comprising the metapopulation must be connected by suitable habitat. Alternatively, discrete habitats could be closely juxtaposed. Roadway crossings should be avoided, although movement can be effected for some species by the use of underpasses (Arnold 1990, Madsen 1990, Soulé 1991). Unsuitable habitat is more likely to be used as a corridor if the transit time between populations is short enough that forage and cover are not required.

Figure 4. Animals find it easier to disperse from one habitat to another if those habitats are closely juxtaposed than if the habitats are widely separated. Close juxtaposition of habitats also facilitates recolonization following local extinction events.

Other species have greater dispersal abilities and broader habitat tolerances. For example, a Florida black bear (*Ursus americanus*) traversed 8 major highways, a dozen other roadways, a river, several canals, fences, farmland and skirted suburban areas in an 11 week journey (Arnold 1990). As another example, McIntyre and Barrett (1992) noted that Australian bird species preferred to move within forested areas, but traversed open areas when necessary. For species such as these, intervening habitat is comprised of various degrees of suitability, each of which can be tolerated for various lengths of time.

Few if any specific guidelines exist for determining the effective distance between populations, or the quality of intervening habitat. In the absence of specific information demonstrating the broader abilities of dispersers, wetland design efforts should focus on establishing corridors of habitat similar to that being used by existing individual populations. Individual populations should be located as closely together as possible so as to minimize transit time, and corridor width should approach home range diameter as transit time increases. This conservative ideal is more likely to ensure long-term survival of species with poor dispersal abilities such as amphibians, reptiles, some bird species and small mammals than other designs. For large mammals and some bird species corridors may consist of dissimilar habitat if species use can be demonstrated. For migratory bird species corridors may be largely unnecessary, and design efforts should focus on providing habitat requirements at breeding and wintering sites, and stopping points in between.

DISTURBANCE

Disturbance is a change in structure caused by factors external to the hierarchical level of the system of interest (Pickett et al. 1989). As it pertains to wildlife, disturbance alters birth and death rates by directly killing individuals or by affecting resources upon which those individuals depend (Petraitis et al. 1989). Generally, small areas are more susceptible to disturbance than larger areas (Norse et al. 1986). Disturbance can be either natural, in that it occurs as part of normal community dynamics, or anthropogenic. Natural disturbance provides for the continued existence of species which use temporary habitats (Soulé and Simberloff 1986), and species richness is generally maximized at moderate frequencies or intensities of disturbance (Connell 1978, Pickett and White 1985). Natural disturbance regimes can be accommodated within a single large reserve (alpha diversity), or by a series of reserves (beta diversity) serving a metapopulation.

By contrast, anthropogenic disturbance is generally detrimental to overall, long-term community health. Anthropogenic disturbance can be intrusive, in which people or domesticated animals enter wetland habitat and normal community processes are disrupted. Intrusive disturbance has been institutionalized at many local, state and federal reserves with the advent of nature trails and viewing areas. Other examples of intrusive disturbance include boat and vehicle incursions, hunting, and lumbering. The primary effect of intrusive disturbance is to disrupt waterfowl, wading bird and raptor nesting and foraging (Pough 1951, Kushlan 1976, Palmer 1976, Kale 1978, U.S. Fish and Wildlife Service 1984, Short and Cooper 1985, Peterson 1986, Ambruster 1987). The simplest way to counter intrusive disturbance is to prevent access to habitats. However, denying public access to natural areas is difficult to justify, and even more difficult to enforce. Moreover, denied access diminishes the educational value of natural areas. Therefore, more realistic efforts will focus on minimizing the disruptive effects of the intrusion by avoiding sensitive areas, restricting access at critical times of the year, and limiting the number of people accessing the area at any one time. Secondary efforts will include avoiding damage to vegetative and water resources by constructing low impact trails and establishing viewing areas at the periphery of the habitat.

Anthropogenic activities external to the wetland also cause disturbances. Classically, the wetland edge was viewed as an area of increased vertebrate biomass and productivity (Leopold 1933, Lay 1938, McAtee 1945, Giles 1978). Certain species of wildlife, notably deer (*Odocoileus spp.*), rabbits (*Sylvilagus spp.*), gamebirds and some raptors appear to benefit from the juxtaposition of wetlands and uplands (Bider 1968, Gates and Hale 1974, Gates and Gysel 1978, Petersen 1979, Wilcove et al. 1986). More recently, there has been recognition of the detrimental effects of this juxtaposition (Figure 5). The edge is subject to an increased frequency and severity of fire, hunting and poaching, nest predation, nest parasitism, and a replacement of the native mammal community by exotic species (Stalmaster and Newman 1978, Robbins 1979, Tremblay and Ellison 1979, Rodgers and Burger 1981, Whitcomb et al. 1981, Brittingham and Temple 1983, Wilcove 1985a and b, Klein in Brown et al. 1989, Soulé 1991). Differences in microclimate, browsing and other disturbances favor weedy plant species and a vegetation community that differs markedly from the interior (Wales 1972, Ranney 1977, Forman and Godron 1986, Lovejoy et al. 1986).

The extent of the edge effect on interior species has only been reasonably quantified for forest bird species. The effect, if any, to emergent and other open wetland systems remains to be addressed. Nevertheless, Wales (1972) and Ranney (1977) determined that major vegetational changes occur from 10 to 30 m into the forest, with the greatest effects occurring along the southerly edge. Lovejoy et al. (1986), working in tropical rainforest, found that microclimate varied up to 100 m into the interior, and that interior birds did not occur within 50 m of the edge. In temperate forest, Whitcomb et al. (1981) found a negative impact from surrounding altered habitats on interior bird species occurrence up to 100 m from the forest edge, whereas Wilcove (1985a) found edge-related nest parasitism and predation to songbirds up to 600 m from the edge.

Whether the edge is viewed as beneficial or detrimental depends upon the objectives of the design. Regardless of objectives, a decrease in the ratio of interior-to-edge will increase the relative number of edge-adapted species and decrease the relative number of interior-adapted species. Isodiametric (round) wetlands will maximize the interior-to-edge ratio, whereas elongated wetlands will minimize the interior-to-edge ratio. Isodiametric wetlands also have a secondary benefit of minimizing internal dispersal distance (Diamond 1975, Wilson and Willis 1975). Maintaining interior-adapted species with a decreasing interior-to-edge ratio will likely require increased protection and management.

Clearly, detrimental edge effects on interior species will be reduced as refuge size increases. If the edge effect extends to 100 m (Lovejoy et al. 1986), then wetland refuges must be larger than 10 ha to accommodate interior species. Wetland refuges must be greater than 100 ha to accommodate interior species if the effect extends 600 m as Wilcove (1985a) has suggested. Wetlands smaller than these minimum sizes will in theory accommodate only edge-adapted or disturbance-tolerant species. However, caution must be exercised in extrapolating effects on temperate and tropical forest bird species to wetland species in general.

Another way in which to minimize disturbance associated with detrimental edge effects is to establish upland buffers to the wetland refuge. Wetlands-related wildlife uses surrounding uplands to fulfill part of their life requisites. Errington (1957) noted rabbits, woodchucks, foxes, ducks, herons, small birds and mammals, skunks, minks and muskrats using the upland areas adjacent to Iowa and South Dakota marshes. Deer (*Odocoileus virginianus*) and pheasants (*Phasianus colchicus*) seek cover in dense upland vegetation surrounding wetlands (Gates and Hale 1974, Linder and Schitosky 1979). Red-tailed hawk (*Buteo jamaicensis*),

Figure 5. Although the edge was historically viewed as an area of increased productivity, recent evidence has illustrated detrimental effects to interior species from exposure to anthropogenic disturbance.

pheasant, northern harrier (*Circus cyaneus*) and leopard frog (*Rana pipiens*) forage in upland borders (Errington and Breckenridge 1936, Dole 1965, Gates and Hale 1974, Petersen 1979). Salamanders, and some frogs and toads, spend the majority of their adult lives in fields and forests (Behler and Find 1979). Waterfowl, turtle, and mammal breed along the upland border of wetlands (Allen and Shapton 1942, Errington 1957, Jahn and Hunt 1964, Pils and Martin 1978, Weller 1978, DeGraaf and Rudis 1986, Kirby 1988). The upland buffer also serves as a travel corridor, a refuge during periods of high water and a shield for wetland species from anthropogenic activity (Meanly 1972, Porter 1981).

In large part, buffers proposed to protect wetlands wildlife are based upon best professional judgment or knowledge of species spatial requirements (Leedy et al. 1978, Roman and Good 1985, Brown et al. 1989, Diamond and Nilson 1988, New Jersey Department of Environmental Regulation 1988, Brady and Buchsbaum 1989). Leedy et al. (1978) determined that a buffer up to 92 m is necessary on either side of a stream to provide required wildlife habitat elements. This opinion has surfaced in several efforts to establish effective wildlife buffers in New Jersey (Roman and Good 1985, Diamond and Nilson 1988, New Jersey Department of Environmental Regulation 1988). Adopting a more rigorous approach, Brown et al. (1989) considered spatial requirements, and using guilds and indicator species determined that buffers should be 98 m to 224 m wide, with a minimum of 15 m of upland included in the buffer. Buffers in the latter study are measured from the waterward edge of forested areas, whereas marsh buffers are measured from the landward edge of the wetland.

There are few studies which note the distance at which wildlife are disturbed by human activity, and much of this information is anecdotal (Short and Cooper 1985, Brady and Buchsbaum 1989, Brown et al. 1989). Disturbance distances for 23 species of birds range from 6 m to 459 m, and average 74 m (Table 1). No strong patterns are apparent which would suggest a relationship between disturbance distance and taxonomic group, body size or ecological niche.

In a less direct manner, buffers protect wetland wildlife by removing sediment, nutrients, salt, bacteria, virus, and chemical pollutants from agricultural and urban surface runoff (Karr and Schlosser 1977, Sullivan 1986, U.S. Department of Agriculture Soil Conservation Service 1986, Potts and Bai 1989). Berger (1989) suggested that one reason for a general decrease in the number of amphibians is excessive agricultural chemical pollution. Vegetated buffers can be effective in preventing or minimizing environmental degradation of wetlands (Dillaha et al. 1989).

Determining appropriate buffer widths adequate to reduce the level of disturbance to wetlands from surface runoff has received more rigorous examination than attempts to establish buffers based upon direct disturbance to wildlife. Recommendations vary considerably and reflect regional and local differences in the aforementioned factors (Table 2). Empirical studies have identified buffer needs of 15 m to 61 m to prevent natural debris and sediment accumulation in streams (Trimble and Sartz 1957, Broderson 1973). Theoretical treatments have identified buffer needs of 15 m to 224 m (New Jersey Department of Environmental Regulation 1988, Brady and Buchsbaum 1989, Brown et al. 1989, Dillaha et al. 1989, Potts and Bai 1989, East Central Florida Regional Planning Council 1991).

Generally, wetland refuges surrounded by natural vegetation, particularly trees and shrubs which serve as visual and auditory screens from anthropogenic activity, will withstand smaller buffers than wetlands surrounded by disturbed habitats. So too, wetlands surrounded by passive recreational areas, such as golf courses and ball fields, will withstand smaller buffers than wetlands surrounded by residential, commercial and industrial development. So too, wetland refuges surrounded by gentle slopes, natural vegetation communities, and course-grained, well-drained soils with a relatively high organic content will withstand smaller buffers than refuges surrounded by steeper slopes, disturbed vegetation communities, and fine-grained, poorly-drained soils with a relatively low organic content. Buffer size should generally increase with an increase in the quality of the wetland, an increase in the size and intensity of surrounding development, and a decrease in surface runoff particle size.

GUIDELINES

Wildlife respond to largely physical characteristics when selecting a habitat. Nevertheless, three factors: 1) the size of the wetland, 2) the relationship of the wetland to other wetlands, and 3) the level and type of disturbance, will largely determine the effectiveness of a wetland for long-term wildlife use. Ideally, a

Table 1. Disturbance tolerance distances for 23 species of birds.

Scientific Name	Common Name	Disturbance Distance (m)
Anas americana	American widgeon	92
Anas clypeata	northern shoveler	92
Anas discors	blue-winged teal	92
Anas fulvigula	mottled duck	37
Anhinga anhinga	anhinga	6
Ardea herodias	great blue heron	100
Buteo jamaicensis	red-tailed hawk	31
Calidra alba	sanderling	73
Calidris alpina	dunlin	92
Calidris mauri	western sandpiper	73
Calidris minutilla	least sandpiper	73
Casmerodius albus	great egret	18
Catoptrophorus semipalmatus	willet	73
Charadrius vociferus	killdeer	55
Egretta caerulea	little blue heron	55
Egretta thula	snowy egret	73
Eudocimus albus	white ibis	73
Fulica americana	American coot	37
Haliaetus leucocephalus	bald eagle	459
Pandion haliaetus	osprey	6
Pelicanus occidentalis	brown pelican	6
Phalacrocorax auritus	double-crested cormorant	6
Podilymbus podiceps	pied-billed grebe	73
AVERAGE		74

SOURCE: Brown et al. (1989), except red-tailed hawk (Brady and Buchsbaum 1989) and great blue heron (Short and Cooper 1985).

wetland designed for wildlife, regardless of whether the design uses preservation, enhancement, restoration or creation to achieve its objectives, should be large enough to support an estimated minimum viable population of the species of interest. In many cases establishment of a wildlife community is the objective, and an indicator species with significant areal requirements should be identified. The

Table 2. Suggested buffer widths to reduce the level of disturbance to wetlands.

Suggested Buffer (m)	Basis	Study
up to 43	Maximum distance sediment transported from logging road to streams	Trimble and Sartz 1957
15-61	Natural control of debris and sediment accumulation in streams	Broderson 1973
15-92	Vegetation interspersion, wetland size, quality of surrounding habitat, potential for impacts	Roman and Good 1985
15-186+	Slope, vegetation, soil characteristics, value of wetland, intensity of development	New Jersey Dept. Environ. Protection 1988
92-122	Water quality maintenance, public health protection, wildlife protection	Brady and Buchsbaum 1989
23-224	Groundwater drawdown, sediment and turbidity control, wildlife.	Brown et al. 1989
20-30	Vegetation	Dillaha et al. 1989
75+	Vegetation and soil characteristics, development intensity and type	Potts and Bai 1989
0-159	Wetland quality, soil type	East Central Florida Regional Planning Council 1991

wetland should be large enough to support at least 50, and ideally 500, breeding individuals, and area requirements can be estimated from knowledge about species home range sizes.

In most instances, wetlands being designed will be large enough to support minimum viable populations of only those species with small areal requirements. For other species, the design should focus on establishment of a metapopulation through ensuring interpopulation movement. Interpopulation movement can be affected for all species by close juxtaposition of individual populations, or provision of habitat corridors. Species with broad habitat tolerances will use a variety of landscape elements as long as barriers such as roads, waterways and waterbodies are avoided. Wetlands for migratory bird species will not require corridors. However, designers of wetlands for migratory bird species should recognize that sustainability of a seasonal population is dependent upon one or more wetlands hundreds or even thousands of kilometers distant.

Wetlands designed to support disturbance-intolerant species should limit intrusions, or protect critical areas during sensitive times of the year. Interior species are unlikely to thrive in wetlands less than 100 ha, or to persist in wetlands less than 10 ha. The effective size of a wetland can be increased by the establishment of an upland buffer. Buffers should be established on a case by case basis through consideration of soil type, vegetation type in the buffer, adjacent land use, slope, runoff particle size, wetland quality and indigenous wildlife. Nevertheless, a buffer of approximately 200 m width appears to be adequate in most instances to minimize direct disturbance to wildlife, and to reduce water quality impacts from contaminated surface runoff (Table 2). An upland buffer also provides life requisites to many wetland wildlife species.

REFERENCES

Allen, A.W. 1986. Habitat suitability index models: mink, revised. U.S. Fish and Wildlife Service Biological Report 82(10.127). 23 pp.

Allen, D.L. and W.W. Shapton. 1942. An ecological study of winter dens, with special reference to the eastern skunk. Ecology 23(1):59-68.

Allendorf, R.W. 1986. Genetic drift and the loss of alleles versus heterozygosity. Zoo Biology 5:181-190.

Allendorf, F.W. and R.F. Leary. 1986. Heterozygosity and fitness in natural populations of animals. Pp. 57-76 In M.E. Soulé (ed.) Conservation Biology: Science of Scarcity and Diversity. Sinauer Associates, Sunderland, Massachusetts.

Ambruster, M.J. 1987. Habitat suitability index models: Greater sandhill crane. U.S. Fish and Wildlife Service Biological Report 82(10.140). 26 pp.

Ambuel, B. and S.A. Temple. 1983. Area dependent changes in the bird communities and vegetation of southern Wisconsin forests. Ecology 64:1057-1068.

Anonymous. 1992. Animal road kill toll tops 2,600 at 64 of Florida's state parks. Florida Environment June 1992, Volume VI, Number VI.

Arnold, C. 1990. Wildlife corridors. California Coast and Ocean summer 1990: 10-21.

Baskett, R.K. 1987. Grand Pass Wildlife Area, Missouri: Modern wetland restoration strategy at work. Increasing Our Wetland Resources pp. 220-224.

Behler, J.L. and F.W. Find. 1979. The Audubon Society Field Guide to North American Reptiles and Amphibians. Alfred A. Knopf, Inc., New York, New York.

Bekele, E. 1980. Island biogeography and guidelines for the selection of conservation units for large mammals. Ph.D. Dissertation, University of Michigan, Ann Arbor.

Berger, L. 1989. Disappearance of amphibian larvae in the agricultural landscape. Ecology International Bulletin 17:65-73.

Bider, J.R. 1968. Animal activity in uncontrolled terrestrial communities as determined by a sand transect technique. Ecological Monographs 38:269-308.

Brady, P. and R. Buchsbaum. 1989. Buffer Zones: The Environment's Last Defense. Report submitted to City of Gloucester, Massachusetts by Massachusetts Audubon: North Shore. 16 pp.

Brittingham, M.C. and S.A. Temple. 1983. Have cowbirds caused forest songbirds to decline? BioScience 33:31-35.

Broderson, J.M. 1973. Sizing Buffer Strips to Maintain Water Quality. M. S. Thesis, University of Washington.

Brown, J.H. 1971. Mammals on mountaintops: Nonequilibrium insular biogeography. American Naturalist 105:467-478.

Brown, J.H. and A. Kodric-Brown. 1977. Turnover rates in insular biogeography: Effect of immigration on extinction. Ecology 58:445-449.

Brown, M., J. Schaefer and K. Brandt. 1989. Buffer Zones for Water, Wetlands, and Wildlife in the East Central Florida Region. Center for Wetlands Publication #89-07, University of Florida, Gainesville, Florida. 78 + pp.

Cain, S. 1938. The species-area curve. American Midland Naturalist 19:573-581.

Chepko-Sade, B.D. and Z.T. Halpin. 1987. Mammalian Dispersal Patterns. University of Chicago Press, Chicago, Illinois.

Cole, R.A. 1978. Habitat Development Field Investigations, Buttermilk Sound Marsh Development Site, Atlantic Intracoastal Waterway, Georgia: Summary Report. Technical Report D-78-26, USACOE Waterways Experiment Station, Vicksburg, Mississippi.

Connell, J.H. 1978. Diversity in tropical rain forests and coral reefs. Science 199:1302-1310.

Crawford, J.A. and D.K. Edwards. 1978. Habitat Development Field Investigations, Miller Sands Marsh and Upland Habitat Development Site, Columbia River, Oregon. Appendix F: Postpropagation Assessment of Wildlife Resources on Dredged Material. Technical Report D-77-38, USACOE Waterways Experiment Station, Vicksburg, Mississippi.

Culver, D.C. 1970. Analysis of simple cave communities I: Caves as islands. Evolution 24:463-474.

Dahl, T.E. 1990. Wetland losses in the United States 1780's to 1980's. U.S. Department of the Interior, Fish and Wildlife Service, Washington, D.C. 13 pp.

DeGraaf, R.M. and D.D. Rudis. 1986. New England wildlife: Habitat, natural history and distribution. U. S. Department of Agriculture Forest Service, Northeast Forest Experiment Station General Technical Report NE-108. 491 pp.

Diamond, J.M. 1973. Distributional ecology of New Guinea birds. Science 179:759-769.

Diamond, J.M. 1975. The island dilemma: lessons of modern biogeographic studies for the design of natural preserves. Biological Conservation 7:129-146.

Diamond, R.S. and D.J. Nilson. 1988. Buffer delineation method for coastal wetlands in New Jersey. In Symposium on Coastal Water Resources. Proceedings of the American Water Resources Association.

Dickerson, G.E., C.T. Blunn, A.B. Chapman, R.M. Kottman, J.L. Krider, E.J. Warwick and J.A. Whatley, Jr., in collaboration with M.L. Baker, J.L. Lush and L.M. Winters. 1954. Evaluation of selection in developing inbred lines of swine. University of Missouri College of Agriculture Research Bulletin 551.

Dillaha, T.A., J.H. Sherrard and D. Lee. 1989. Long-term effectiveness of vegetative filter strips. Water Environment and Technology, November.

Dole, J.W. 1965. Summer movements of adult leopard frogs, *Rana pipiens*. Ecology 46(3):236-255.

East Central Florida Regional Planning Council. 1991. ECFRPC Wetland Buffer Criteria and Procedures Manual. Draft # 7, East Central Florida Regional Planning Council, Winter Park, Florida.

Errington, P.L. 1957. Of Men and Marshes. MacMillan Co., New York, New York.

Errington, P.L. and W.J. Breckenridge. 1936. Food habits of marsh hawks in the glaciated prairie region of north-central United States. The American Midland Naturalist 17:831-848.

Forman, R.T.T. and M. Godron. 1986. Landscape Ecology. John Wiley & Sons. 619 pp.

Franklin, I.R. 1980. Evolutionary change in small populations. Pp. 135-149 In Soulé, M.E. and M.E. Wilcox (eds.) Conservation Biology: An Evolutionary - Ecological Perspective. Sinauer Associates, Sunderland, Massachusetts.

Fritz, R.S. 1979. Consequences of insular population structure: Distribution and extinction of spruce grouse populations. Oecologia 42:57-65.

Galli, A.E., E.C.F. Leck and R.T.T. Forman. 1976. Avian distribution patterns within different sized forest islands in central New Jersey. Auk 93:356-364.

Garbisch, E.W., Jr. 1977. Recent and planned marsh establishment work throughout the contiguous United States: A survey and basic guidelines. Technical Report D-77-3. USCOE Waterways Experiment Station, Vicksburg, Mississippi.

Garrett, M.G. and W.L. Franklin. 1988. Behavioral ecology of dispersal in the black-tailed prairie dog. Journal of Mammals 69:236-250.

Gates, J.M. and J.B. Hale. 1974. Seasonal Movement, Winter Habitat Use, and Population Distribution of an East Central Wisconsin Pheasant Population. Wisconsin Department of Natural Resources Technical Bulletin # 76, Madison, Wisconsin.

Gates, J.E. and L.W. Gysel. 1978. Avian nest dispersion and fledgling success in field-forest ecotones. Ecology 59:871-883.

Gavareski, C.A. 1976. Relation of park size and vegetation to urban bird populations in Seattle, Washington. Condor 78:375-382.

Gilbrook, M.J. 1989. Spatial and Size Distribution of Wetland Habitats in the East Central Florida Region. East Central Florida Regional Planning Council, Winter Park, Florida. 78 pp.

Giles, R.H., Jr. 1978. Wildlife Management. W.H. Freeman, San Francisco, California.

Gilpin, M.E. 1987. Spatial structure and population vulnerability. Pp. 125-139 In Soulé, M.E. (ed.) Viable Populations for Conservation. Cambridge University Press, Cambridge, Massachusetts. 189 pp.

Goodman, D. 1987. The demography of chance extinction. Pp. 11-34 In Soulé, M.E. (ed.) Viable Populations for Conservation. Cambridge University Press, Cambridge, Massachusetts. 189 pp.

Gottfried, B.M. 1979. Small mammal populations in woodlot islands. American Midland Naturalist 102:105-112.

Graves, B.M. and S.H. Anderson. 1987. Habitat suitability index models: bullfrog. U.S. Fish and Wildlife Service Biological Report 82(10.138). 22 pp.

Grinnell, J. and H.S. Swarth. 1913. An account of the birds and mammals of the San Jacinto area of southern California, with remarks upon the behavior of geographic races on the margins of their habitats. University of California Publications Zoology 10:197-406.

Harris, L.D. 1984. The Fragmented Forest: Island Biogeography Theory and the Preservation of Biotic Diversity. University of Chicago Press, Chicago, Illinois. 211 pp.

Holekamp, K.E. 1984. Natal dispersal in Belding's ground squirrel (*Spermophilus beldingi*). Behavioral Ecology and Sociobiology 16:21-30.

Jahn, L.R. and R.A. Hunt. 1964. Duck and coot ecology and management in Wisconsin. Wisconsin Department of Natural Resources Technical Bulletin # 33, Madison, Wisconsin.

Kale, H.W. 1978. Rare and Endangered biota of Florida: Birds. University Presses of Florida, Gainesville, Florida. 121 pp.

Karr, J.R. and I.J. Schlosser. 1977. Impact of nearstream vegetation and stream morphology on water quality and stream biota. United States Environmental Protection Agency Report #600/3-77-097.

Kiester, A.R., C.W. Schwartz and E.R. Schwartz. 1982. Promotion of gene flow by transient individuals in an otherwise sedentary population of box turtles (*Terrapene carolina triunguis*). Evolution 36(3):617-619.

Kirby, R.E. 1988. American black duck breeding habitat enhancement in the northeastern United States: A review and synthesis. U. S. Fish and Wildlife Service Biological Report 88(4). 50 pp.

Kushlan, J.A. 1976. Site selection for nesting colonies by the American white ibis (*Eudomicus alba*) in Florida. Ibis 118:590-593.

Kushlan, J.A. 1979. Design and management of continental wildlife reserves: Lessons from the Everglades. Biological Conservation 15:281-290.

Kusler, J.A. and M.E. Kentula. 1990. Wetland Creation and Restoration: The Status of the Science. Island Press, Washington, D. C. 594 pp.

Landin, M.C., E.J. Clairain, Jr. and C.J. Newling. 1989. Wetland habitat development and long term monitoring at Windmill Point, Virginia. Wetlands 9(1):13-26.

Landin, M.C. and J.W. Webb. 1989. Wetland development and restoration as part of Corps of Engineer programs: Case studies. Pp. 388-391 In Proceedings: National Wetland Symposium - Mitigation of Impacts and Losses. Association of State Wetland Managers, Inc. 460 pp.

Lay, D. 1938. How valuable are woodland clearings to birdlife. Wilson Bulletin 50:254-256.

Leedy, D.L., R.M. Maestro and T.M. Franklin. 1978. Planning for Wildlife in Cities and Suburbs. Urban Wildlife Research Center, Inc., Ellicott City, Maryland.

Leopold, A. 1933. Game Management. Scribners, New York, New York.

Linder, R.L. and F. Schitosky, Jr. 1979. Use of wetlands by upland wildlife. Pp. 307 In Greeson, P.E., J.R. Clark and J.E. Clark (eds.) Wetland Functions and Values: The State of Our Understanding. Proceedings of the National Symposium on Wetlands.

Lopez, B. 1992. Implacable corridors of death. Los Angeles Times 4 October, 1992.

Lovejoy, T.E., R.O. Bierregaard, Jr., A.B. Rylands, J.R. Malcolm, C.E. Quintela, L.H. Harper, K.S. Brown, Jr., A.H. Powell, G.V.N. Powell, H.O.R. Schubart and M.B. Hays. 1986. Edge and other effects of isolation on Amazon forest fragments. Pp. 257-285 In Soulé, M. E. (ed.) Conservation Biology: The Science of Scarcity and Diversity. Sinauer Associates, Inc., Sunderland, Massachusetts.

Lunz, J.D., R.J. Diaz and R.A. Cole. 1978a. Upland and Wetland Habitat Development with Dredged Material: Ecological Considerations. USACOE Waterways Experiment Station, Vicksburg, Mississippi. 50 pp.

Lunz, J.D., T. Ziegler, R.T. Huffman, B.R. Wells, R.J. Diaz, E.J. Clairain, Jr. and L.J. Hunt. 1978b. Habitat Development Field Investigations, Windmill Point, Marsh Development Site, James River, Virginia: Summary Report. Technical Report D-77-23. USACOE Waterways Experiment Station, Vicksburg, Mississippi.

Lynch, J.F. and D.F. Whigham. 1984. Effects of forest fragmentation on breeding bird communities in Maryland, USA. Biological Conservation 28:287-324.

MacArthur, R.H. and E.O. Wilson. 1963. An equilibrium theory of insular zoogeography. Evolution 17:373-387.

MacArthur, R.H. and E.O. Wilson. 1967. The Theory of Island Biogeography. Princeton University Press, Princeton, University.

Madsen, A.B. 1990. Otters, *Lutra lutra,* and traffic. Flora and Fauna 96(2):39-46.

McAtee, W.L. 1945. The Ring-Necked Pheasant and its Management in North America. American Wildlife Institute, Washington, D.C.

McIntyre, S. and G.W. Barrett. 1992. Habitat variegation, an alternative to fragmentation. Conservation Biology 6(1):146-147.

Meanly, B. 1972. Swamps, River Bottoms and Canebrakes. Barre Publishers, Barre, Massachusetts.

Michael, E.D. and L.S. Smith. 1988. Creating Wetlands Along Highways in West Virginia. West Virginia Department of Highways and U. S. Department of Transportation, FHWA/WV-85/001. 261 pp.

Moore, N.W. and M.D. Hooper. 1975. On the number of bird species in British woods. Biological Conservation 8:239-250.

Nei, M., T. Maruyama and R. Chakraborty. 1975. The bottleneck effect and genetic variability in populations. Evolution 29:1-10.

New Jersey Department of Environmental Regulation. 1988. Wetland Buffer Delineation Method. State of New Jersey Division of Coastal Resources, Trenton, New Jersey.

Newling, C.J. 1990. Restoration of bottomland hardwood forests in the lower Mississippi Valley. Restoration and Management Notes 8(1):23-28.

Norse, E.A., K.L. Rosenbaum, D.W. Wilcove, B.A. Wilcox, W.H. Romme, D.W. Johnston and M.L. Stout. 1986. Conserving Biological Diversity In Our National Forests. The Wilderness Society. 116 pp.

Noss, R.F. 1987. Corridors in real landscapes: A reply to Simberloff and Cox. Conservation Biology 1(2):159-164.

Oxley, D.J., M.B. Fenton and G.R. Carmody. 1974. The effects of roads on populations of small mammals. Journal Applied Ecology 11(1):51-59.

Palmer, R.S. (ed.) 1976. Handbook of North American Birds. Volume 2. Yale University Press, New Haven, Connecticut.

Peterken, G.F. 1974. A method of assessing woodland flora for conservation using indicator species. Biological Conservation 6:239-245.

Peterken, G.F. 1977. Habitat conservation priorities in British and European woodlands. Biological Conservation 11:223-236.

Petersen, L. 1979. Ecology of Great Horned Owls and Red-tailed Hawks in Southeastern Wisconsin. Wisconsin Department of Natural Resources, Madison, Wisconsin.

Peterson, A. 1986. Habitat suitability index models: Bald eagle (breeding season). U.S. Fish and Wildlife Service Biological Report 82(10.126).

Petraitis, R.S., R.E. Latham and R.A. Niesenbaum. 1989. The maintenance of species diversity by disturbance. Quarterly Review of Biology 64:393-418.

Pickett, S.T.A. and J.N. Thompson. 1978. Patch dynamics and the design of nature reserves. Biological Conservation 13:27-37.

Pickett, S.T.A. and P.S. White (eds.). 1985. The ecology of natural disturbance and patch dynamics. Academic Press, Orlando, Florida.

Pickett, S.T.A., J. Lolasa, J.J. Armesto and S.L. Collins. 1989. The ecological concept of disturbance and its expression at various hierarchical levels. Oikos 54:129-136.

Pils, C.M. and M.A. Martin. 1978. Population dynamics, predatory-prey relationships and management of the red fox in Wisconsin. Wisconsin Department of Natural Resources Technical Bulletin # 107, Madison, Wisconsin.

Porter, B.W. 1981. The wetland edge as a community and its value to wildlife. Pp. 15-25 In Richarson, B. (ed.) Selected Proceedings of the Midwest Conference on Wetland Values and Management. 660 pp.

Potts R.R. and J.L. Bai. Establishing variable width buffer zones based upon site characteristics and development type. 1989. Pp. 6A-3-6A-14 In Water: Laws and Management. Proceedings of the American Water Resources Association.

Pough, R.H. 1951. Audubon Water Bird Guide. Doubleday and Co., Garden City, New Jersey.

Ralls, K., P.H. Harvey and A.M. Lyles. 1986. Inbreeding in natural populations of birds and mammals. Pp. 35-56 In M. E. Soulé (ed.) Conservation Biology: Science of Scaricity and Diversity. Sinauer Associates, Sunderland, Massachusetts.

Ranney, J.W. 1977. Forest island edges - their structure, development and implication to regional forest ecosystem dynamics. EDFB/IBP-77-1, Oak Ridge National Laboratory, Oak Ridge, Tennessee.

Ray, D.K. and W.O. Woodroof. 1988. Mitigating impacts to wetlands and estuaries in California's coastal zone. Pp. 106-108 In Proceedings: National Wetland Symposium - Mitigation of Impacts and Losses. Association of State Wetland Managers, Inc. 460 pp.

Reimold, R.J. and D.A. Thompson. 1988. Wetland mitigation effectiveness. Pp. 259-262 In Proceedings: National Wetland Symposium - Mitigation of Impacts and Losses. Association of State Wetland Managers, Inc. 460 pp.

Rey, J.R., J. Shaffer, D. Tremain, R.A. Crossman and T. Kain. 1990. Effects of reestablishing tidal connections in tow impounded subtropical marshes on fishes and physical conditions. Wetlands 10(1):27-45.

Robbins, C.S. 1979. Effect of forest fragmentation on bird populations. Pp. 198-212 In DeGraaf, R. M. and K. E. Evans (eds.) Management of North Central and Northeastern Forests for Nongame Birds. U. S. Department of Agriculture, Forest Service General Technical Report NC-51.

Rodgers, J.A., Jr. and J. Burger. 1981. Concluding remarks: Symposium on human disturbance and colonial waterbirds. Colonial Waterbirds 4:69-70.

Roman, C.T. and R.E. Good. 1985. Buffer Delineation Model for New Jersey Pinelands Wetlands. Center for Coastal and Environmental Studies, Division of Pinelands Research, New Jersey Pinelands Commission. 73 pp.

Seidensticker, J.C., M.G. Hornocker, W.V. Wiles and J.P. Messick. 1973. Mountain lion social organization in the Idaho Primitive area. Wildlife Monographs 35:1-60.

Senner, J.W. 1980. Inbreeding depression and the survival of zoo populations. Pp. 209-244 In Soulé, M.E. and M.E. Wilcox (eds.) Conservation Biology: An Evolutionary - Ecological Perspective. Sinauer Associates, Sunderland, Massachusetts.

Shaffer, M.L. 1981. Minimum population sizes for conservation. Bioscience 31:131-34.

Shaw, S.P. and C.G. Fredine. 1956. Wetlands of the United States. U.S. Fish and Wildlife Service Circular 39. 67 pp.

Shirer, H.W. and J.F. Downhower. 1968. Radio tracking of dispersing yellow-bellied marmots. Transactions of the Kansas Academy of Science 71:463-479.

Short, H.L. and R.J. Cooper. 1985. Habitat suitability index models: Great blue heron. U.S. Fish and Wildlife Service Biological Report 82(10.99). 23 pp.

Soulé, M.E. 1980. Thresholds for survival: Maintaining fitness and evolutionary potential. Pp. 151-169 In Soulé, M.E. and M.E. Wilcox (eds.) Conservation Biology: An Evolutionary - Ecological Perspective. Sinauer Associates, Sunderland, Massachusetts.

Soulé, M.E. 1991. Land use planning and wildlife maintenance: Guidelines for conserving wildlife in an urban landscape. American Planning Association Journal 57(3):313-323.

Soulé, M.E. and D. Simberloff. 1986. What do genetics and ecology tell us about the design of nature reserves? Biological Conservation 35:19-40.

Stalmaster, M.V. and J.R. Newman. 1978. Behavioral responses of wintering bald eagle to human activity. Journal of Wildlife Management 42:506-513.

Steinhart, P. 1986. Artificial species. Audubon 88:8-9.

Stephenson, A.B, A.J. Wyatt and W.W. Nordskog. 1953. Influence of inbreeding on egg production in the domestic fowl. Poultry Science 32:510-517.

Storm, G.L., R.D. Andrews, R.L. Phillips, R.A. Bishop, D.B. Siniff and J.R. Tester. 1976. Morphology, reproduction, dispersal, and mortality of Midwestern red fox populations. Wildlife Monographs 49:5-82.

Sullivan, J.K. 1986. Using buffer zones to battle pollution. Environmental Protection Agency Journal 12(4):8-10.

Talbot, C.W., K.W. Able and J.K. Shisler. 1986. Fish species composition in New Jersey salt marshes: Effects of marsh alterations for mosquito control. Transactions of the American Fisheries Society 115:269-278.

Thompson, L.S. 1978. Species abundance and habitat relations of an insular montane avifauna. Condor 80:1-14.

Tremblay, J. and L.N. Ellison. 1979. Effects of human disturbance on breeding of black-crowned night herons. Auk 96:364-369.

Trimble, G.R., Jr. and R.S. Sartz. 1957. How far from a stream should a logging road be located? Journal of Forestry pp. 339-341.

U.S. Army Corps of Engineers. 1976. Third Annual Report of the Material Research Program. Environmental Effects Laboratory, Waterways Experiment Station, Vicksburg, Mississippi. 78 pp.

U.S. Army Corps of Engineers. 1989. Evaluation of Freshwater Wetland Replacement Projects in Massachusetts. New England Division, Waltham, Massachusetts.

U.S. Department of Agriculture, Soil Conservation Service. 1986. Urban Hydrology for Small Watersheds. Technical Release Number 55.

U.S. Fish and Wildlife Service. 1984. Management Guidelines for the Bald Eagle in the Southeast Region.

Vuilleumier, F. 1970. Insular biogeography in continental regions. I. The northern Andes of South America. American Naturalist 104:373-388.

Vuilleumier, F. 1973. Insular biogeography in continental regions. II. Cave faunas from Tesoin, southern Switzerland. Systematic Zoology 22:64-76.

Wales, B.A. 1972. Vegetation analysis of north and south edges in a mature oak-hickory forest. Ecological Monograph 42:451-471.

Webb, J.W., M.C. Landin, and H.H. Allen. 1988. Approaches and techniques for wetlands development and restoration of dredged material disposal sites. Pp. 132-134 In Proceedings: National Wetland Symposium - Mitigation of Impacts and Losses. Association of State Wetland Managers, Inc. 460 pp.

Weller, M.W. 1978. Wetland habitats. Pp. 210 In Greeson, P.E., J.R. Clark and J.E. Clark (eds.) Wetland Functions and Values: The State of Our Understanding. Proceedings of the National Symposium on Wetlands.

Weller, M.W. 1987. Freshwater Marshes: Ecology and Wildlife Management. University of Minnesota Press, Minneapolis. 150 pp.

Weller, M.W. 1990. Waterfowl management techniques for wetland enhancement, restoration and creation useful in mitigation procedures. Pp. 517-528 In Kusler, J.A. and Kentula, M.E. (eds.) Wetland Creation and Restoration: The Status of the Science. Island Press, Washington, D.C. 594 pp.

Weller, M.W., G.W. Kaufmann and P.A. Vohs, Jr. 1991. Evaluation of wetland development and waterbird response at Elk Creek Wildlife Management Area, Lake Mills, Iowa, 1961 to 1990. Wetlands 11(2):245-262.

Whitcomb, R.F. 1977. Island biogeography and habitat islands of eastern forest. American Birds 31:3-5.

Whitcomb, R.F., J.F. Lynch, M.K. Klimkiewicz, C.S. Robbins, B.L. Whitcomb and D. Bystrak. 1981. Effects of forest fragmentation on avifauna of the eastern deciduous forest. Pp. 125-206 In Burgess, R. (ed.) Forest Island Dynamics in Man-Dominated Landscapes: Ecological Studies 41. Springer-Verlag, New York.

Wiggett, D.R. and D.A. Boag. 1989. Intercolony natal dispersal in the Columbian ground squirrel. Canadian Journal of Zoology 67:42-50.

Wilcove, D.S. 1985a. Forest Fragmentation and the Decline of Migratory Songbirds. Ph.D. thesis, Princeton University, Princeton, New Jersey.

Wilcove, D.S. 1985b. Nest predation in forest tracts and the decline of migratory songbirds. Ecology 66:1211-1214.

Wilcove, D.S., C.H. McLellan and A.P. Dobson. 1986. Habitat fragmentation in the temperate zone. Pp. 237-256 In Soulé, M.E. (ed.), Conservation Biology: The Science of Scarcity and Diversity. Sinauer Associates, Inc., Sunderland, Massachusetts.

Wilcox, B.A. 1980. Insular ecology and conservation. Pp. 95-117 In Soulé, M.E. and B.A. Wilcox (eds.) Conservation Biology, an Evolutionary Ecological Perspective. Sinauer Associates, Inc., Sunderland, Massachusetts.

Wilson, E.O. and E.O. Willis. 1975. Applied biogeography. Pp. 522-534 In Cody, M. and J.M. Diamond (eds.) Ecology and Evolution of Communities. Belknap Press of Harvard University, Cambridge, Massachusetts. 545 pp.

Wright, S. 1969. Evolution and the Genetics of Populations, Vol. 2. The Theory of Gene Frequencies. University of Chicago Press, Chicago. 511 pp.

CHAPTER 14

MANAGING WETLANDS FOR WILDLIFE

Donald M. Kent

Environments can, by the judicious use of those tools employed in gardening and landscaping or farming, be built to order with assurance of attracting the desired bird (Leopold 1933).

Prior to the 1930's, wildlife management had been concerned primarily with the administration and regulation of waterfowl and furbearer harvests. It was about this time that wildlife managers, as well as the public, recognized that wildlife resources were not limitless. Aldo Leopold crystallized this emerging perspective in his book *Game Management* (1933), and gave birth to the scientific management of wildlife populations and wildlife habitats.

Wetlands are especially critical habitats for wildlife, and exceed all other land types in wildlife productivity (Vaught and Bowmaster 1983, Cowardin and Goforth 1985, Payne 1992). A majority of wildlife species use wetlands on either a permanent or transitory basis for breeding, food or shelter (Pandit and Fotedar 1982, Rakstad and Probst 1985). Wetlands provide critical habitat for 80 of 276 threatened and endangered species. Approximately 64 percent of the wildlife in the Great Lakes region inhabit or are attracted to wetlands, including 62 percent of the birds, 69 percent of the mammals and 71 percent of the amphibians and reptiles (Rakstad and Probst 1985). From 67 to 90 percent of commercial fish and shellfish species are either directly or indirectly dependent upon wetlands (Peters et al. 1979, Vaught and Bowmaster 1983, Radtke 1985). Wetlands are also the principal habitat for furbearers and waterfowl. Ten to 12 million ducks breed in the lower 48 states, and 45 million ducks depend on wetlands throughout the United States and Canada for their existence (Vaught and Bowmaster 1983, Radtke 1985).

Wetlands wildlife has a quantifiable economic value. Hundreds of millions of dollars are spent annually on birdwatching and other wildlife observations. Freshwater fisherman spent $7.8 billion dollars in 1980, and waterfowl hunters spent $950 million in 1975 (Radtke 1985). In 1975-1976, more than 8.5 million furbearer pelts with a value in excess of $35.5 million were harvested (Chabreck 1979).

Valuable wetland habitats are being lost and degraded at an alarming rate. Almost 0.5 million acres of wetlands are lost per year (Low in Payne 1992). Annual losses to agriculture range from 1 to 4 percent (Weller 1981). Prairie potholes in the United States and Canada are lost at a rate of 1 to 2 percent per year, and 75 percent of northern central United States wetlands were lost between 1850 and 1977 (U. S. Department of Agriculture 1980, Radtke 1985, Melinchuk and MacKay 1986). Bottomland hardwood forests were cleared at a rate of 165,000 acres per year between 1940 and 1980, reducing forested wetlands in some states by 96 percent (Korte and Fredrickson 1977, Radtke 1985). Coastal wetlands have also suffered dramatic losses, with almost 25,000 acres per year being lost from Gulf Coast wetlands (Chabreck 1976, Gagliano 1981). Many of the wetlands which remain are degraded from channelization, damming, and agricultural and urban surface runoff. As well, these remaining wetlands are typically fragmented or isolated, and occur on private land.

Coincident with the loss and degradation of wetlands is a decline in continental waterfowl populations. Breeding mallard populations have declined at a rate of up to 19 percent since 1970 (Melenchuk and MacKay 1986). Weller (1981) estimates that 90 million waterfowl nests were lost in the north central United States between 1850 and 1977, and wintering waterfowl populations declined by 70 percent between the mid 1950s and 1983 (Whitman and Meredith 1987). The effect of wetlands loss and degradation on other wildlife remains largely undetermined.

Management of wetlands wildlife is necessitated by the exceptional value of wetlands as wildlife habitat, and the continued loss and degradation of wetlands. Preservation of existing high quality wetlands is a management priority. However, the majority of high quality wetland habitats, those uninfluenced by extrinsic disturbances, are already preserved in parks and refuges. Many other wetland habitats offer less than optimal habitat. For the latter, application of management techniques can increase productivity.

MANAGEMENT APPROACHES

Passive Versus Active Management

Management of wildlife ranges from passive approaches exemplified by preservation of self-regulating habitat, to semiactive approaches such as the installation of nest boxes, to active approaches such as impoundments which require periodic water, soil and vegetation manipulation. As management schemes become more active, and intrinsically more complex, monetary and labor costs increase, and the chances for sustainability and success decrease.

To many, purchase of existing habitats is the only feasible way of protecting unique areas for bird nesting or migration stopovers (Weller 1987). This approach is illustrated by the actions of the National Audubon Society, Ducks Unlimited, The Nature Conservancy and the U. S. Fish and Wildlife Service. The latter group has been purchasing Waterfowl Production Areas since the 1960's in an effort to maintain continental waterfowl populations.

It is vainglorious to expect that managers can improve on the complex dynamic processes of natural undisturbed wetlands.   Active management will by necessity enhance habitat for some species while degrading habitat for other species. Management may fail because of inadequate or inaccurate information, imprecise water control, colonization and modification by nuisance species, or even political or public pressure to terminate or modify management techniques or goals (Fredrickson 1985). Therefore, it seems reasonable to reserve active management for wetlands known to be degraded and created wetlands.

Single Species Versus Diversity

Historically, wildlife management overwhelmingly emphasized waterfowl, and other species were managed incidentally if at all (Figure 1, Payne 1992). In large part, management was applied to game species (Graul and Miller 1984). Single species (or in some cases guild) management typically included prioritization of wildlife species, determining the requirements of these species, obtaining information on local environmental conditions, and determining the wildlife value and growth requirements of local plants (Chabreck 1976).

The 1970s gave rise to regulations which required the management of wildlife for diversity, as well as to an increased public interest in managing species other than waterfowl (Rundle and Fredrickson 1981). Both consumptive and nonconsumptive species were to be preserved (Odom 1977, Martin 1979). The Colorado Nongame Act of 1973 required that all native species be perpetuated, and the 1976 National Forest Management Act required the maintenance of animal diversity. Managers recognized early that the single species approach was inadequate for ensuring the maintenance of diversity, and yet it was impossible to manage for all species (Wagner 1977).

Graul and Miller (1984) reviewed several approaches to managing for diversity. The management indicator approach is intended to benefit a featured species, typically game but sometimes threatened or endangered species or a species of public interest (Gould 1977). The relationship between the featured species and other species must be understood to ensure maintenance of diversity. The ecological indicator approach manages for stenotopic species and assumes that eurytopic species will have their requirements satisfied indirectly (Graul et al. 1976). The habitat diversity approach manages vegetation stand type and age class rather than individual wildlife species (Siderits and Radtke 1977). The success of the approach is sensitive to the size of habitat blocks set aside. Finally, the special features approach emphasizes habitat features such as snags, edges, perches, etc. (Graul 1980). None of these approaches have been tested in any rigorous manner which would permit determination of their effectiveness. Another approach applicable to active management, and also untested, is to mimic the soil, hydrology or vegetation of a natural, undisturbed wetland which has the desired species.   The previous chapter discusses other factors in addition to habitat structure that may affect a wetland's wildlife species composition.

Figure 1. Until recently, wildlife management efforts were focused largely on preservation and creation of waterfowl production areas.

MANAGEMENT TECHNIQUES

There are a number of techniques used to manage wetlands for wildlife (e.g. Weller 1987, Whitman and Meredith 1987, Payne 1992). The majority of these techniques are directed at managing vegetation. Other techniques are directed at providing nonvegetative structural requirements such as feeding opportunities and breeding sites. Selected vegetation management techniques, including burning, grazing, herbicide application, mechanical manipulation, propagation and water level manipulation are discussed herein. Also discussed are artificial breeding and loafing sites.

Vegetation Management

Burning

It is likely that fires were a naturally occurring event in many palustrine emergent wetlands prior to mankind's intervention. Fire functioned to eliminate accumulated plant material, and to return nutrients to the soil. Burning is widely used for marsh management, particularly in Gulf Coast marshes (Hoffpauer 1968, Chabreck 1976, Wright and Bailey 1982). Wetland wildlife managers use fire to promote the growth of green shoots, roots and rhizomes of grasses and sedges which are then available to foraging geese. Fallen seeds are exposed to ducks. Dead plant material is eliminated which increases the value of the habitat to ducks, muskrats (*Ondatra zibethica*) and nutria (*Myocastor coypus*). Burning creates deep pools and edge for nesting and feeding waterfowl, and controls or eliminates undesirable vegetation.

There are three types of burns (Payne 1992). Cover (surface, wet) burns are conducted when the water level is at or above the root horizon and are used to convert monotypic stands of reed (*Phragmites communis*), cattail *(Typha spp.)* or unproductive sedge to plants that provide food and cover to nutria, muskrat, duck and geese. This is accomplished by releasing plants with an earlier growing season than the undesirable plant species. Root burns are used to control or eliminate climax vegetation or other plants of low wildlife value. Hotter than cover burns, root burns are conducted when the soil is dry to a depth of 8-15 cm. The roots of undesirable plants are burned, and more desirable plants, which have roots extending to greater depths, are spared. Peat burns are conducted during droughts in an effort to convert marsh into aquatic habitat. The fire burns a hole in the peat, which then fills with water. Peat burns are more common in freshwater marshes where there is sufficient organic material in the soil to support the fire than in coastal marshes, .

Timing of the burn depends upon the type of burn (cover, root or peat) and the intended objective. Patchy late summer cover burns expose insects to migrating shorebirds (Bradbury 1938). Root and peat burns at this time of year can be used to eliminate reed and cattail, and are most effective if reflooding can be accomplished (Mallik and Wein 1986). Late summer or early fall cover burning will decrease muskrat populations by decreasing the availability of den-building material (Daiber 1986). Elimination of undesirable woody vegetation is also accomplished in late summer or early fall through a root burn (Linde 1985). Patchy winter cover burns increase edge and access for waterfowl nesting the following spring, and provide for control of reed and cattail (Ward 1942, Beule 1979). Olney threesquare (*Scirpus olneyi*), American bulrush (*Scirpus americanus*), and saltmarsh bulrush (*Scirpus robustus*) benefit by late winter cover burning and reflooding of saltmarsh cordgrass (*Spartina patens*). Spring cover burns will increase muskrat populations by stimulating the production of succulent shoots (Daiber 1986).

Fires are difficult to direct and extinguish. This is particularly true in bog wetlands, where fires may burn for days or weeks (Payne 1992). A means for extinguishing the fire, either an auxiliary water supply which can be sprayed on the marsh, or a means for reflooding, should be provided. Burns should not be conducted during the waterfowl breeding season as ducklings are particularly susceptible to fast fires through dead vegetation. Nor should burns be conducted in areas with a high erosion potential, or in drought years unless a root or peat burn is intended.

Burning has some short-term adverse impacts on wetlands wildlife. Inevitably, cover is reduced, forcing ducks to concentrate in unburned areas, which increases their susceptibility to predation. Winter cover is reduced, which has an ancillary effect of reducing the wetland's ability to trap and retain snowfall. This latter effect can be significant in precipitation deficit regions (Ward 1968). Burning also results in a short-term reduction in the insect prey base (Opler 1981). These short-term impacts are overshadowed by the long-term benefits to wildlife.

Grazing

Weller (1987) has suggested that bison grazing on northern prairies may have benefited certain wildlife species by opening up dense stands of vegetation (Figure 2). Grazing in wetlands arrests plant community succession, and tends to reduce undesirable perennials and increase annuals (Chabreck et al. 1989). Grazing animals may create openings in dense vegetation bordering riparian areas (Krueger and Anderson 1985). In uplands bordering wetlands, grazing reduces cover for predators and fuel for fires, and inhibits grassland invasion by brush.

Today, cattle are the primary agent for habitat management through grazing, although sheep, horses and even muskrat can be effective. Sheep are more easily controlled than cattle, and tend to be more effective at removing undesirable plants through their close grazing (Ermacoff 1968). Horses are better at controlling woody vegetation (Pederson et al. 1989). Muskrats are very effective at reducing persistent emergent cover, but are very difficult to control (Weller 1987).

Waterfowl are the primary beneficiaries of controlled grazing, although shorebirds and furbearers may occasionally benefit. Openings in otherwise impenetrable cover allow access by nesting ducks. Migrating and wintering waterfowl, especially snow geese (*Chen caerulescens*), forage on exposed seeds, sprouts stimulated by grazing, and roots and rhizomes exposed by hoove impacts (Chabreck 1968, Daiber 1986). Snipe (*Gallinago gallinago*) concentrate in overgrazed marshes, and upland sandpiper (*Bartramia longicauda*) are attracted by shorter cover (Chabreck 1976, Weller 1987). O'Neil (1949) has suggested that marshes managed for furbearers be subjected to grazing.

Grazing intended to improve waterfowl nesting habitat should occur in the winter or early spring (Rutherford and Snyder 1983), and no more than 50 percent of the forage plants should be removed annually (Payne 1992). To increase the value of marsh habitat for migratory and wintering waterfowl, grazing may continue through spring and early summer. Cattle (or other grazing agent) should

Figure 2. Controlled grazing may benefit certain wildlife species by opening up dense vegetation and increasing annuals. Nevertheless, grazing, especially overgrazing, can reduce habitat value.

be excluded July through September to promote the growth and seed production of annual grasses and sedges (Chabreck 1976, Chabreck et al. 1989). The marsh can be flooded in fall and winter to attract ducks, or left unflooded to attract geese.

There are drawbacks to the management of wetlands through grazing in that management for one guild, waterfowl, inevitably reduces habitat value for other species such as small birds and mammals. Cattle trample dens and cause underground tunnels to collapse. Domestic grazers also compete with wildlife herbivores for food. Overgrazing has more serious consequences in that it negatively impacts habitat for all wildlife species. Ducks and other waterfowl will not nest or feed in exposed areas, and overgrazing of riparian areas can destroy the streambank, increase erosion and decrease plant vigor.

Herbicide Application

Application of herbicides should in most instances be used as a last resort. Herbicides exhibit lethal and sublethal effects on plants and animals, a problem exacerbated by the difficulty needed to apply and control their application.

Therefore, application is likely to affect nontarget species, including desired species. Furthermore, some herbicides will persist in the soil and could potentially contaminate surface water and ground water. In aquatic environments, treatment of large areas can result in oxygen depletion from decaying plants.

To minimize the potential negative impacts of herbicide application, herbicides should only be used where other methods are ineffective, treat the smallest area possible, apply the herbicide when hazards to wildlife are lowest and follow the manufacturers instructions. The safest herbicides are organophospates and carbamates, as they persist in the environment for relatively short periods compared with organochlorine herbicides.

Despite the drawbacks of herbicide use, they can be effective in eliminating or reducing dense emergent vegetation. Herbicides have successfully been used to create open water areas in dense marsh vegetation, thus increasing habitat value for waterfowl (Weller 1987, Payne 1992). Water hyacinth (*Eichornia crassipes*) has for the most part been impossible to control without the use of herbicides. Extensive efforts have been undertaken along the eastern seaboard to rid coastal marshes of common reed, and to encourage revegetation with native species (Jones and Lehman 1987). Application of herbicide to reed is generally required in two successive years, with burning off of dead reed following treatment.

Annuals, biennials and perennials are most easily controlled at the seedling stage. Summer annuals are best treated in spring, and winter annuals in fall, during germination. Biennials are most effectively treated as rosettes in the fall. The vegetative stage of annuals and perennials are moderately to poorly controlled with herbicides, and control during seed set is largely ineffective (Hansen et al. 1984). Aquatic vegetation is best treated when the foliage is above the water and is rain or dew free (Gangstad 1986). Herbicides injected into the ground are most effective when applied in the spring.

Cattail and common reed are treated somewhat differently. Existing growth should be cut so that herbicide can be applied to seedlings. If cutting is not feasible, an herbicide that will be translocated from the leaves to the rhizome should be applied to the foliage in the fall immediately prior to senescence.

For the most part, woody perennials can be effectively treated at any time of the year. Herbicide can be applied to cuts or notches in the trunk or to stumps after cutting. Herbicides applied to the canopy should occur after the leaves have fully formed, but prior to the development of a heavy cuticle.

Mechanical Management

Mechanical methods of management include blasting, bulldozing, crushing, cutting, disking, draglining and dredging. They are applied to decrease the density of existing vegetation, or to ensure the maintenance of open areas in newly created or restored habitat. This is achieved by scraping vegetation, deepening basins, breaking up sod-forming grasses or organic deposits, and retarding woody growth. The result is the creation of open water areas for feeding and breeding waterfowl in deeper marshes, and for ground-dwelling wildlife in shallow marsh and wet meadow habitat (Linde 1985, Weller 1987, Payne 1992). Also, mechanical management is used to prepare newly constructed wetlands or dewatered wetlands for planting. Prior to management through mechanical methods, an evaluation should be conducted which compares long-term expected benefits to short-term impacts to plant and animal communities.

Blasting

Blasting is a relatively inexpensive method for creating openings in dense, emergent vegetation. It is best applied in mineral soils, when the water level is at the soil surface to 20 cm below the surface (Hopper 1971). Historically, dynamite was used to blast openings, but beginning in the 1960s ammonium nitrate fertilizer mixed with fuel oil has been used. The latter is approximately 1/10th the cost of dynamite and much safer to handle.

Basins created by blasting are generally bowl or cone-shaped with steep sides. Although blasting is the least expensive way to create open water areas and reduce the density of vegetation, there is very little control over the shape and size of the areas created. Sloughing of the sides will likely occur over the first year or two following blasting, and basins in organic soils may refill requiring a repeat of the process.

Bulldozing, Draglining and Dredging

Bulldozing, draglining and dredging are used to open up dense marsh by scraping or basin deepening (Figure 3). To prevent regrowth, basins must be excavated below the photic zone. The methods can be applied to dewatered impoundments, or to newly constructed wetlands that have not been flooded, to the benefit of waterfowl and muskrat (Linde 1985, Weller 1987). The abrupt edges and absence of vegetation which characterize newly constructed basins cause them to appear unnatural for a few years, and will likely be less attractive to water birds (Provost 1948, Strohmeyer and Fredrickson 1967, Weller 1987). Bulldozing is the more economical of the methods, and allows the most accurate contouring. It can be effective in dry to moist mineral soils. Draglines are used in wetter situations, and are more difficult to control than a bulldozer. The dragline can be operated from upland, on mats placed on the wetland, or on a frozen surface during winter in temperate regions. Dredging is the most costly of the methods and is generally impractical except for very large projects.

Crushing

Crushing is an effective method for creating openings in dense, persistent emergent vegetation such as cattail and common reed. It is most effective when treatment takes advantage of the plant's natural carbohydrate translocation cycle (Beule 1979). Carbohydrate reserves stored in the rhizome are used in spring to

Figure 3. Bulldozers can be used to open up dense vegetation or create new basins.

produce new leaves. When fully formed, the new leaves will replace the drawn-down reserves by translocating the products of photosynthesis. Crushing of the new leaves when they are nearly fully formed interrupts the cycle and rhizome reserves are depleted. Regrowth will be stunted or precluded. Successive annual treatments are generally sufficient to permit establishment of more desirable plants species from the seed bank, or from planting or seeding.

Cutting

Plants vary in their response to cutting. For example, cattail abundance is generally decreased by cutting, especially successive annual cutting. By contrast, marsh smartweed (*Polygonum coccineum*) appears to tolerate, and perhaps even to be stimulated by cutting (Pederson et al. 1989).

Cutting is effective because rhizomes are dependent upon oxygen, which is supplied to the rhizomes by way of aerenchyma in the leaves. If the leaves are cut, and the substrate remains inundated, oxygen flow to the rhizomes is cut off and the plant will die (Laing 1941). An effective strategy is to cut the leaves as close to the ground surface as possible in the fall or winter. In colder climates, the leaves can be on ice. Cutting success can be improved by inundating the stubble to a depth of 10-20 cm, and repeating the operation for 2 or 3 consecutive years.

Disking

Disking is used to break up sod-forming grasses too dense to be used by ground-dwelling wildlife. Also, disking helps to prevent growth of woody vegetation, retards filling in of wetlands by breaking up organic deposits, prepares new marsh for planting by breaking up sod and peat, and aerates dewatered soils thereby stimulating plant growth upon reflooding (Griffith 1948, Burger 1973). The technique is most effective if repeated every 3-4 years. Disking is largely ineffective for control of cattail and other plants with rhizomes due to resprouting (Payne 1992).

Managers of marshes in the southeastern United States use disking to stimulate the growth of Asiatic dayflower (*Aneilema keisak*), which is a valuable duck food. Disking occurs in February or March. Fall disking creates habitat for common snipe (*Gallinago gallinago*).

Propagation

Propagation of wetland plants was relatively common in the 1940's, and was used primarily to establish emergents and marsh edge species (Linduska 1964). The technique has experienced a revival of sorts with the advent of wetlands mitigation. However, propagation is a relatively expensive operation, and failures are still common. In most instances, the natural seed bank will be more than adequate provided water levels are appropriate (Ermacoff 1968, Weller 1987).

Nevertheless, propagation can be an effective strategy (Figure 4). For example, three-cornered rush (*Scirpus olneyi*) can be successfully established in southeastern salt marsh, where it enhances habitat for snow geese and muskrat (Ross and Chabreck 1972). It can also be an effective strategy for small areas, and for newly created areas if invasion by undesirable species is likely (Weller 1990, Payne 1992). Payne and Copes (1986) suggest that propagation is a reasonable strategy if less than 15 percent of an existing wetland is vegetated with desirable perennial species.

Propagules come in many forms including seed, rootstock, rhizome, tuber, cutting, seedling and transplant. Seeding constitutes the least expensive propagation method, and is especially appropriate for large areas. It is also the most convenient method, in that seeds can be stored until needed. Note however that the cost and logistical difficulty of seeding increase dramatically if seeds are collected rather than purchased. Mulch can increase the success of most seeding efforts. The mulch shelters seedlings, helps to retain soil moisture, and reduces soil erosion until such time as the propagules establish soil-holding root systems.

Tubers can also be broadcast, and in some instances can be harvested mechanically. Tubers, roots and rhizomes make maximum use of the plant material, but are susceptible to washout if flooding occurs after propagating (Payne 1992). Cuttings are a relatively low cost option, and can be collected fairly quickly. As with tubers, roots and rhizomes, cuttings are susceptible to washout. Survival is generally lower than that of rooted propagules (Payne 1992).

Figure 4. Although the natural seed bank will generally be more than adequate under appropriate conditions, propagation can be effective for small or newly created wetlands.

Seedlings and transplants are the most expensive propagules, but exhibit generally greater survival than other propagules (Payne 1992). Seedlings generally permit greater flexibility as they can be stored and planted over longer periods of time than transplants. Transplanting should ideally occur during dormancy.

Propagules should be selected on a site-specific basis. Undoubtedly, the wildlife manager will select plants that maximize use of the wetland by target wildlife. For example, summer annuals for ducks, or annual and perennial grasses and legumes for geese. However, care should be taken to ensure that the selected species are appropriate for the intended soil and hydrological conditions. Water quality, especially turbidity and pH, can also be important factors contributing to propagation success. Whenever possible, propagules should originate from a nearby site, or the site itself, as native, locally adapted species are most successful (Coastal Zone Resources Division 1978). Other factors, such as the ability to withstand waves or ice, can be important in certain situations.

Annuals are propagated by seed. Broadcasting is less reliable than row cropping, and requires 50-55 percent more seed (Crawford and Bjugstad 1967). Perennials are propagated by seed or vegetative propagule. The latter should consist of vigorous stock, and care should be taken not to damage roots and the stem during removal, transport and planting. The propagule should be pruned after planting to decrease transpiration, and a support should be provided to prevent wind damage. Large, dense plantings are more resistant to herbivory than small, sparse plantings (Payne 1992).

To ensure that adequate nutrients are available, the propagule can be fertilized as planted, or the surrounding soil can be fertilized just prior to planting. Slow release fertilizer is generally best, and can be placed in or near the planting hole. If placed in the planting hole, the fertilizer should be well below the root ball so as not to burn the roots. As a maintenance measure, fertilizer can be applied during or just before the growing season. Warm season species are best fertilized in the spring or early summer, whereas cool season species are best fertilized at the time of seeding, and again at midwinter (Carpenter and Williams 1972).

Water Level Manipulation

Water depth and hydroperiod determine wetland plant community composition. As such, water level manipulation encourages or discourages particular types and species of plants. Wildlife is attracted to the type of vegetation community which develops in response to water level manipulation, and is directly affected through the creation or elimination of aquatic habitat.

Manipulating water levels in wetlands can be an effective management technique for increasing wildlife productivity (Wilson 1968, Chabreck 1976, Rundle and Fredrickson 1981, Fredrickson and Taylor 1982, Knighton 1985). Although, the degree of substrate drainage, surface relief and substrate composition are equally important in determining vegetation interspersion (Knighton 1985). Water level manipulation can be accomplished relatively inexpensively if a small, simple water control structure can be constructed at the downstream end of a naturally occurring basin. The water control structure should be capable of effecting a complete drawdown, and be able to manipulate water levels with a precision of 5-7 cm. Costs increase dramatically if levee construction is required. Water level manipulation can also be affected by pumping water in or out of an impoundment, although this option is more costly than periodic adjustment of a water control structure. Deep organic soils (>15 cm) generally preclude management through water level manipulation, as reflooding results in floating mats (Knighton 1985).

Manipulation includes both drawdown and flooding (Figure 5). Drawdowns reduce or eliminate undesirable plant species, facilitate decomposition of vegetation and the return of nutrients to the soil, allow desirable plant species to germinate or recover from flood stress, concentrate prey for wildlife, and reduce or eliminate nuisance fish and wildlife (Kadlec 1960, Linde 1969, Weller 1987). Flooding decreases the density of emergent vegetation and increases the production of invertebrates, thereby enhancing the suitability of the wetland for waterfowl breeding and brood rearing.

The response of plants to the manipulation of water levels depends on the timing and extent of drawdown and flooding, and the plant community successional stage. Consequently, wildlife community composition and productivity is dependent on these factors. For example, late summer flooding will freeze-proof

Figure 5. Management for wildlife through water level manipulation includes drawdown and flooding. Drawdown will encourage germination of desirable plants, whereas flooding will increase the production of invertebrates.

emergent marshes thereby increasing muskrat numbers. Conversely, a partial late fall drawdown will expose invertebrates and minnows to migrant ducks. A fall drawdown conducted over several years will reduce or eliminate muskrat, carp (*Cyprinus carpio*) and bullhead (*Ictalurus spp.*).

Management techniques using water level manipulation have been developed largely to benefit waterfowl. Generally, achievement of a 1:1 ratio of emergent vegetation to open water will maximize waterfowl use (Weller and Spatcher 1965, Weller 1975, Bookhout et al. 1989, Pederson et al. 1989). Maintenance of pool level will encourage perennial emergent vegetation suitable for dabbler duck nesting and brood rearing. A partial drawdown in spring will encourage regrowth of perennials if more cover is needed (Weller 1987, Bookhout et al. 1989).

Management for migratory and wintering waterfowl requires periodic drawdown and reflooding. Typically, drawdown occurs in spring or early summer so that soils can drain, thereby stimulating germination and growth of grasses, sedges and other seed-producing plants from the seed bank (Kadlec 1960). The seed bank may be inadequate in saline wetlands and impounded bays (Pederson and Smith 1988). Floating-leaved and submergent vegetation density will be reduced by drawdown, and perennial emergent vegetation growth will be suppressed. Complete drawdown exposes mudflats and encourages dense stands of moist soil plants, primarily nonpersistent annual and biennial emergents, which are prolific seed producers. Seed viability declines if mudflat emergents are continuously flooded for several years (Knighton 1985). Complete drawdown also encourages undesirables such as willow (*Salix spp.*) and purple loosestrife (*Lythrum salicaria*); therefore, periodic inspections will be required to prevent invasion by these species. The wetland is reflooded in fall to make moist soil plant seeds available to waterfowl. Reflooding should be gradual to avoid flotation of emergents, scouring and mortality from turbidity (Weller 1987). Dabbling ducks are benefited by reflooding to a depth of 10-25 cm, whereas diving ducks and coots (*Gallinula chloropus*) benefit from deeper depths (Payne 1992).

Water levels can also be manipulated to manage wetlands for rails and shorebirds (Neely 1959, Griese et al. 1980, Rundle and Fredrickson 1981, Payne 1992). Rails are attracted to wetlands with robust emergent vegetation and water depths less than 50 cm deep, and preferably less than 15 cm deep. When spring rail use is desired, wetlands vegetated with annual grasses and smartweeds must by dewatered over winter to protect vegetation from ice and waterfowl. Fall use can be encouraged by reflooding drawndown wetlands in late summer (Johnson and Dinsmore 1986). Shorebirds are attracted to gradual drawdowns creating extremely shallow water (0-5 cm) interspersed with exposed, saturated soil. Spring flooding or disking can be used to suppress growth of aquatic plants. Rail management will, to some extent, also benefit dabbling ducks, whereas shorebird management will provide some benefit to geese.

Manipulation of water levels in hardwood forests can encourage oaks and other mast producing species to the benefit of ducks, beaver (*Castor canadensis*), woodcock (*Scolopax minor*), mink (*Mustela vison*), squirrel (*Sciurus spp.*), raccoon (*Procyon lotor*) and muskrat. These so-called green tree reservoirs are flooded to a mean depth of approximately 40 cm in fall and drawndown in late winter or early spring prior to the start of the growing season. Complete drawdown is essential to preclude development of undesirable vegetation (Hunter 1978, Vaught and Bowmaster 1983). Waterfowl are especially benefited by the described manipulations, as flooding makes mast available to wintering waterfowl, and the drawdown concentrates invertebrate prey items for migrating waterfowl.

Water level manipulation is a long-term management strategy, and some type of managed disturbance will be required at intervals of 5 years or less (Payne 1992). Drawdowns for waterfowl should reasonably be accomplished every 2-4 years in marshes. Moreover, the cyclic nature of water level manipulation management, variation inherent in the seed bank, and seed bank response to edaphic conditions will likely result in differing marsh vegetation communities from year to year. In green tree reservoirs, the soil should be allowed to dry out every few years so as to discourage the establishment of vegetation characteristic of wetter habitats. Flooding should be withheld for a 2-3 year period after acorn production to encourage seedling establishment. Annual differences in wildlife community composition and use can be expected.

Artificial Nesting and Loafing Sites

Development of upland areas adjacent to wetlands and overharvesting of trees have contributed to a decline in waterfowl populations. Artificial nesting and loafing sites can be an effective means for increasing the local habitat carrying capacity for waterfowl when upland nesting sites are limited (Johnson et al. 1978, Lokemoen et al. 1984). Artificial areas increase the shoreline to wetland surface area ratio, thereby increasing the amount of sites available to breeding pairs. Islands provide an increased measure of security from predators, as well as reducing the level of anthropogenic disturbance. Nevertheless, artificial areas are less likely to be accepted by wildlife, and are more likely to be used by upland nesting waterfowl such as Canada goose (*Branta canadensis*) than by marsh edge birds such as American coot (*Fulica americana*) (Weller 1987).

Artificial nesting and loafing sites should be located in areas frequented by waterfowl, but where natural nesting and loafing sites are limited. Depending upon local predators and the nature of proximal disturbance, islands should be located 9-170 m from the mainland, and closer to leeward than to the windward side of the mainland (Jones 1975, Giroux 1981, Ohlsson et al. 1982). Water depth around the island should be 0.5-0.75 m (Hammond and Mann 1956). Islands should be 0.5-5 ha in size, as smaller areas are too small to support a breeding pair, and larger areas might support predators ((Duebbert 1982, Higgins 1986). Long, narrow, rectangular islands will maximize the number of breeding pairs. A low profile will be less attractive to predators, although the area should be high enough to prevent flooding of ground nesters (Hoffman 1988). Groundcover should be fairly dense (> 50 percent), and consist of grass, legumes and forbs. Coots and grebes like vegetation extending into the water (Swift 1982). Dense grass will encourage nesting by gulls, whereas woody shrubs will provide habitat for herons (Soots and Parnell 1975). Trees should be avoided as they provide perches for raptors and crows.

Fisheries

Managing wetlands for wildlife while at the same time maintaining a fisheries is difficult, and in many instances the two are mutually exclusive. The typical freshwater wetland managed for wildlife is subject to variations in water level, whereas management for fisheries requires a relatively stable water level. Water level fluctuation increases turbidity, and periodically increases water temperature and decreases dissolved oxygen levels. Freezing to the bottom is also likely to occur in shallow impoundments. As such, only those species tolerant of fluctuating and relatively extreme conditions, such as carp and bullhead, will persist, and then only if deeper areas which can serve as refugia are provided. Carp and bullhead are typically discouraged in managed freshwater wetlands, as they dislodge vegetation and increase turbidity.

Wetlands connected to deepwater habitats provide a better opportunity for reconciling the conflict between wildlife and fisheries. Diked wetlands on Lake Erie, which are managed for waterfowl, allow for fish movement through water control pipes (Petering and Johnson 1991). Relatively many species have been recorded in the wetlands (Johnson 1989), but for the most part these species are tolerant of extreme conditions and do not use the diked wetlands for spawning.

The effectiveness of the diked wetlands for fisheries is limited by the number of access points, and the provision of sufficiently deep water only during the fall waterfowl migration period.

Limited access and lowered water levels also limit the fisheries value of impounded coastal wetlands. Waterfowl management requires the closure of impoundments during peak fish recruitment periods. As such, marsh nursery use by estuarine dependent saltwater species decreases, whereas use by freshwater species increases (Herke et al. 1987, Batzer and Resh 1992, Herke et al. 1992). Nevertheless, use of impounded coastal wetlands by marine transient species can be increased by providing deepwater refugia such as perimeter ditches, and by reducing impediments to movement between the impoundment and deepwater areas. The latter can be accomplished by increasing the number of water control structures, and by using water control structures such as variable crest weirs or slotted weirs which permit fish movement.

REFERENCES

Batzer, D.P. and V.H. Resh, 1992. Macroinvertebrates of a California seasonal wetland and responses to experimental manipulation. Wetlands 12(1):1-7.

Beule, J.D. 1979. Control and management of cattails in southeastern Wisconsin wetlands. Wisconsin Department Natural Resources Technical Bulletin 112. 40 pp.

Bookhout, T.A., K.E. Bednarik and R.W. Kroll. 1989. The Great Lakes marshes. Pp. 131-156 In Smith, L.M., R.L. Pederson and R.M. Kaminski (eds.) Habitat Management for Migrating and Wintering Waterfowl in North America. Texas Technical University Press, Lubbock, Texas.

Bradbury, H.M. 1938. Mosquito control operations on tide marshes in Massachusetts and their effect on shore birds and waterfowl. Journal Wildlife Management 2:49-52.

Burger, G.V. 1973. Practical Wildlife Management. Winchester Press, New York. 218 pp.

Carpenter, L.H. and G.L. Williams. 1972. A literature review on the role of mineral fertilizers in big game range management. Colorado Game, Fish and Parks Department Special Report 28. 25 pp.

Chabreck, R.H. 1968. Weirs, plugs and artificial potholes for the management of wildlife in coastal marshes. Proceedings Marsh Estuary Management Symposium 1:178-192.

Chabreck, R.H. 1979. Wildlife harvest in wetlands of the United States. Pp. 618-631 In Greeson, P. E., J. R. Clark and J. E. Clark (eds.) Wetland Functions and Values: The State of Our Understanding. American Water Resources Association, Minneapolis, Minnesota.

Chabreck, R.H. 1976. Management of wetlands for wildlife habitat improvement. Pp. 226-233 In Wiley, M. (ed.) Estuarine Processes Volume I. Academic Press, New York.

Chabreck, R.H. 1979. Wildlife harvest in wetlands of the United States. Pp. 618-631 In Greeson, P.E., J.R. Clark and J.E. Clark (eds.) Wetland Functions and Values: The State of Our Understanding. American Water Resources Association, Minneapolis, Minnesota.

Chabreck, R.H., T. Joanen and S.L. Paulus. 1989. Southern coastal marshes and lakes. Pp. 249-277 In Smith, L.M., R.L. Pederson and R.M. Kaminski (eds.) Habitat Management for Migrating and Wintering Waterfowl in North America. Texas Technical University Press, Lubbock, Texas.

Coastal Zone Resources Division. 1978. Handbook for Terrestrial Wildlife Habitat Development on Dredged Material. U. S. Army Corps of Engineers Technical Report D-78-37. 388 pp.

Cowardin, L.M. and W.R. Goforth. 1985. Summary and research needs. Pp. 135-136 In Water Impoundments for Wildlife: A Habitat Management Workshop. General Technical Report NC-100, United States Department of Agriculture, Forest Service, North Central Forest Experimental Station.

Crawford, H.S., Jr. and A.J. Bjugstad. 1967. Establishing grass range in the southwest Missouri Ozards. U.S. Forest Service Research Note NC-22. 4 pp.

Daiber, F.C. 1986. Conservation of tidal marshes. Van Nostrand Reinhold, New York. 341 pp.

Duebbert, H.F. 1982. Nesting of waterfowl on islands in Lake Audubon, North Dakota. Wildlife Society Bulletin 10:232-237.

Ermacoff, N. 1968. Marsh and habitat management practices at the Mendota Wildlife Area. California Department Fish Game Leaflet 12. 10 pp.

Fredrickson, L.H. 1885. Managed wetland habitats for wildlife: Why are they important? Pp. 1-8 In Water Impoundments for Wildlife: A Habitat Management Workshop. General Technical Report NC-100, United States Department of Agriculture, Forest Service, North Central Forest Experimental Station.

Fredrickson, L.H. and T.S. Taylor. 1982. Management of seasonally flooded impoundments for wildlife. U. S. Fish and Wildlife Service Resource Publication 148.

Gagliano, S.M. 1981. Special report on marsh deterioration and land loss in the deltaic plain of coastal Louisiana. Coastal Environments, Inc., Baton Rouge, Louisiana. 14 pp.

Gangstad, E.O. 1986. Freshwater Vegetation Management. Thomas Publishers, Fresno, California. 380 pp.

Giroux, J. 1981. Use of artificial islands by nesting waterfowl in southeastern Alberta. Journal Wildlife Management 45:669-679.

Givens, L.S., M.C. Nelson and V. Ekedahl. 1964. Farming for waterfowl. Pp. 599-610 In J.P. Linduska (ed.) Waterfowl Tomorrow. U.S. Fish and Wildlife Service, Washington, D. C.

Gould, N.E. 1977. Featured species planning for wildlife on southern national forests. Transactions North American Wildlife and Natural Resources Conference 42:435-437.

Graul, W.D. 1980. Grassland management practices and bird communities. Pp. 38-47 In DeGraaf, R.M and N.G. Tilghman (compilers) Workshop proceedings; management of western forests and grasslands for nongame birds. U. S. Department Agriculture, Forest Service General Technical Report INT-86, Intermountain Forest and Range Experiment Station, Ogden, Utah.

Graul, W.D. and G.C. Miller. 1984. Strengthening ecosystem management approaches. Wildlife Society Bulletin 12:282-289.

Graul, W.D., J. Torres and R. Denney. 1976. A species-ecosystem approach for nongame programs. Wildlife Society Bulletin 4:79-80.

Griese, H.J., R.A. Ryder and C.E. Braun. 1980. Spatial and temporal distribution of rails in Colorado. Wilson Bulletin 92:96-102.

Griffith, R. 1948. Improving waterfowl habitat. Transactions North American Waterfowl Conference 13:609-617.

Hammond, M.C. and G.E. Mann. 1956. Waterfowl nesting islands. Journal Wildlife Management 20:345-352.

Hansen, G.W., F.E. Oliver and N.E. Otto. 1984. Herbicide Manual. U.S. Bureau Reclamation, Denver, Colorado. 346 pp.

Herke, W.H., E.E. Knudson and P.A. Knudsen. 1987. Effects of Semi-Impoundment on Estuarine-Dependent Fisheries. Pp. 403-423 In Whitman, W. R. and W. H. Meredith (eds.) Waterfowl and Wetlands Symposium: Proceedings of a Symposium on Waterfowl and Wetlands Management in the Coastal Zone of the Atlantic Flyway. Delaware Coastal Management Program, Delaware Department of Natural Resources and Environmental Control, Dover, Delaware. 522 pp.

Herke, W.H., E.E. Knudsen, P.A. Knudsen and B.D. Rogers, 1992. Effects of semi-impoundment of Louisiana on fish and crustacean nursery use and export. North American Journal of Fisheries Management. 12:151-160.

Higgins, K.F. 1986. Further evaluation of duck nesting on small manmade islands in North Dakota. Wildlife Society Bulletin 14:155-157.

Hoffman, R.D. 1988. Ducks Unlimited's United States construction program for enhancing waterfowl production. Pp. 109-113 In Zelazny, J. and J.S. Feierabend (eds.) Proceedings of a conference on Increasing our Wetland Resources. National Wildlife Federation, Washington, D.C.

Hoffpauer, C.M. 1968. Burning for coastal marsh management. Proceedings Marsh Estuary Management Symposium 1:134-139.

Hopper, R.M. 1971. Use of ammonium nitrate-fuel oil mixtures in blasting potholes for wildlife. Colorado Department Natural Resources Game Information Leaflet 85. 4 pp.

Hunter, C.G. 1978. Managing green tree reservoirs for waterfowl. International Waterfowl Symposium 3:217-223.

Johnson, D.L. 1989. Lake Erie wetlands: Fisheries considerations. Pp. 257-273 In Krieger, K.A. (ed.) Lake Erie and its Estuarine Systems: Issues, Resources, Status and Management. NOAA Estuary of the Month Seminar Series 14, 4 May, 1988. Washington, D.C.

Johnson, R.F., R.O. Woodward and L.M. Kirsch. 1978. Waterfowl nesting on small manmade islands in the prairie wetlands. Wildlife Society Bulletin 6:240-243.

Johnson, R.R. and J.J. Dinsmore. 1986. Habitat use by breeding Virginia rails and soras. Journal Wildlife Management 50:387-392.

Jones, J.D. 1975. Waterfowl nesting island development. U.S. Bureau Land Management Technical Note 260. 17 pp.

Jones, W.L. and W.C. Lehman. 1987. Phragmites control and revegetation following aerial applications of glyphosate in Delaware. Pp. 185-196 In Whitman, W.R. and W.H. Meredith (eds.) Waterfowl and Wetlands Symposium: Proceedings of a Symposium on Waterfowl and Wetlands Management in the Coastal Zone of the Atlantic Flyway. Delaware Coastal Management Program, Delaware Department of Natural Resources and Environmental Control, Dover, Delaware. 552 pp.

Kadlec, J.A. 1960. The effect of drawdown on the ecology of a waterfowl impoundment. Michigan Department Conservation Report 2276. 181 pp.

Knighton, M.D. 1985. Vegetation management in water impoundments: Water level control. Pp. 39-50 In Knighton, M.D. (ed.) Water Impoundments for Wildlife: A Habitat Management Workshop. General Technical Report NC-100. St. Paul, Minnesota: U.S. Department of Agriculture, Forest Service, North Central Forest Experiment Station. 136 pp.

Korte, P.A. and L.H. Fredrickson. 1977. Loss of Missouri's lowland hardwood ecosystem. Transactions North American Wildlife Natural Resources Conference 42:31-41.

Krueger, H.O. and S.H. Anderson. 1985. The use of cattle as a management tool for wildlife in shrub-willow riparian systems. Pp 300-304 In Johnson, R.R., C.D. Ziebell, D.R. Patton, P.F. Folliott and R.H. Hamre (eds.) Riparian Ecosystems and Their Management: Reconciling Conflicting Uses. U.S. Forest Service General Technical Report RM-120.

Laing, H.E. 1941. Effect of concentration of oxygen and pressure or water upon growth of rhizomes of semisubmerged water plants. Botanical Gazette 102:712-724.

Leopold, A. 1933. Game Management. Charles Scribner's Sons, New York. 481 pp.

Linde, A.F. 1969. Techniques for wetland management. Wisconsin Department of Natural Resources Report 45. 156 pp.

Linde, A.F. 1985. Vegetation management in water impoundments: Alternatives and supplements to water-level control. Pp. 51-60 In M.D. Knighton (ed.) Water Impoundments for Wildlife: A Habitat Management Workshop. U.S. Forest Service General Technical Report NC-100.

Linduska, J.P. (ed.). 1964. Waterfowl Tomorrow. Fish and Wildlife Service, Washington, D.C.

Lokemoen, J.T., F.B. Lee, H.F. Duebbert and G.A. Swanson. 1984. Aquatic habitats-waterfowl. Pp. 161B-176B In Henderson, F.R. (ed.) Guidelines for increasing wildlife on farms and ranches. Kansas State University Cooperative Extension Service, Manhattan, Kansas.

Mallik, A.V. and R.W. Wein. 1986. Response of a *Typha* marsh community to draining, flooding and seasonal burning. Canadian Journal Botany 64:2136-2143.

Martin, E.M.   1979.   Hunting and harvest trends for migratory game birds other than waterfowl: 1964-1976.   U. S. Fish and Wildlife Service Special Scientific Report - Wildlife 218.   37 pp.

Melinchuk, R. and R. MacKay.   1986.   Prairie Pothole Project: Phase 1 Final Report.   Wildlife Technical Report 86-1, Wildlife Branch, Saskatchewan Parks and Renewable Resources, Regina, Saskatchewan.   32 pp.

Neely, W.W.   1959.   Snipe field management in the southeastern states. Proceedings Annual Conference Southeast Association Game Fish Commission 14:30-34.

Odom, R.R.   1977.   Sora (*Porzana carolina*).   Pp. 57-65 In G.C. Sanderson (ed.) Management of migratory shore and upland game birds in North America. International Association Fish and Wildlife Agencies, Washington, D.C.

Ohlsson, K.E., A.E. Robb, Jr., C.E. Guindon, Jr., D.E. Samuel and R.L. Smith. 1982.   Best current practices for fish and wildlife on surface-mined land in the northern Appalachian coal region. U.S. Fish Wildlife Service FWS/OBS-81/45.   305 pp.

O'Neil, T.   1949.   The muskrat in the Louisiana coastal marshes.   Louisiana Department Wildlife and Fisheries, New Orleans.   152 pp.

Opler, P.A.   1981. Management of prairie habitats for insect conservation. Journal Natural Areas Association 1:3-6.

Pandit, A.K. and D.N. Fotedar.   1982.   Restoring damaged wetlands for wildlife. Journal of Environmental Management 14:359-368.

Payne, N.F.   1992.   Techniques for Wildlife Habitat Management of Wetlands. McGraw-Hill, Inc., New York.   549 pp.

Payne, N.F. and F. Copes.   1986.   Wildlife and Fisheries Habitat Improvement Handbook.   U.S. Forest Service, Washington, D.C.   402 pp.

Pederson, R.L., D.G. Jorde and S.G. Simpson.   1989.   Northern Great Plains.   Pp. 281-310 In Smith, L.M., R.L. Pederson and R.M. Kaminski (eds.) Habitat Management for Migrating and Wintering Waterfowl in North America. Texas Tech University Press, Lubbock, Texas.

Pederson, R.L. and L.M. Smith.   1988.   Implications of wetland seed bank research: A review of Great Basin and prairie marsh studies. Pp. 81-98 In Wilcox, D.A. (ed.) Interdisciplinary Approaches to Freshwater Research. Michigan State University Press, East Lansing.

Petering, R.W. and D.L. Johnson.   1991.   Distribution of fish larvae among artificial vegetation in a diked Lake Erie wetland. Wetlands 11(1):123-138.

Peters, D.S., D.W. Ahrenhoz and T.R. Rice.   1979.   Harvest and value of wetland associated fish and shellfish.   Pp. 606-617 In Greeson, P.E., J.R. Clark and J.E. Clark (eds.) Wetland Functions and Values: The State of Our Understanding.   American Water Resources Association, Minneapolis, Minnesota.

Provost, M.W.   1948.   Marsh blasting as a wildlife management technique. Journal Wildlife Management 12:350-387.

Radtke, R.E.   1985. Wetland management: public concern and government action. Pp. 9-14 In Water Impoundments for Wildlife: A Habitat Management

Workshop. General Technical Report NC-100, United States Department of Agriculture, Forest Service, North Central Forest Experimental Station.

Rakstad, D. and J. Probst. 1985. Wildlife occurrence in water impoundments. Pp. 80-94 In Water Impoundments for Wildlife: A Habitat Management Workshop. General Technical Report NC-100, United States Department of Agriculture, Forest Service, North Central Forest Experimental Station.

Ross, W.M. and R.H. Chabreck. 1972. Factors affecting the growth and survival of natural and planted stands of *Scirpus olneyi*. Proceedings 26th Annual Conference of the Southeastern Association of Game and Fish Commissions, New Orleans. 329 pp.

Rundle, W.D. and L.H. Fredrickson. 1981. Managing seasonally flooded impoundments for migrant rails and shorebirds. Wildlife Society Bulletin 9(2):80-87.

Rutherford, W.H. and W.D. Snyder. 1983. Guidelines for habitat modification to benefit wildlife. Colorado Division Wildlife, Denver. 194 pp.

Siderits, K. and R.E. Radtke. 1977. Enhancing forest wildlife habitat through diversity. Transactions North American Wildlife and Natural Resources Conference 42:425-434.

Soots, R.F., Jr. and J.F. Parnell. 1975. Introduction to the nature of dredge islands and their wildlife in North Carolina and recommendations for management. Pp. 1-34 In Parnell, J.F. and R.F. Soots (eds.) Proceedings of a conference on Management of Dredge Islands in North Carolina Estuaries. University of North Carolina Sea Grant Collaborative Program Publication UNC-SG-75-01.

Strohmeyer, D.L. and L.H. Fredrickson. 1967. An evaluation of dynamited potholes in northwest Iowa. Journal Wildlife Management 31:525-532.

Swift, J.A. 1982. Construction of rafts and islands. Pp. 200-203 In Scott, D.A. (ed.) Managing Wetlands and Their Birds. Proceedings 3rd Technical Meeting on Western Palearctic Migratory Bird Management, International Waterfowl Research Bureau, Slimbridge, Gloucester, England.

U.S. Department of Agriculture. 1980. Soil and Water Resources Conservation Act (Appraisal). Washington, D.C.

Vaught, R. and J. Bowmaster. 1983. Missouri Wetlands and Their Management. Conservation Commission of the State of Missouri, Jefferson City, Missouri. 23 pp.

Wagner, F.H. 1977. Species vs. ecosystem management: concepts and practices. Transactions North American Wildlife and Natural Resources Conference 42: 14-24.

Ward, E. 1942. Phragmites management. Transactions North American Wildlife Conference 7:294-298.

Ward, P. 1968. Fire in relation to waterfowl habitat of the Delta Marshes. Proceedings Annual Tall Timbers Fire Ecology Conference 8:255-268.

Weller, M.W. 1975. Studies of cattail in relation to management for marsh wildlife. Iowa State Journal Science 49:383-412.

Weller, M.W. 1981. Estimating wildlife and wetland losses due to drainage and other perturbations. Pp. 337-346 In Richardson, B. (ed.) Selected Proceedings of the Midwest Conference on Wetland Values and Management. Minnesota

Water Planning Board, Water Resources Research Center, University of Minnesota, Upper Mississippi River Basin Commission and Great Lakes Basin Commission.

Weller, M.W. 1987. Freshwater Marshes: Ecology and Wildlife Management (2nd ed.). University of Minnesota Press, Minneapolis, Minnesota. 150 pp.

Weller, M.W. 1990. Waterfowl management techniques for wetland enhancement, restoration, and creation useful in mitigation procedures. Pp. 517-528 In J. A. Kusler and M.E. Kentula (eds.) Wetland Creation and Restoration: The Status of the Science. Island Press, Washington, DC.

Weller, M.W. and C.E. Spatcher. 1965. Role of habitat in the distribution and abundance of marsh birds. Iowa State University Agriculture and Home Economics Experiment Station Special Report 43. 31 pp.

Whitman, W.R. and W.H. Meredith. 1987. Introduction. Pp. 5-10 In Whitman, W.R. and W.H. Meredith (eds.) Waterfowl and Wetlands Symposium: Proceedings of a Symposium on Waterfowl and Wetlands Management in the Coastal Zone of the Atlantic Flyway. Delaware Coastal Management Program, Delaware Department of Natural Resources and Environmental Control, Dover, Delaware. 522 pp.

Wilson, K.A. 1968. Fur production on southeastern coastal marshes. Proceedings Marsh Estuary Management Symposium 1:149-162.

Wright, H.A. and A.W. Bailey. 1982. Fire ecology: United States and southern Canada. Wiley Press, New York, New York. 501 pp.

CHAPTER 15

COASTAL MARSH MANAGEMENT

Robert Buchsbaum

HISTORY OF COASTAL MARSH MANAGEMENT

When European settlers first arrived in the northeast they often settled around salt marshes (Nixon 1982). Marshes were valued as a source of food for livestock because there was little open grazing land. Marshes had traditionally been used for grazing sheep and cattle in Europe (Jensen 1985), thus it was not surprising that they would be similarly valued in the New World.

As more and more farm land was cleared for pasture, attitudes toward coastal wetlands changed for the worse. Marshes were at best ignored and at worst were perceived as worthless land that bred mosquitoes and other pestilence. The best use of the coastal wetlands was in being reclaimed, and put to some useful purpose. Up until about the 1970s, the two most widespread management activities in coastal wetlands were outright destruction and mosquito abatement.

Outright Destruction

Coastal wetlands have been filled and degraded to create more land area for homes, industry, and agriculture. Estimates of wetlands lost since colonial times have not distinguished coastal from inland wetlands, so we must rely on figures for all wetlands to get some idea of coastal losses. Dahl (1990) estimated that the United States has lost 30 percent of its original wetlands acreage (53 percent if Alaska and Hawaii are excluded). An estimated 46 percent of the original wetlands area of Florida and Louisiana, the two states with the largest acreage of coastal wetlands (almost 17 million acres combined), have been lost (Watzin and Gosselink 1992). About 90 percent of California's original area of wetlands has been destroyed (Figure 1, Watzin and Gosselink 1992).

Over half our original salt marshes and mangrove forests have been destroyed, much of it between 1950 and the mid-1970s (Watzin and Gosselink 1992). Between the mid-1950s and mid-1970s, the coterminous United States lost an estimated 373,300 acres of vegetated estuarine wetlands, a 7.6 percent loss (Frayer et al. 1983). Such losses and modifications have been particularly acute in San

Figure 1. Salt marsh dominated by pickleweed *(Salicornia virginica)* near Stinson Beach, California. Over 90 percent of California's wetlands, including most of its original coastal marshes, have been destroyed.

Francisco Bay. Most of the bay's tidal marshes have been filled by the activities of gold miners, agriculture, and salt production. Hydrologic changes caused by dams, reservoirs, and canals have reduced the freshwater flow to only about 60 percent of its original volume. Similar activities have occurred in other urban areas. Major airports were built on filled tide lands in New York City, Boston, and New Orleans. The upscale Back Bay section of Boston was once a shallow embayment fringed with salt marshes. Old maps of this city show extensive areas of water that are now dry land. Similarly, the original shoreline of Manhattan was irregular with bays and inlets, a far cry from the present almost linear expanse of piers and highways. Marshland, with its rich, peaty soil was often reclaimed for agriculture in Europe. Both mangrove swamps and salt marshes in Florida have also been destroyed to create waterfront homes and marinas and for the construction of the Intracoastal Waterway (Florida Department of Natural Resources 1992a, b). Over 40 percent of the salt marshes and mangroves in Tampa Bay have been lost since 1940 (Florida Department of Natural Resources 1992a, b). Lake Worth in Palm Beach County has lost 87 percent of its mangroves and 51 percent of its salt marshes.

Estimating the extent of loss of coastal marshes in the United States is tricky because of the lack of historical information, absence of uniformity of methods, and the difficulty of estimating acreage of smaller marshes by aerial photography. The historical destruction of salt marshes is likely to be underestimated for at least two reasons: 1) small marshes most susceptible to filling are ignored in estimates, and 2) new marshes are constantly being created either naturally or through alterations of hydrology by human activity. At Good Harbor Beach in Gloucester, Massachusetts, for example, a salt marsh developed within the past 200 years after a channel was blocked.

Mosquito Control

Mosquito control activities in coastal wetlands have involved both physical alteration of the habitat, to make it less suitable for mosquito breeding (source reduction), and the use of chemical and/or biological agents to directly kill adult and larval mosquitoes. Although the use of pesticides often receives the most public attention, habitat alteration is ultimately of more concern because of its potential to irreversibly alter coastal wetlands.

Biology of Salt Marsh Mosquitoes

Mosquito breeding areas on salt marshes and mangrove forests typically occur at the irregularly flooded upper edges of these habitats (Figure 2). Sites may include spring tides associated with the new and full moons. They also may breed among sporadically inundated tufts of high marsh plants, such as salt marsh hay (*Spartina patens*) in east coast marshes. Eggs of most species, such as *Aedes solicitans*, the most common nuisance mosquito in the northeast, are laid on the surface of a marsh typically in shallow depressions or along the edges of drying salt pannes at least several days after the last spring tide. The eggs incubate in the air and hatch only after the subsequent spring tide or rain refills depressions on the marsh surface. The larvae, known as wrigglers because of their corkscrew-like movements, undergo four feeding stages (instars) and a nonfeeding, but active, pupal stage. Adults emerge anywhere from several days to several weeks after the eggs hatch depending on the temperature.

Salt marsh mosquitoes typically produce several broods per year and are said to be multivoltine. Because they are tied to the lunar tidal cycle, the emergence of adults from marshes tends to be synchronized. Coastal residents experience this as periodic waves of mosquitoes, which may occur every two to four weeks depending on the height of the spring tide and weather conditions.

The success of mosquito breeding on a salt marsh depends on a number of factors. If the pool dries out before the larvae can complete all stages and emerge as adults, the larvae will die. Similarly, permanent pools that support predatory fish such as *Fundulus spp.* and *Gasterosteus spp.* will not support mosquito larvae and are not a suitable habitat for eggs. Low marsh areas that are flooded daily by

Figure 2. Typical habitat of salt marsh mosquito larvae during a spring tide. The pools are within a short form smooth cordgrass *(Spartina alterniflora)* marsh, and will usually dry up prior to the next spring tide precluding a permanent fish population.

tides are not sites of mosquito breeding because they do not provide the prolonged period of air incubation the eggs require, and they are accessible to predatory fish.

Habitat Alteration by Grid Ditching

Although the most radical habitat alteration for mosquito control is filling the marsh, most mosquito control activities have involved water management of some kind. Habitat alteration for mosquito control in coastal wetlands reached the zenith of activity in the United States during the Depression (Provost 1977). Both the Civilian Conservation Corps and the Works Progress Administration had programs to reclaim marshes by digging ditches at regular intervals on the marsh surface. Although these ditches were ostensibly intended to remove standing water from the marsh surface and to lower the water table, they usually were built without regard for where pannes existed and where mosquitoes actually bred. Thus many marshes or sections of marshes that did not breed mosquitoes were ditched. At the time, such considerations were not considered significant, as a major purpose of the ditching projects was to put people to work. The grid-ditching pattern, estimated to occur in over 95% of New England marshes, is now familiar to anyone who flies over the east coast.

The effectiveness of grid ditching of marshes in controlling mosquitoes and its impacts on marsh processes have been debated for the last forty years (Bourne and Cottam 1950, Lesser et al. 1976, Provost 1977). The debate was largely initiated by the publication of observations that waterfowl use of a marsh in Kent County, Delaware had declined after the marsh was subjected to grid ditching (Bourne and Cottam 1950). Bourne and Cottam noted declines in invertebrate populations in the ditched portion of this marsh compared to an unditched section. They also noted the dominance of high marsh shrubs, groundsel tree (*Baccharis halmifolia*) and salt marsh elder (*Iva frutescens*), along the edges of ditches. Bourne and Cottam predicted that these high marsh shrubs would continue to spread onto the ditched marsh at the expense of the previously existing smooth cordgrass, *Spartina alterniflora*, as long as the ditches remained functional. This initiated a long standing debate about grid ditching between wildlife managers whose goal was to manage salt marshes for waterfowl, and mosquito control agencies whose goal was to reduce mosquito populations. In retrospect, there really is very little evidence on either side about the harmful effects of grid ditching on marsh wildlife (Provost 1977).

The debate centered around the purported lowering of water tables and gradual drying out of marshes in response to ditching. Clearly, a ditch that drains a panne will negatively affect wildlife that depend on that panne. But because marsh peat has such a strong affinity for water, the water table itself may only be lowered in the immediate vicinity (ca. 1 m) of the ditch (Balling and Resh 1982). Thus ditches are not likely to cause an overall lowering of the marsh water table. Lesser et al. (1976) reexamined the Kent County, Delaware marsh in the 1970s and found that, contrary to the prediction of Bourne and Cottam, smooth cordgrass still dominated much of the ditched marsh even though the ditches were maintained in good working order. After the cessation of navigational dredging in the channel, which had caused a general lowering of the water table in the marsh, the area of high marsh shrubs had actually declined, and smooth cordgrass had increased (Provost 1977). Studies of the effects of ditching are often complicated by dredging of navigable waters adjacent to marshes (Lesser et al., 1976).

The most intensive studies of the effects of ditching on marsh vegetation and marsh organisms have been carried out in San Francisco Bay, New Jersey and Delaware marshes. Putting aesthetic considerations aside, ditching a marsh obviously increases the amount of tidal water flowing into the high marsh, creating narrow bands where low marsh habitats penetrate into high marsh. Strips of smooth cordgrass penetrate salt marsh hay habitat along ditches in east coast marshes. Ditching allows the tall ecophenotype of smooth cordgrass (Valiela et al. 1978), which dominates the lower part of the intertidal zone along the edges of tidal creeks, to extend into the high marsh. Increased productivity of marsh vegetation and invertebrates can result (Shisler et al. 1975, Lesser et al. 1976, Balling and Resh 1983). Such factors, however, are compromised by the improper placement of dredge spoils and other structural alterations of the habitat that result from ditching.

Ditching increases the heterogeneity of the marsh, both in terms of physical characteristics and the biota. The banks of the mosquito ditches are characterized by lowered salinities compared to the adjacent high marsh because regular tidal flushing prevents the buildup of hypersaline conditions (Balling and Resh 1982). In addition, the substratum along the edges of ditches is likely to be better oxygenated than areas further back because of the lowered water table at low tide (Mendelssohn et al. 1981, Howes et al. 1981, Balling and Resh 1982). In San Francisco Bay, pickleweed (*Salicornia virginica*), a low marsh species, tends to have higher productivity along ditches than elsewhere on the marsh (Balling and Resh 1983). Balling and Resh attribute this higher productivity to the tendency of the immediate vicinity of ditches to have lower salinities than the surrounding marsh. In less saline marshes, the tendency of pickleweed to be outcompeted by baltic rush (*Juncus balticus*), a brackish water species, is also attributed to lower average salinities along ditches.

The response of invertebrates to ditching in San Francisco Bay varies seasonally. Diversity of arthropods decreased away from ditches during the dry season in San Francisco Bay salt marshes (Balling and Resh 1982). The reverse was true during the wet season except in a natural channel and an old ditch that had relatively greater biomass of vegetation and more complex structure than most of the ditches present (Balling and Resh 1982). Balling and Resh conclude that the arthropod community adjacent to mosquito ditches will eventually resemble that adjacent to natural channels.

Along the east coast, a number of studies indicate that ditching has no marked affect on invertebrate populations of salt marshes (Shisler and Jobbins 1975, Lesser et al. 1976, Clarke et al. 1984). Lesser et al., for example, found an increase in populations of fiddler crabs (*Uca* spp.) and the salt marsh snail (*Melampus bidentatus*) in ditched marshes compared to controls. Ditching may very likely enhance fish populations of salt marshes. Fish density and diversity increased in ponds when these were connected to a ditching system ((Resh and Balling 1983). As long as ponds are not drained, ditching increases the amount of available marsh habitat to fish by increasing the amount of open water at high tide. It also allows the fish access to parts of marsh that are normally not available to them. The ditches serve as corridors by which fish may enter the vegetated surface of the marsh at high tides (Rozas et al. 1988). This movement of fish, particularly the mummichog (*Fundulus heteroclitus*), is important to the productivity of marsh fish in that it allows the fish to feed on invertebrates of the marsh surface, resulting in more rapid growth rates (Weissberg and Lotrich 1982). Ecologically, it is a mechanism by which the productivity of the vegetated surface of the marsh is transported into the surrounding estuarine habitats, as these fish themselves become prey for larger fish or birds. Using flume nets, more than three times as many individual fish and 14 times the fish biomass per area were caught in intertidal rivulets of tidal freshwater marshes than in larger creekbanks (Rozas et al. 1988). These intertidal rivulets are structurally similar to mosquito ditches.

Ditching of salt marshes has historically been considered harmful to populations of salt marsh birds (Urner 1935, Bourne and Cottam 1950). Clarke et al. (1984) found lower numbers of shorebirds, waders, terns and swallows on ditched marshes compared to adjacent control marshes that had substantial areas of pannes. Since there were no differences in invertebrate populations, they attributed this observation to difficulty of foraging along ditches, possibly because of their steep sides. Other than swallows, the number of passerines (songbirds) was unaffected.

Perhaps the most destructive aspect of ditching to salt marsh ecosystems has been related to the placement of dredge spoils. In many cases, spoils have simply been left along the side of the excavated ditch where they form levees that are rarely, if ever, inundated by the tides. These levees are typically colonized by species of plants normally found at the upland edge of the marsh, such as the salt marsh elder in east coast marshes. If the levees are high enough, the normal flow of high tides over the surface of the marsh is impeded, and cannot access the marsh surface.

The negative impact of dredge spoil disposal can be avoided by proper management procedures designed to insure that the spoils do not form levees along the border of mosquito ditches. A rotary ditcher, for example, spreads dredge spoils thinly over the marsh surface and has a temporary fertilizing effect (Burger and Shisler 1983). Using the dredge spoils from ditches to create small islands that do not impede the general sheet flow of water over a marsh during a high tide may actually be beneficial to wildlife that require a mixture of upland and wetland habitats. Shisler et al. (1978) found that clapper rails (*Rallus longirostris*) frequently nested on spoil islands in New Jersey marshes.

Pesticides and Bacterium

Pesticides are still used to control salt marsh and mangrove mosquitoes. Broad spectrum pesticides, such as organophosphates (e.g. malathion) or pyrethroids (e.g. resmethrin) are sprayed on marshes in an attempt to kill emerging adults as they fly off the marsh. In Essex County, Massachusetts, malathion use has been timed to coincide with the emergence of adults from the marsh before they have had a chance to disperse to upland habitats (pers. comm. W. Montgomery, Essex County Mosquito Control Project). These pesticides break down relatively quickly in the environment compared to those in wide use twenty years ago, such as organochlorines (e.g. DDT, dieldrin). However, organophosphates are toxic to nontarget organisms, particularly aquatic invertebrates and fish, and need to be applied repeatedly.

Bacillus thuringiensis israelensis (Bti) is a bacterium that produces a protein toxin that affects mosquito larvae. Bti may be spread by hand or aerially over salt pannes that contain mosquito larvae. Although more specific than pesticides, Bti may still have some impact on nontarget dipterans that may occur in marshes, particularly chironomids (Lacey and Undeen 1986). Chironomid larvae are an important item in the diet of sticklebacks (Ward and Fitzgerald 1983). Bti treatment of salt marsh pools may potentially impact the food sources of these fish, which themselves are essential in the trophic structure of salt marshes, as they are consumed by other fish, birds and mammalian predators. Bti is less toxic to chironomid larvae than to mosquito larvae (Lacey and Undeen 1986), thus avoiding nontarget affects on chironomids requires judicious measurement of final concentrations.

Exploitive Uses of Coastal Wetlands

When coastal wetlands were not being destroyed outright, or ditched for mosquito control, they were sometimes managed to provide useful products for humans. The vegetation itself has been used by humans, and salt marsh hay is still cut from New England marshes. Although not the most ideal fodder for livestock, it has the advantage of containing virtually no weed seeds, thus it is much sought after by gardeners for mulch. Nixon (1982) cited a 19th Century survey which showed that farmers in Rhode Island cut 1,717 tons of salt marsh hay from 2,506 acres of marsh in 1875. The salt marsh hayers benefited from the creation of mosquito ditches that drained pannes and created a regular grid pattern on the marsh, making it easier to move equipment around on the marsh. As the hayers were primarily interested in the salt marsh hay, a high marsh species, they would sometimes build dikes or other barriers to restrict regular tidal inundation.

In tropical regions, tannins are extracted from mangrove bark, and the wood is used for charcoal. Mangrove swamps, however, have not historically been managed to the extent that salt marshes have.

In many drier regions of the world, extensive irrigation projects for traditional agricultural species have led to increasing soil water salinities. Halophytes have been researched as potential food crops because of their tolerance for relatively high soil water salinities. The glasswort, *Salicornia bigelovii* Torr., has shown particular promise, as it can be irrigated directly with seawater (Glenn et al. 1991).

Marshes have also been managed to provide wildlife for hunting. Typically, impoundments have been created on salt marshes to provide open water habitat for waterfowl. In the northeast, freshwater impoundments, thought to be the most suitable nesting habitat for waterfowl, were created out of salt marshes at the Parker River National Wildlife Refuge in Massachusetts and Brigantine National Wildlife Refuge in New Jersey. Impoundments often create a new set of problems, most notably invasion by aggressive alien plant species such as common reed (*Phragmites australis*) and purple loosestrife (*Lythrum salicaria*) that are more tolerant of brackish conditions. Impoundments may reduce the exchange of tidal water into the marsh, and thus reduce the ability of coastal wetlands to export organic matter into surrounding coastal waters (Montague et al. 1987). They also act as barriers to the movement of marsh fish, as well as anadromous fish that may be passing through marshes.

Impacts of Marsh Diking

Diking of marshes has been carried out to create impoundments for wildlife, for flood control, to create pleasure boating and swimming areas, and for the construction of causeways for roads and railroads. Often this causes habitat degradation behind the dike, as tidal flushing is reduced and the water stagnates.

Diking can have drastic effects on marsh vegetation and, by extension, seriously alter populations of marsh fauna. If salinities behind the dike are diminished due to reduced tidal flushing, aggressive brackish water species, such as the common reed and cattails (*Typha spp.*) will replace the natural salt marsh vegetation (Figure 3, Niering and Warren 1980, Roman et al. 1984, Beare and Zedler 1987). Overall productivity of the vegetation may increase in response to lowered salinities, or decrease if the tidally restricted area becomes hypersaline (Zedler et al. 1980).

Often, marsh creeks behind dikes have poorer water quality than those seaward. Portnoy (1991) measured lower dissolved oxygen and higher than normal levels of sulfides behind a dike on the Herring River in Wellfleet, Massachusetts. This area is plagued by periodic fish kills and high numbers of mosquitoes, both consequences of stagnation. In the past, road construction on fill over marshes did not plan for maintenance of adequate tidal flushing in their design. Sheet flow of tidal water over the marsh surface is blocked by roads, and culverts for tidal creeks are often too small to maintain the normal tidal range and flushing. Flood and ebb tides behind a road across a marsh in Ipswich, Massachusetts are delayed several hours by an inadequately sized culvert compared to that seaward of the road, and the tidal range is reduced by about 25 percent.

Restoring the normal tidal circulation to a formerly diked area can reverse these negative effects. Slavin and Shisler (1983) noted substantial increases in waders, waterfowl, shorebirds and gulls in a marsh when the dike of a tidally restricted salt marsh hay farm was breached. Conversely, the number of passerines declined. They also observed increases in smooth cordgrass and declines in salt marsh hay. Recent studies of Connecticut salt marshes have documented a striking decline in brackish species and expansion of the natural salt marsh with removal of dikes (Sinicrope et al. 1990). Simply removing a dike, however, does not always lead to the return of the natural salt marsh vegetation. If the peat has been oxidized or eroded behind the dike, the wetland surface may be lowered and the area remain unvegetated and flooded when tidal restrictions are removed (pers. comm. J. Portnoy, S. Warren).

Marshes are dynamic systems that may move up or down the shoreline in response to changes in sea level. Diking along the upland edge of marshes, a common flood control measure in urban areas, prevents the normal migration of the marsh. The future of such marshes is dubious if rising sea levels occur as a result of increases in atmospheric carbon dioxide and other greenhouse gasses.

CURRENT MANAGEMENT ISSUES

Federal and state laws and regulations reflect a new appreciation by the general public for the function and value of coastal wetlands. The outright legal destruction of large areas of coastal marshes and mangrove swamps is, hopefully,

Figure 3. Common reed *(Phragmites australis)* encroaching on salt marsh cordgrass in Long Island, New York. Such scenes are common along the upland edge of east coast marshes, particularly where tidal flow has been restricted.

a thing of the past. Nonetheless, several significant management issues still remain including recent wetlands losses caused by direct or indirect human impacts, effects of activities that are still permitted by federal and state wetlands regulations, such as mosquito control procedures and the construction of docks and piers over marshes, cumulative impacts of activities in watersheds surrounding the coastal wetland, including activities in wetlands buffers, and restoration of degraded coastal wetlands.

Recent Trends in Wetlands Acreage

Losses of coastal wetlands still occur, albeit at a slower rate than prior to 1970. The U.S. Fish and Wildlife Service's National Wetlands Inventory Project estimated that a net loss of 70,799 acres of vegetated estuarine wetlands occurred in the coterminous United States between 1974 and 1983 (Tiner 1991). This is about 1.5 percent of the total existing wetland acreage in 1973, and represents a decline in the rate of loss from the mid-1950s through the mid-1970s.   An increase of 11,534 acres occurred in nonvegetated estuarine wetlands, such as tidal flats.

Recent losses have been more subtle than those of the past, consisting primarily of a transformation of estuarine vegetated wetlands to deepwater habitat rather than conversion to urban or agricultural land (Tiner 1991). The majority of the recent losses have occurred in the Mississippi delta and the Florida Everglades (Field et al 1991). Studies along the northern Gulf of Mexico have implicated rapid shoreline subsidence, and the inability of marshes to keep up with this subsidence due to relatively low accretion rates as major factors (DeLaune et al. 1989, Turner and Rao 1990). Localized alteration of hydrology caused by the building of canals and levees for flood control has increased surface water levels on marshes, stressing and killing the vegetation (DeLaune et al. 1989, Mendelssohn and McKee 1988). The breakup of a vegetated wetland into smaller, and then larger, ponds can occur several kilometers from a canal (Turner and Rao 1990). Louisiana lost 2.9 percent (57,000 acres) of its wetlands from 1974-1983, largely through these processes.

During this same period, Texas experienced a loss of about 10,000 acres and New Jersey and South Carolina lost over 1,000 acres (Tiner 1991). For the entire country, 45,000 acres were lost to urban development, about 2500 of which were mangrove swamps in Florida. Another 4000 acres were converted to agricultural use.

Mosquito Control by Open Marsh Water Management

In response to the concern over grid ditching, a technique called Open Marsh Water Management (OMWM) was developed in the mid-Atlantic states in the 1950s (Ferrigno et al. 1975). OMWM consists of a system of reservoirs and canals in mosquito breeding areas that allow predatory fish, generally *Fundulus sp.*, access to waterlogged areas of high marsh where mosquito larvae develop. Often, the reservoirs are several-acre champagne ponds and are at least three feet deep to provide an adequate refuge for the fish during the several weeks of neap tide. In New England, old mosquito ditches are converted into reservoirs by deepening them and plugging up their junction with natural tidal creeks (Hruby and Montgomery 1986). Canals are dug from reservoirs to mosquito breeding areas to allow passage of fish (Figure 4). The success of OMWM in controlling salt marsh mosquitoes has been documented (Ferrigno et al. 1975, Hruby et al. 1985).

OMWM Versus Grid Ditching

The chief advantage of OMWM compared to grid ditching is that it does not drain the pannes and pools on the marsh surface. An OMWM system, unlike grid ditches, is not connected to tidal creeks. Sea water enters reservoirs and canals by sheet flow directly over the marsh surface during spring high tides, in the same manner as it floods nearby mosquito breeding habitats. The water is then trapped in the reservoirs and canals and does not drain out at low tide, as connections with

Figure 4. Small reservoir pool and two shallow radial canals in a salt marsh in Gloucester, Massachusetts. These are part of an OMWM system. The reservoir is 1 m deep, and the canals 0.3 m deep. The reservoirs of OMWM systems of mid-Atlantic and southern marshes may be as large as 3-4 acres.

the tidal creeks have been eliminated. Where water flows through old mosquito ditches, a sill at the junction of the ditch with a tidal creek, or at a relatively high point in marsh topography, achieves the same goal.

OMWM systems are site specific. This is more a function of the recent overall enlightened management of salt marshes than of OMWM in particular, as grid ditching could also have been site specific. For an OMWM system to function properly, managers must identify the mosquito breeding sites through a monitoring program and then integrate the OMWM design into the hydrology of the area.

Another advantage of OMWM systems compared to grid ditching is that they are easier to maintain. Grid ditches periodically have to be cleaned out or redug, as the steep banks become scoured by tidal action and often collapse. This is particularly true in northern New England where large tidal ranges occur. Ironically, this often creates more of a mosquito problem than was initially present, as a clogged ditch that no longer allows passage to fish is an excellent mosquito breeding habitat. Portnoy (1982) found higher numbers of mosquitoes (*Aedes cantator*) in mosquito ditches, even those treated with a larvicide, than in natural surface pools untreated with larvicides in a diked river basin on Cape Cod. As OMWM systems are not subject to the scouring action of tidal water rushing through creeks they should require less maintenance, although no one has actually compared the two types of systems for any length of time.

Finally, the development of rotary ditchers has allowed dredge spoils to be placed over marshes in a thin layer, thus reducing impacts to vegetation. In OMWM systems in Massachusetts, vegetation visibly recovers the second growing season after deposition of spoils by rotary ditchers (pers. comm. W. Montgomery, T. Hruby). In New Jersey, thin deposition of dredge had little visible effect on vegetation even in the initial year of deposition (Burger and Shisler 1983).

Effect of OMWM on Mosquitoes

Monitoring the success of OMWM efforts in terms of its ability to control mosquitoes is complex, as mosquito numbers and the numbers of broods produced per year vary substantially both temporally and spatially on the marsh depending on the timing of rainfall, spring tide events and temperature. Nevertheless, Hruby et al. (1985) determined that the number of mosquito larvae declined by 75-99 percent in the OMWM marsh compared to the numbers in the same site the year before it was altered, while larval numbers in adjacent control areas remained roughly the same.

In addition to allowing fish predation on the mosquito larvae, OMWM systems are likely to interfere with the hatching cycle of mosquito eggs, which need to incubate for a period in air. The number of larval mosquitoes that survive to pupate as adults on the marsh surface is negatively correlated with both tidal inundation and with fish numbers (Figures 5 and 6). In a comparison of mosquito emergence in an unaltered marsh, a grid-ditched marsh and an OMWM system, significantly fewer mosquitoes were observed emerging from the OMWM system than the unaltered marsh (unpublished data). No mosquitoes emerged from the ditched marsh. Fish were present in the ditches of the grid ditched marsh, and on the unaltered and OMWM system marsh surfaces during the spring tide.

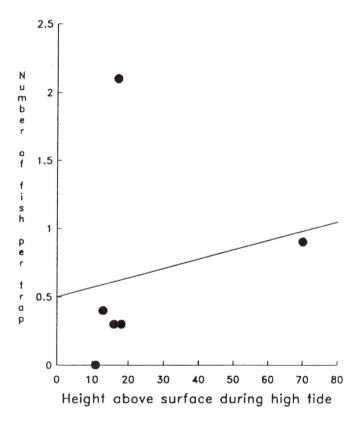

Figure 5. Relationship between tide height above the marsh surface during a spring high tide and the number of mosquitoes that survive to emerge as adults. Sampling occurred in a ditched marsh, an OMWM marsh, and an unaltered control marsh.

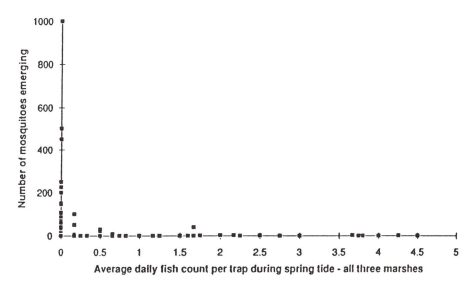

Figure 6. Relationship between fish numbers on the salt marsh and emerging mosquito numbers. Sampling occurred in a ditched marsh, an OMWM marsh, and an unaltered control marsh.

Effect of OMWM on Marsh Processes

There have been few studies of the long-term impacts of OMWM. Brush et al. (1986) concluded that OMWM had little impact on bird numbers on a Massachusetts salt marsh. In the first year after a ditched marsh was converted to an OMWM system, shorebird (sandpipers, plovers, etc.) numbers increased presumably because the spoils provided easily accessible foraging for invertebrates. Their numbers declined in subsequent years as the vegetation grew up through the spoils. Marsh passerines declined at first and then increased to pre-OMWM levels, and other groups of birds were unaffected. Brush et al suggest that bird numbers were more closely related to the number of pannes on a marsh than to whether it was altered by OMWM, ditched, or remained unaltered.

Peat cores from a northern Massachusetts salt marsh were examined for invertebrates using Berlese funnels. No significant difference in types of organisms was detected between unaltered marsh and an OMWM marsh (unpublished data).

Occasionally, the reservoir pools of an OMWM system become stagnant during periods of hot weather and neap tides. Mosquito ditches that have been appropriated for OMWM seem especially prone to this, as they tend to be long, narrow, and relatively deep. A thick layer of algae may form on the surface, with hypoxic or anoxic water beneath. In Massachusetts marshes, fish kills have not been observed, but in some Florida counties stagnation and declining water quality of pools has occurred, leading to declines in fish populations. The latter has been mitigated by constructing channels to increase tidal exchange.

Other Potential Management Uses of OMWM

In the future, OMWM techniques may be useful as part of an integrated approach to restoring degraded marshes. A decline in the invasive brackish water common reed is evident in OMWM marshes that include a perimeter ditch at the border of the upland. The perimeter ditch prevents freshwater surface and groundwater flow from moving out over the natural salt marsh, and salinity levels are maintained at those appropriate for salt marsh vegetation. Dredge spoils created by OMWM may be potentially useful for enhancing nesting areas for marsh birds. Dredge spoils may be colonized by wildlife such as rails that forage on marshes but which require drier nest sites (Shisler et al. 1978). Judicious placement of dredge spoils into small islands rather than levees may enhance wildlife habitat of the marsh without impeding circulation. Another way to enhance wildlife habitat values is to design the reservoir pools so that they function as foraging areas for shorebirds and waders. This can be accomplished by creating a gently sloping edge to reservoir pools.

Recommendations for Mosquito Control

OMWM is an alteration of a salt marsh with ecological consequences and should only be used if there is a documented, compelling reason. Ideally, OMWM would comprise one facet of an integrated approach to mosquito control. At a minimum, an Integrated Pest Management (IPM) approach to controlling nuisance mosquitoes on salt marshes is a useful theoretical framework for determining when and how to intervene. An IPM approach should begin by determining a threshold level at which control measures are necessary. Although many residents near marshes might assert that even one mosquito bite per hour is too many, a threshold level of annoyance that is specifically related to salt marsh mosquitoes needs to be determined before proceeding with any action. If the predetermined annoyance threshold has been exceeded, then trapping of adult mosquitoes is carried out to determine if mosquitoes from salt marshes are a significant cause of the nuisance. If disease is an issue, the threshold will be some indicator of the presence of the disease.

If salt marsh mosquitoes are shown to be the cause of the nuisance, then monitoring of levels of larval mosquitoes in salt marshes is initiated. Standards developed by the Massachusetts Audubon Society and the Essex County (Massachusetts) Mosquito Control Project include one full year of monitoring of mosquito larvae before any alterations are considered (Hruby and Montgomery 1986). Monitoring includes weekly sampling with a standard mosquito larvae dipper throughout the growing season at stations where mosquito larvae are suspected of occurring. Control is considered appropriate if three broods of mosquitoes containing at least five larvae per dip occur during the growing season. If two broods are observed with greater than five mosquitoes per site, then the site is remonitored for another season before a decision is made. Potential impacts of alterations on other marsh organisms, such as breeding and foraging birds, are also taken into consideration before proceeding further.

When control is justified, the ecologically most benign procedures or combination of procedures are selected. A proactive approach includes public education to reduce the risk of public exposure and review of drainage plans for developments, subdivisions, and roads, by appropriate local and state agencies to insure that they do not contribute to reduced tidal flushing in coastal wetlands. A source reduction approach for disturbed habitats that support large populations of larval mosquitoes includes OMWM (3-4 acres) areas, or selective ditching in smaller areas. Although commonly employed, pesticide application for nuisance control is less cost effective than source reduction in the long run. Pesticide use requires repeated application, affects nontarget organisms, and will likely require increasing application frequencies and doses as mosquitoes develop resistance (Ofiara and Allison 1986).

Any control mechanism should be evaluated on a regular basis, and modified as necessary. Both mosquitoes and nontarget organisms should be monitored.

Impacts of Docks and Piers

Outright destruction of coastal marshes by private landowners is no longer allowed in many parts of the country, however building docks and piers over salt marshes is still a permitted activity. Although there is little data on this subject, there has been some recent concern on the part of resource managers over the impacts of these structures on marshes, particularly the effects of shading on vegetation and the potential for scouring around support posts. The Massachusetts Office of Coastal Zone Management has recently sponsored research on the effects of small docks and piers on tidal flats and eelgrass beds adjacent to salt marshes. The U.S. Army Corps of Engineers, which is responsible for permitting these structures under Section 404b(1) of the Federal Water Pollution Control Act of 1972, requires that marsh vegetation beneath a dock not be impacted by the dock, but does not give design standards to insure that light does reach the plants. The Massachusetts Executive Office of Environmental Affairs has adopted pier guidelines (Table 1).

Table 1. Summary of guidelines for the construction of piers in critical areas (Massachusetts Office of Coastal Zone Management 1988).

- T-docks and floats at the end of a pier are preferred in that they allow a vessel to be parallel to the shoreline and in deeper water. In any case, the stem of a vessel at a pier must be facing toward deeper water.
- Seasonal docks are preferred over permanent ones.
- Pile driving is preferred over jetting.
- Docks should be as small in scope as possible, with as few pilings as possible. The goal is to allow unimpeded transport of sediment.
    - Spacing of pilings should be no closer than 20 X the diameter of the pilings when the dock is located in a salt marsh, and never any closer together than 10 feet.
    - Piers in salt marshes should be of wood construction. If treated lumber is used, only nonleaching types.
    - No stabilization structures should be proposed, even if the pier is adjacent to an eroding bank.
    - Piers located in an anadromous fish run should be designed so that there is no change in the rate of flow within the run. Construction should not be allowed at times when the fish are running.

Table 1. (cont.)

- The pier and the uses of the pier should not encroach upon navigational channels and mooring areas.
- A dock should not be designed to end in water that is too shallow to float the vessel at all tides, however very long elevated walkways are likely to have environmental impacts.
- No pier should extend beyond the length of existing piers used for similar purposes and in no case should the length extend more than 1/3 of the way across a water body.
- A higher level of environmental review is required for proposed walkways wider than 4 feet, proposed floats greater than 300 sq ft, and any structure proposed to be less than 4 feet above a salt marsh. The review should clearly demonstrate that the structure will have no adverse impact on the marsh, adjacent mudflats, or submerged aquatic vegetation.
- The dock should be designed so that the approach path of vessels is at least 50 feet from the edge of any salt marsh.
- The primary factor in determining the location of the dock on a lot is the avoidance of sensitive resources.
- The necessity for mitigation should be avoided to the maximum extent possible.

The docks themselves may be less of a problem than the human and boat traffic they encourage (Figure 7). Marshes are very susceptible to trampling, as scientists who have repeatedly sampled in the same area of marsh are well aware. A catwalk over a marsh is probably a less harmful way to reach a boat in the long run than walking directly on the marsh. Footsteps gradually erode a path into the marsh surface leading to loss of vegetation and subsidence of peat. The boats themselves, maneuvering in the shallow water of marsh creeks, resuspend sediment which may affect submerged vegetation and shellfish beds. Boats also contribute to the erosion and slumping of peat along the banks of marshes. Community piers are an alternative to a plethora of individual ones.

Figure 7. Boats left high and dry at low tide at a marina adjacent to an Essex, Massachusetts, salt marsh. Resuspension of sediments in shallow water, erosion of marsh peat and disturbance of wildlife are associated with boating activity around salt marshes.

Buffer Zones Around Coastal Wetlands

Wetland laws typically regulate activities within a wetland and in a small buffer strip surrounding a wetland. The rationale for regulatory authority over a fringe area around the wetland is that activities within that fringe potentially impact the wetland. A number of states have adopted wetlands buffers ranging from 45 to 300 meters from the edges of wetlands in critical areas (Table 2). Buffer distances such as these represent a compromise between protecting the private property interest of the landowner and the public interest in the wetland. In the ideal world, each buffer distance would be determined after a case by case analysis of soils, hydrology, slope, vegetational cover, the characteristics of the wetlands resources being protected, and the nature of the development being proposed. Some states have adopted such a case by case approach using a ranking system based on the above criteria (Diamond and Nilson 1988). Unfortunately, political realities, the expense of evaluating individual properties, and scientific uncertainty about the relationship of the buffer to wetlands functions often make a case by case approach unrealistic in practice.

Table 2. Buffer distances around critical areas adopted by various states. (Brady and Buchsbaum 1989).

State	Buffer Zone Width (m)
Maine	45-90
Maryland	90
New Jersey	90
Rhode Island	60
Washington	60
Wisconsin	300

The rationale for protecting coastal and inland wetlands is that they perform certain functions that are valuable to the public. Coastal wetlands enhance the water quality of coastal waters by removing potentially harmful constituents before they reach open water. Coastal wetlands are also rich habitats for fish, shellfish, and wildlife. Water quality and habitat values are related, because poor water may result in increased incidence of disease in marine organisms as well as rendering some, such as shellfish, unharvestable. In addition to these two values, buffers also help to maintain the aesthetics of a wetland by preserving scenic qualities.

Water Quality Aspects of Buffers

Pathogenic Microorganisms

Buffers can play a key role in a management strategy to reduce pollution before it reaches a coastal wetland (Abernathy et al. 1985; Lawrence et al. 1985,

1986). Particulate pollutants, such as microorganisms, suspended solids, and pollutants associated with sediments, are slowed down and entrained by physical processes within the soil. Soluble nutrients may be removed by biological activity within the buffer.

Pollution by bacteria and viruses from domestic wastes is a particular concern in coastal wetlands because they render shellfish unfit for human consumption and create a public health threat at swimming beaches. The percentage of potentially harmful microorganisms in sewage and stormwater that reaches a wetlands area is largely a function of time. The longer it takes for these organisms to be transported from their source to a resource area, the smaller their numbers will be at that resource area. Bacteria and viruses are adsorbed to soil particles in the buffer, downgradient movement is slowed, and they eventually die (Hemond and Benoit 1988).

The velocity of water in surface runoff is much faster than that of groundwater. Therefore, stormwater runoff will potentially pose a greater risk of microbial contamination than domestic sewage. Stormwater runoff was the major source of high fecal coliform counts in Buttermilk Bay and many other water bodies in Wareham, Massachusetts (Heufelder 1988). Fecal coliform bacteria were almost completely attenuated within 7 m of leaching fields.

Characteristics of the buffer itself determine to what extent microorganisms are detained. Important characteristics include the permeability of the soil, slope gradient, extent of vegetation, and degree of channelization. Occasionally, water percolating through the soil will reach an impermeable layer, such as bedrock or clay, and then be channelized rapidly into coastal wetlands.

The distance a microorganism will travel from its source and still remain viable depends not only on characteristics inherent to the buffer but also on the type of organism. Survival times of up to 27 weeks in soils were reported for some microorganisms, and the distances these organisms traveled from their source ranged from two to 837 meters (Hagedorn 1984). Viruses may travel as far as 400 meters in groundwater from sewage infiltration basins (Keswick and Gerba 1980). In seawater and freshwater, average times for 90 percent decline in concentrations of coliform bacteria were 2.2 hr and 57 hr respectively (Mitchell and Chamberlain 1978). The lateral distance microorganisms will travel from a leaching field can be estimated based on these die-off rates and the groundwater flow rate.

It is clear that under certain conditions, viruses and bacteria which are potentially harmful to humans may travel further from their source than buffer distances of even 100 m, particularly in poorly functioning or antiquated wastewater treatment systems or in stormwater runoff. Buffer distance is a difficult parameter to isolate in the field because it usually correlates with overall density of development. In one study there was a significant correlation between concentrations of fecal coliform bacteria in a salt marsh creek and the number of houses within 30 m of the edge of the marsh (Bochman 1991). The number of houses within 30 m, however, correlated with the total number of houses within the watershed.

Nitrogen

Nitrogen is of particular concern to coastal waters because it stimulates blooms of phytoplankton and seaweeds. These blooms harm the aesthetics of coastal waters and threaten other marine plant and animal life due to the lowering of light and dissolved oxygen levels. The efficacy of removing nitrogen with a large buffer depends on whether the nitrogen enters the wetland through surface runoff or in groundwater. Nitrate in surface runoff will slowly percolate through the root zone of plants where it will be taken up by vegetation both in the buffer and the wetland (Woodwell 1977, Ehrenfeld 1987, Knight et al. 1987). However, if the surface runoff is very rapid, vegetation in the buffer may be ineffective in taking up nitrogen before it reaches a wetland.

Nitrogen in groundwater will only be removed when it nears the surface of a wetland or a water body either through plant and algal uptake or through denitrification. When in the groundwater, nitrogen in groundwater is thought to be conservative and not subject to attenuation by biological or physical processes (Valiela and Costa 1988). If this is the case, a large buffer will not protect a coastal embayment or salt pond from potential eutrophication unless it provides sufficient area for dilution. Instead, nitrogen can be managed by land use controls that limit the amount of nitrogen entering the watershed or by wastewater treatment procedures (denitrification systems, tertiary treatment) that remove nitrogen before it can enter groundwater.

Wildlife Habitat Aspects of Buffers

There are several reasons why a buffer is thought to enhance the wildlife habitat value of a coastal wetland. Many wildlife species that live in wetlands depend on the surrounding upland for cover, for nesting, as an alternate foraging area, and as a migration route. According to the classic ecological notion of the ecotone, wildlife will be more abundant at the wetland-buffer boundary because two habitat types coexist in close proximity. Many birds and mammals forage on the abundant invertebrates and fish of a salt marsh, but require the surrounding upland for nesting or as a refuge during high tides. In addition to providing an alternative source of essential resources, a buffer also insulates the animals of the wetland from disturbance from developments located around its periphery, a characteristic that is particularly valuable for smaller wetlands. Human activity brings not only noise, but also domestic animals and those native and alien wildlife (e.g., starlings, cowbirds) that are well-adapted to human habitation and often compete with native wildlife in transitional areas. A 90-meter distance has been recommended to provide a buffer against disturbance around state and federal wildlife refuges and conservation areas (Diamond and Nilson 1988).

Determining the actual impact of different sized buffers on wetlands wildlife in the field is difficult, primarily because wildlife respond to a number of different aspects of the landscape that cannot be controlled. A reasonable approach is to infer the role of the wetlands buffer by examining the life history of different animals. In their Habitat Evaluation Procedures, the U.S. Fish and Wildlife Service typically considers buffer size as one component in calculating habitat suitability indices. Rogers et al. (1975) also evaluated adequacy of aquatic buffers in their ranking of biotic natural areas of the Eastern Shore of Maryland.

Several examples illustrate the importance of buffers to wildlife species that occur in coastal salt and freshwater marshes. The area immediately surrounding coastal wetlands is important for providing the seclusion a number of species of waterfowl need to nest free from predation and disturbance. Kirby (1988) states that, *It has long been recognized that lands adjacent to areas managed for waterfowl play a major role in the entire management scheme.* Black ducks (*Anas rubripes*) nest either on islands in wetlands or in areas immediately adjacent to wetlands up to 1.2 km from the wetland (Kirby 1988). An ideal nesting habitat of black ducks is heavily vegetated on at least one side to provide them with concealment from predators. Gadwalls (*Anas strepera*) typically nest in drier shoreline areas within 30.5 m of water (DeGraaf and Rudis 1986). Canada geese (*Branta canadensis*) also nest near the water's edge in fresh and salt marshes (DeGraaf and Rudis 1986).

Short and Cooper (1985) state that human disturbance and the resulting loss of nesting sites have probably been the most important factors contributing to declines in some great blue heron (*Ardea herodias*) populations in New York and British Columbia. Although nesting sites for great blue herons may be many miles from their foraging areas (DeGraaf and Rudis 1986), creeks and shallow ponds on marshes are typical feeding sites. A suitable foraging area must be at least 100 m from human activities and habitation (Short and Cooper 1985). In Essex Bay, Massachusetts, there is a significant inverse correlation between the extent of development in the buffer along the marsh edge and the number of foraging wading birds (Figure 8).

The U.S. Fish and Wildlife Service's Habitat Suitability Index (HSI) for mink (*Mustela vison*) assumes that a minimum strip of 100 m along the edge of a wetland enhances the value of that wetland to mink (Allen 1986). In a study in Alaska, 68 percent of all observations of mink were either in a wetland itself or within 100 m of the wetland (Allen 1986). Rapid declines of mink populations along the shores of Lake Ontario were associated with small increases in the human population along the shoreline. River otter (*Lutra canadensis*) also occur along the edges of coastal ponds, salt marshes and estuaries (DeGraaf and Rudis 1986).

Some Examples of Buffer Protection Programs

A number of states and regions have investigated the question of setting buffer distances around coastal and inland wetlands. These include Maryland (Rogers et al. 1975), Rhode Island (Rhode Island Coastal Resource Management Program 1984), New Jersey pinelands (Roman and Good 1985), central Florida (Brown and Shaeffer 1990) and Washington (Castelle et al. 1992a, b). In addition

Phillips and Phillips have independently proposed a method for the estimation of shoreline buffer zones. Two approaches are illustrated herein.

Phillips and Phillips (1988) proposed a simple physical model to enable managers to predict the size of a buffer necessary to attenuate pollutants from stormwater. The delivery of pollutants to a wetland or water body is related to the energy of surface flow generated during a storm event. This energy can be predicted from (1) the permeability of the soil within the buffer, (2) the slope gradient down which the stormwater runs, (3) the width of the area over which the

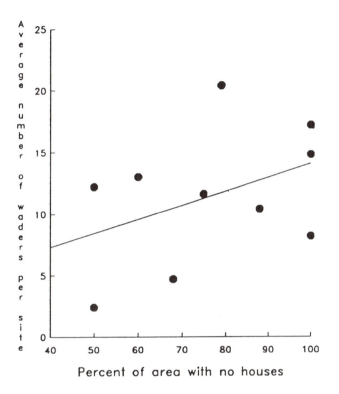

Figure 8. Relationship between the number of houses along the marsh edge and number of foraging wading birds in a salt marsh in Essex Bay, Massachusetts.

runoff flows (i.e. the buffer size), and (4) the surface roughness of the buffer (Phillips and Phillips 1988). The first is measured as hydraulic conductivity or infiltration rate and is generally available from the U.S. Soil Conservation Service for the various soil types. The second is obtained during the planning stages of any project. Vegetation is a major part of surface roughness, which is factored into the model as a roughness coefficient. The use of this model requires field measurement of a nearby reference buffer to which the buffer under consideration is compared. Because this model is based on transport of suspended particles, it is valid for bacteria and pollutants associated with particles, but not for soluble pollutants.

The New Jersey Pinelands Commission, which has regulatory authority over development in the globally-significant pine barrens of southeastern New Jersey, has set up a systematic decision-making flow chart for establishing buffer distances from wetlands resources (Roman and Good 1985). Major variables taken into account are the perceived value of the wetland resource and the nature of the activity being proposed near the wetland.

The New Jersey Pinelands model includes several special cases in which the buffer is set at 90 meters (300 feet). These include proposed activity in a designated preservation district (the core area of the Pinelands), within 90 m of an Atlantic white cedar (*Chamaecyparis thyoides*) swamp, activities in which resources are extracted, and activities which include onsite wastewater treatment systems. Atlantic white cedar swamps in the Pine Barrens are afforded particularly strong protection because they contain a large assemblage of rare species, and because they have historically been severely impacted by human activity. Activities proposed within existing densely developed areas are required to have a minimum buffer of 15 meters.

If none of the special cases apply, then the proposed activity is subject to a three step procedure to establish the buffer distance. The evaluator of the project develops a numerical index of wetlands value based on comparing the site-specific wetland to a standard evaluation scheme. Criteria comprising value index include presence or absence of endangered species, vegetation quality, surface water quality, water quality maintenance, wildlife habitat, and sociocultural values. Potential localized, cumulative, and watershed-wide impacts, and the slope, are all considered in deriving this index. Lastly, the values and impacts indices are averaged to derive a buffer delineation index. Wetlands that have a high value index and a high potential for impacts are assigned the largest buffer distance, 90 meters.

Restoration of Degraded Wetlands with Particular Emphasis on Introduced Species

One aspect of degradation of salt marshes that is often associated with human activities is the invasion of salt marshes by introduced species, many of which are exotics (nonnative). The danger is that these species, lacking natural controls in their adopted habitats, will outcompete the native vegetation and cause overall changes in the flora and fauna of the marshes.

Introduced low marsh species tend to colonize bare substrate at the lower margins of marshes (Frenkel and Boss 1988), and do not generally compete with the native salt marsh species. A number of species of low marsh plants have been introduced into Pacific Coast marshes. These include the cordgrasses *Spartina alterniflora* and *Sp. anglica* and the eelgrass *Zostera japonica* (Frenkel and Boss 1988, Calloway and Josselyn 1992). The cordgrass *Sp. densiflora*, a native of Chile, has been accidentally introduced into Humboldt Bay, California. Modes of introduction of these species include transport of seeds in oyster culture and in ballast from ships. An example of an introduced low marsh plant competing with a native species is *Sp. x townsendi*, a vigorous hybrid of the native cordgrass, *Sp. anglica*, and the introduced North American species, *Sp. alterniflora*. This hybrid is outcompeting *Sp. anglica* in British salt marshes.

The mid- and high-marshes are less susceptible to colonization by introduced species than the bare substrate of the low marsh (Frenkel and Boss 1988). In most salt marsh systems, these sections tend to be densely vegetated already, requiring a strong competitive ability if an introduced species is to be successful. In addition, the suspected source of marsh plant introductions, oyster culture and ballast, are generally in direct contact only with low marsh. Nonetheless, the *changing mosaic* nature of salt marsh vegetation (Miller and Egler 1950) insures that bare areas are always present for opportunistic species. Salt marsh hay, a native of East Coast marshes, has colonized a number of marshes in the Pacific Northwest. In Siuslaw Estuary, Oregon, this species is limited to the relatively open *Scirpus maritimus-Deschampsia caespitosa* community of the mid-marsh, where its dense growth habit has crowded out native marsh plants. The plants were likely introduced accidentally in packing material (Frenkel and Boss 1988). Management of mid- and high-marsh invasive species requires mechanical removal of plants, an increasingly difficult process as the population expands.

The vegetation of the upper edge of coastal wetlands, where saltwater grades into freshwater is particularly prone to alteration through changes in hydrology. As mentioned earlier, invasions of common reed along the east coast and cattails in southern California have been associated with reduction in salinities (Niering and Warren 1980, Roman et al. 1984, Beare and Zedler 1987). Mitigation requires restoration of the historical tidal exchange by removing barriers and, where fill has been deposited, regrading the marsh to elevations suitable for salt marsh species. As common reed tolerates a wide range of salinities up to about 25 ppt, mechanical removal of this species may be required to allow reestablishment of salt marsh species. Although salt marsh managers often view common reed dominated habitats as less valuable to wildlife than the native salt marsh (Roman et al. 1984), it is important to consider the context in which this plant occurs within a landscape that includes both native salt marsh vegetation, common reed, and a surrounding upland. In urban marshes, dense stands of common reed adjacent to salt marshes may provide the only cover for wildlife species in an otherwise exposed environment. The common reed dominated zone of Belle Isle Marsh in metropolitan Boston was the part of the marsh where muskrats (*Ondatra zibethica*) and redwinged blackbirds (*Agelaius phoenicus*) nested, and black-crowned night herons (*Nycticorax nycticorax*) roosted (Buchsbaum and Hall 1992). The American bittern (*Botaurus lentiginosus*), an uncommon shy species, occurs primarily within the common reed zone at this marsh.

FUTURE CONSIDERATIONS

In the last twenty years the amount of legal protection given coastal wetlands by the federal government and many coastal states has increased substantially. This, combined with an increase in our scientific understanding of the ecology of these habitats offers hope for the future protection and management of these habitats. Nevertheless, large-scale destruction of marshes for development is being replaced by other potential threats. Mosquito control will continue to be a major factor in the management of salt marshes and mangrove swamps, thus evaluation of the effects of OMWM and related procedures on various marsh processes is still required. The projected expanding population along the coast (Culliton et al. 1990) will continue to place pressure on coastal wetlands even when the actual wetland is protected. More people means development of buffer areas and surrounding watersheds, incremental losses of small pieces of coastal habitats, and increases in recreational-related structures and activities that impact coastal wetlands.

Management and mismanagement of coastal wetlands in the past and present have occurred on a local scale, in which decisions have focused on protecting (or not protecting) individual wetlands. The most significant future management issues will be related to global and regional changes that occur on a scale beyond which there is any precedent for management. The transformation of marshes to deepwater habitats in Louisiana due to shoreline subsidence, the largest single cause of recent vegetated wetland loss in the United States, is not a problem amenable to traditional localized solutions. Similarly, addressing the threat to coastal wetlands caused by projected rising sea levels will require global cooperation.

REFERENCES

Abernathy, A.R., J. Zirschy, and M.P. Borup. 1985. Overland flow wastewater treatment at Easley, S.C. Journal Water Poll. Control Fed. 57:291-299.

Allen, A.W. 1986. Habitat suitability models: mink, revised. U.S. Fish and Wildlife Service Biol. Rep. 82(10.127). 23 pp.

Allen, A.W. and R.D. Hoffman. 1984. Habitat suitability index models: muskrat. U.S. Dept. Int. Fish and Wildlife Serv. FWS/OBS-82/10.46. 27 pp.

Balling, S.S. and V.H. Resh. 1982. Arthropod community response to mosquito control recirculation ditches in San Francisco Bay salt marshes. Environmental Entomology 11: 801-808.

Balling, S.S. and V.H. Resh. 1983. The influence of mosquito control recirculation ditches on plant biomass, production, and composition in two San Francisco Bay salt marshes. Estuarine and Coastal Shelf Science 16: 151-161.

Beare, P.A. and J.B. Zedler. 1987. Cattail invasion and persistence in a coastal salt marsh: the role of salinity reduction. Estuaries 10: 165-170.

Bochman, A. 1991. Buffer zones and water quality in the Parker River/Essex Bay Area of Critical Environmental Concern: a correlational study. Unpublished M.S. thesis. Harvard University Extension School.

Bourne, W.S. and C. Cottam. 1950. Some biological effects of ditching tidewater marshes. Res. Rep. 19. U.S. Fish and Wildlife Service. 30 pp.

Brady, P. and R. Buchsbaum. 1989. Buffer zones: the environment's last defense. Massachusetts Audubon Society, Wenham, MA.

Brown, M. and J. Shaefer. 1990. Buffer zones for wetlands and wildlife in east central Florida. University of Florida Center for Wetlands, Publ. 89-07. Gainesville.

Brush, T., R.A. Lent, T. Hruby, B.A. Harrington, R.M. Marshall, and W.G. Montgomery. 1986. Habitat use by salt marsh birds and response to open marsh water management. Colonial Waterbirds 9: 189-195.

Buchsbaum, R. and M. Hall. 1992. Baseline assessment of Belle Isle Marsh in anticipation of restoration. Report to the Massachusetts Environmental Trust. 46 pp.

Burger, J. and J. Shisler 1983. Succession and productivity on perturbed and natural *Spartina* salt marsh areas in New Jersey. Estuaries 6: 50-56.

Calloway, J.C. and M. Josselyn. 1992. Introduction and spread of *Spartina alterniflora* in south San Francisco Bay. Estuaries 15: 218-226.

Castelle, A.J., C. Conolly, M. Emers, E.D. Metz, S. Meyer, and M. Witter. 1992a. Wetlands buffers: an annotated bibliography. Adolfson Associates. Prepared for Shorelands and Coastal Zone Manage. Progr. Washington State Dept. of Ecol. Olympia. Publ. No. 92-10.

Castelle, A.J., C. Conolly, M. Emers, E.D. Metz, S. Meyer, M. Witter, S. Mauermann, T. Erickson, and S.S. Cooke. 1992b. Wetlands buffers: uses and effectiveness. Adolfson Associates. Prepared for Shorelands and Coastal Zone Manage. Progr. Washington State Dept. of Ecol. Olympia. Publ. No. 92-10.

Clarke, J.A., B.A. Harrington, T. Hruby, and F. Wasserman. 1984. The effect of ditching for mosquito control on salt marsh use by birds in Rowley, Massachusetts. Journal of Field Ornithology 55: 160-180.

Culliton, T.J., M.A. Warren, T.R. Goodspeed, D.G. Remer, C.M. Blackwell, and J.J. McDonough III. 1990. Second report of coastal trends: 50 years of population change along the nation's coast, 1960-2010. U.S. Dept. of Commerce, National Oceanic and Atmospheric Admin., NOAA Strategic Assessment Branch, Rockville, MD.

Dahl, T.E. 1990. Wetlands losses in the United States 1780's to 1980's. U.S. Department of the Interior. Fish and Wildlife Service, Washington, D.C. 21 pp.

DeGraaf, R.M. and D.D. Rudis. 1986. New England Wildlife: Habitat, natural history, and distribution. U.S. Dept. Agr. Forest Service. N.E. Forest Exp. Sta. Gen. Tech. Report NE-108, Broomall, PA 461 pp.

DeLaune, R.D., J.H. Whitcomb, W.H. Patrick, J.H. Perdue, and S.R. Pezeshki. 1989. Accretion and canal impacts in a rapidly subsiding wetland. I. ^{137}Cs and ^{210}Pb techniques. Estuaries 12: 247-259.

Diamond, R.S. and D.J. Nilson. 1988. Buffer delineation method for coastal wetlands of New Jersey. Symp. Coast. Water Res. Amer. Water Res. Assoc. May 1988.

Dzierzeski, M. 1991. Vegetation changes in a hydrologically altered salt marsh ecosystem. Unpublished Masters thesis. University of New Hampshire, Durhham, NH.

Ehrenfeld, J.G. 1987. The role of woody vegetation in preventing groundwater pollution by nitrogen from septic tank leachate. Water Res.1 21:605-614.

EOEA. 1991. Report of the Living Resources Committee of the Technical Advisory Group on Marine Affairs. Commonwealth of Massachusetts, Executive Office of Environmental Affairs.

Ferrigno, F., P. Slavin, and D.M. Jobbins. 1975. Salt marsh water management for mosquito control. Proceedings of the 62nd Annual Meeting of the New Jersey Mosquito Control Association: 32-38.

Field, D.W., A.J. Reyer, P.V. Genovese, and B.D. Shearer. 1991. Coastal wetlands of the United States, an accounting of a valuable national resource. Strategic Assessment Branch, National Oceanic and Atmospheric Administration and U.S. Fish and Wildlife Service. 13 pp.

Florida Department of Natural Resources. 1992a. Florida salt marshes. 4 pp. (pamphlet).

Florida Department of Natural Resources. 1992b. Florida's mangroves. 4 pp. (pamphlet).

Frayer, W.E., T.J. Monahan, D.C. Bowden, and F.A. Graybill. 1983. Status and trends of wetlands and deepwater habitats in the coterminous United States. Dept. of Forest and Wood Science, Colorado State Univ. Ft. Collins, CO. 32 pp.

Frenkel, R.E. and T.R. Boss. 1988. Introduction, establishment and spread of Spartina patens on Cox Island, Siuslaw Estuary, Oregon. Wetlands 8: 33-49.

Glenn, E.P., J.W. O'Leary, M.C. Watson, T.L. Thompson, and R.O. Kuehl. 1991. Salicornia bigelovii Torr.: an oilseed halophyte for seawater irrigation. Science 251:1065-1067.

Hagedorn, C. 1984. Microbiological aspects of groundwater pollution due to septic tanks. In Bitton, G. and C.P. Gerba (Eds.), Groundwater Pollution Microbiology. John Wiley and Sons, Inc. New York. pp 181-196.

Hemond, H.F. and J. Benoit 1988. Cumulative impacts on water quality functions of wetlands. Environmental Management. 12:639-653.

Heufelder, G. 1988. Bacterial monitoring in Buzzards Bay. EPA 503/4-88-001. U.S. Env. Prot. Agency, Region 1, Boston, MA.

Howes, B, R. Howarth, J.M. Teal, and I. Valiela. 1981. Oxidation-reduction potentials in a salt marsh. Spatial patterns and interactions with primary production. Limnol. Oceanogr. 26:350-360.

Hruby, T., W.G. Montgomery, R.A. Lent, and N. Dobson. 1985. Open marsh water management in Massachusetts: adapting the technique to local conditions and its impact on mosquito larvae during the first season. Journal of the American Mosquito Control Association 1:85-88.

Hruby, T., and W.G. Montgomery. 1986. Open marsh water management manual for open tidal marshes in the Northeast. Massachusetts Audubon Society. Ecosystems Management Series.

Jensen, A. 1985. The effect of cattle and sheep grazing on salt marsh vegetation at Skallingen, Denmark. Vegetation 60:37-48.

Jordan, T.E., D.L. Correll, W.T. Peterjohn, and D.E. Weller. 1986. Nutrient flux in a landscape: the Rhodes River watershed and receiving waters. In D.

Correll (ed.) Watershed Research Perspectives. Smithsonian Env. Research Center. Edgewater, MD. pp. 57-76.

Keswick, B. and C. Gerba. 1980. Viruses in groundwater. Environmental Science and Technology. 14:1290-1297.

Kirby, R.E. 1988. American black duck breeding habitat enhancement in the northeastern United States: a review and synthesis. U.S. Fish Wildl. Serv. Biol. Report 88(4). 50 pp.

Knight, R.L., T.W. McKim, and H.R. Kohl. 1987. Performance of a natural wetland system for wastewater management. Journal Water Poll. Control. Fed. 59:746-754.

Lacey, L.A. and A.H. Undeen. 1986. Microbial control of mosquitoes and blackflies. Annual Review of Entomology 11:265-296.

Lawrence, R.R., R. Leonard, and J.M. Sheridan. 1985. Managing riparian ecosystems to control nonpoint pollution. Journal of Soil and Water Cons. 40:87-91.

Lawrence, R.R., J.K. Sharpe, and J.M. Sheridan. 1986. Long term sediment deposition in the riparian zone of a coastal plain watershed. Journal of Soil and Water Cons. 41:266-271.

Lesser, C.R., F.J. Murphy, and R.W. Lake. 1976. Some effects of grid system mosquito control ditching on salt marsh biota in Delaware. Mosquito News 36: 69-77.

Massachusetts Office of Coastal Zone Management. 1988. Guidelines for dock and pier construction in ACECs and ocean sanctuaries. Mass. Coastal Zone Management, Exec. Off. of Environmental Affairs, Boston, Massachusetts.

Mendellsohn, I.A., K.L. McKee, and W.H. Patrick. 1981. Oxygen deficiency in *Spartina alterniflora* roots: metabolic adaptations to anoxia. Science 214:439-441.

Mendelssohn, I.A. and K.L. McKee. 1988. *Spartina alterniflora* die-back in Louisiana: time course investigation of soil waterlogging effects. Journal of Ecology 76: 509-521.

Miller, W.B. and F.E. Egler. 1950. Vegetation of the Weqetequock-Pawcatuck tidal-marshes, Connecticut. Ecological Monographs 20:143-172.

Mitchell, R. and C. Chamberlain. 1978. Survival of indicator organisms. In G. Berg (ed.) Indicators of viruses in water and food. Ann Arbor Science Publ., Inc., MI. pp. 15-38.

Montague, C.L., A.V. Zale, and H.F. Percival. 1987. Ecological effects of coastal marsh impoundments: a review. Environmental Management 11: 743-756.

Niering, W.A. and R. Scott Warren. 1980. Vegetation patterns and processes in New England salt marshes. BioScience 30: 301-307.

Nixon, S. 1982. The ecology of New England high salt marshes. U.S. Fish and Wildlife Serv., Office of Biol. Serv., Washington, D.C. FWS/OBS-81-55. 70 pp.

Ofiara, D.D. and J.R. Allison. 1986. A comparison of alternative mosquito abatement methods using benefit-cost analysis. Journal of the American Mosquito Control Assoc. 2:522-528.

Phillips, J.D. and L.R. Phillips 1988. Delineation of shoreline buffer zones for stormwater pollution control. Symp. on Coastal Water Res. Amer. Water Res. Assoc.

Portnoy, J.W. 1982. Report on the mosquitoes of the Herring River Basin, Cape Cod National Seashore, 1982. U.S. Dept. of Interior, National Park Service. North Atlantic Region Water Resource Program, Report #12.

Portnoy, J.W. 1991. Summer oxygen depletion in a diked New England estuary. Estuaries 14: 122-129.

Provost, M.W. 1977. Source reduction in salt marsh mosquito control: past and future. Mosquito News 37: 689-698.

Resh, V.H. and S.S. Balling. 1983. Tidal circulation alteration for salt marsh mosquito control. Environmental Management 7: 77-84.

Rhode Island Coastal Resources Management Program. 1989. A Special Area Management Plant for the Salt Pond Region. November, 113 pp.

Rogers, J., S. Syz, and F. Golden. 1975. Maryland upland natural areas study. Volume 1: Eastern Shore. Prepared for Maryland Coastal Zone Management Program, Annapolis, MD. 67 pp.

Roman, C.T. and R. Good. 1985. Buffer delineation model for New Jersey Pinelands wetlands. Div. of Pinelands Res. Center for Coastal and Env. Studies, Rutgers Univ, New Brunswick, NJ. 73 pp.

Roman, C.T., W.A. Niering, and R.S. Warren. 1984. Salt marsh vegetation change in response to tidal restriction. Environmental Management 8: 141-150.

Rozas, L.P., C.C. McIvor, and W.E. Odum. 1988. Intertidal rivulets and creek-banks: corridors between tidal creeks and marshes. Marine Ecology Progress Series 47: 303-307.

Shisler, J. and D.M. Jobbins. 1975. Aspects of biological productivity in mosquito ditched salt marshes. Proc. N.J. Mosq. Exterm. Assoc. 62: 48-49.

Shisler, J., F.H. Lesser, and T.L. Schultze. 1975. Re-evaluation of some effects of water management on the Mispillion Marsh, Kent County, Delaware. Proc. N.J. Mosq. Exterm. Assoc. 66: 276-278.

Shisler, J., T.L. Schultze and B.I. Howes. 1978. The effect of the marsh elder (*Iva frutescens*) on the standing crop biomass of *Spartina patens* and associated wildlife. Biol. Conserv, 14:159-166.

Short, H.J. and R.J. Cooper. 1985. Habitat suitability index models: great blue heron. U.S. Fish Wildl. Serv. Biol. Rep. 82(10.99). 23 pp.

Sinicrope, T.L., P.G. Hine, R.S. Warren, and W.A. Niering. 1990. Restoration of an impounded salt marsh in New England. Estuaries 13: 25-30.

Slavin, P. and J.K. Shisler. 1983. Avian utilization of a tidally restored salt hay farm. Biological Conservation 26: 271-285.

Tiner, R.W. 1991. Recent changes in estuarine wetlands of the coterminous United States. Coastal Wetlands Coastal Zone '91 Conference-ASCE. pp. 100-109.

Turner, R.E., and Y.S. Rao. 1990. Relationships between wetlands fragmentation and recent hydrologic changes in a deltaic coast. Estuaries 13: 272-282.

Urner, C.A. 1935. Relation of mosquito control in New Jersey to bird life of the salt marshes. Proceedings of the 22nd Annual Meeting of the New Jersey Mosquito Extermination Association: 130-136.

Valiela, I. and J. Costa 1988. Eutrophication of Buttermilk Bay, a Cape Cod coastal embayment: concentrations of nutrients and watershed nutrient budgets. Environ. Manage. 12:539-553.

Valiela, I., J.M. Teal, and W.G. Deuser. 1978. The nature of growth forms of the salt marsh grass, *Spartina alterniflora*. American Naturalist 112:461-470.

Ward G. and G.J. Fitzgerald. 1983. Fish predation on the macrobenthos of tidal salt marsh pools. Can. J. Zool. 61:1358-1361.

Watzin, M.C. and J.G. Gosselink., 1992. The fragile fringe: coastal wetlands of the continental United States. Louisiana Sea Grant College Program, Louisiana State University, Baton Rouge, LA; U.S. Fish and Wildlife Service, Washington, D.C.; and Notional Oceanic and Atmospheric Administration, Rockville, MD.

Weissberg, S.B. and V.A. Lotrich. 1982. Importance of an infrequently flooded intertidal marsh surface as an energy source for the mummichog, *Fundulus heteroclitus*: an experimental approach. Marine Biology 66:307-310.

Woodwell, G.M. 1977. Recycling sewage through plant communities. American Scientist 65:556-562.

Zedler, J.B., T. Winfield, and P. Williams. 1980. Salt marsh productivity with natural and altered tidal circulation. Oecologia 44: 236-240.

CHAPTER 16

WETLANDS EDUCATION

Kathleen Drake and Mario Del Vicario

Twenty years after enactment of the Clean Water Act, wetlands are continuing to decline in acreage and functional quality. The general public is aware of wetlands as a controversial issue, but for the most part cannot identify wetlands in their local area, and most likely do not understand how a local wetland contributes to the proclaimed values of flood protection, pollutant attenuation, aquifer protection, food chain support, and fisheries and shell fisheries production. Nor, generally, does the public understand that the loss of these values can result in local economic burdens in the form of natural disasters, sanitary engineering, water supply and quality, and lost revenue. Those who are concerned about the status of remaining wetlands have looked to the role of wetlands education as an effective, nonregulatory means towards wetlands protection. Given the extent of remaining wetlands in private ownership (74%, Ducks Unlimited 1990 Annual Report), there is emphasis among educators to impart an understanding of the resource as valuable and to foster the growth of a personal desire to protect and manage wetlands properly. The emphasis on wetlands education in affecting wetlands protection is gaining momentum. Within the past few years agencies and organizations have begun to inventory what materials exist, assess the focus of their education and outreach efforts and assess how organizational energies should be directed. There has been a surge in the development of new wetlands educational materials. Cooperative efforts among agencies and organizations have been undertaken. Educational efforts have become one of the priority tools in the wetlands protection strategy.

There are a number of agencies, organizations, and associations that function as educators in that they produce wetlands informational materials or are otherwise involved in wetlands education efforts (Figure 1). There are a wide array of existing wetlands educational materials. This chapter identifies the wetlands educators, briefly summarizes their educational and outreach efforts and synthesizes the array of existing materials. This synthesis provides the basis for development of situation-specific education programs. To facilitate the discussion, wetlands informational materials are divided into four categories: 1) wetlands awareness, 2) wetlands regulations and legislation, 3) wetlands conservation and planning, and

Figure 1. A number of agencies and organizations are directly or implicitly involved in wetlands education, and produce a variety of educational materials. Photo courtesy of the U.S. Fish and Wildlife Service.

4) wetlands science and management. Within each category the focus of information and the primary target audiences are identified. Informational materials pertaining to these categories are listed in appendices A through D, and Appendix E provides contact information enabling inquiry and acquisition of the materials. Please note that these listings are extensive but not exhaustive, and new materials are planned and are being produced. Therefore, the tracking of informational materials is an ongoing effort. Following examination of the educators and the informational materials, apparent gaps in educational efforts are noted. Finally, the chapter provides general guidance in implementing a wetlands educational effort. Hopefully, this information will assist in planning a wetlands education program and avoiding redundancy in such effort, and will guide an educator over the initial hurdles of identifying contacts and acquiring materials.

WETLANDS EDUCATORS

Federal Agencies

There are a variety of federal laws that directly or indirectly involve and affect wetlands. *Wetlands Overview: Federal and State Policies, Legislation, and*

Programs, Statewide Wetlands Strategies: A Guide to Protecting and Managing the Resource, and the *Wetlands Protection Workbook* written by the Environmental Law Institute in conjunction with the United States Environmental Protection Agency provide an overview. A number of federal agencies are authorized to implement the various laws, or are authorized to oversee public lands that include extensive wetland acreage. Many of these federal agencies have generated, or are beginning to generate, outreach materials on wetlands.

United States Department of Commerce

National Oceanic and Atmospheric Administration

NOAA has established an extensive outreach network within their Sea Grant program. Academic institutions are involved in program research, and Sea Grant personnel coordinate the research information in their outreach activities and publications. There is a National Sea Grant Depository at the University of Rhode Island, Narragansett, serving as a centralized library of all Sea Grant publications. The depository provides index searches, and loans written materials.

National Oceanic and Atmospheric Administration also administers the National Estuarine Research Reserve System and the National Marine Sanctuaries. There are currently twenty three estuaries within the system, whose focus is research and education. Most reserves offer interpretive programs, tours, and workshops. Many have produced guides and workshops for the application of the curriculum to other audiences. For example, Old Woman Creek National Estuarine Research Reserve has developed wetlands informational materials offered in realtor certification courses. There are eight marine sanctuaries and, although they are used by the public, organized outreach programs and materials generally have not been developed. According to personnel from the Sanctuaries and Reserve Division, the sanctuaries are soon likely to place greater emphasis on outreach. National Oceanic and Atmospheric Administration's Sanctuaries and Reserves Division is in the process of centralizing information on their educational materials and programs, and should be kept in mind for future contact.

United States Department of Agriculture

Cooperative Extension Service, Agricultural Stabilization and Conservation Service, Soil Conservation Service

The Cooperative Extension Service has established an extensive outreach network within their Land Grant program. As with the Sea Grant program, academic institutions are involved in program research, and Cooperative Extension Service personnel coordinate those efforts within their outreach area. The Cooperative Extension Service does not have a centralized depository, but they recently inventoried the state offices and assembled a listing of wetlands informational materials available through those offices. This listing can be attained by contacting the Cooperative Extension Service office of Natural Resources and Rural Development in Washington, D.C.

The Agricultural Stabilization and Conservation Service administers the Swampbuster, Conservation Reserve, and Wetland Reserve provisions of the Food Security Act. The Swampbuster provision directly affects agricultural producers' eligibility for other federal benefits, as violations of the provision result in farmers losing eligibility for commodity program benefits, crop insurance, disaster payments, and other federal benefits. The Conservation Reserve Program authorizes the federal government to enter into contract with agricultural producers to remove highly erodible cropland from production for a period of ten to fifteen years in return for annual rental payments. Cropped wetlands were eligible for enrollment beginning in 1989. The Wetlands Reserve Program was established in 1990 with the goal of restoring farmed wetlands and riparian corridors on private lands by means of lump sum payments for thirty-year and permanent easements, and implementation of restoration plans. The Soil Conservation Service provides technical expertise in the implementation of these programs, and although its focus has not been to develop educational materials, its outreach has been extensive through field work, personal contact with landowners, and appearances at local information sessions. There exists an extensive network, at a very local level, of technical support through the state offices of the Soil Conservation Service and their partnerships with the Conservation Districts. These agencies provide informational materials on the Department of Agriculture programs and provide local technical support to landowners potentially involved in the programs.

National Association of Conservation Districts

A conservation district is a local subdivision of state government that utilizes and coordinates assistance from a variety of private and public sources including local, state and federal agencies and organizations. The Conservation Districts work to balance the conservation and development of land, water, forest, wildlife and related resources. The National Association of Conservation Districts was organized in 1946 by districts and their state associations to serve as a national voice. National Association of Conservation Districts develops national policies, maintains relationships with organizations and government agencies, and publishes information about districts.

The Conservation Districts have established the Envirothon program, a competitive, problem-solving event for high school level students which teaches natural resource information. The local Conservation District coordinator establishes an Envirothon resource committee comprised of soils, wildlife, and fisheries experts, environmental organizations, and interested civic groups who then coordinate with teachers and advisors in schools, 4-H, and other environmental clubs that wish to work with a natural resource training program. The Envirothon resource committee and the teachers and advisors plan training activities in five topics of natural resources, one of which is aquatic resources, which includes educational efforts on wetlands. There are county, state, and ultimately national competitions with apparently great appeal to high school students. The program is unique in its synthesis of federal, state, and local agencies and organizations and offers great opportunity for the dissemination of outreach materials and efforts.

United States Department of the Interior

United States Fish and Wildlife Service

The Fish and Wildlife Service has a varied involvement in wetlands research and outreach. They administer the National Wetlands Inventory program which provides status reports on the acreage and rate of loss of the nation's wetlands, and they offer a telephone information service that provides assistance in acquiring National Wetlands Inventory maps (1-800-USA-MAPS). The Division of Biological Services researches natural wetlands systems, and the results are reported in Biological Services Program publications. The Fish and Wildlife Service serves as technical consultant on wetlands restoration and creation efforts of the Conservation Reserve and Wetland Reserve programs administered by the Agricultural Stabilization and Conservation Service. The Division of Refuges administers the National Wildlife Refuge system, which often has interpretive centers and programs addressing wetland habitats.

The Fish and Wildlife Service also administers the Duck Stamp program whose revenues go towards the acquisition of additional lands for the National Wildlife Refuge System. A Junior Duck Stamp program has been established in California which includes wetlands activities for youths. The Service administers the North American Waterfowl Management Plan, an agreement signed by the United States and Canada which sets goals for duck, goose, and swan populations and identifies habitat conservation needs in specific regions of the country. The plan has served as a framework for federal agencies such as the Bureau of Land Management, Bureau of Reclamation, and Bureau of Indian Affairs, through cooperative agreements with the FWS, to restore degraded wetlands within the public lands those agencies oversee. Similarly, Fish and Wildlife Service has entered into cooperative agreement with Ducks Unlimited, Inc., in efforts towards private lands restoration.

FWS has a Publications Unit (703-358-1711) and an Audiovisual Department (202-208-5611) which collectively serve as a centralized clearinghouse for documents, posters, slides, and videos. These offices provide free copies or loan of FWS materials. Documents which are no longer in print generally can be acquired through the Government Printing Office (202-783-3238) or National Technical Information Service (703-487-4780).

United States Forest Service

The U.S. Forest Service has identified 12 million acres of wetlands within the 191 million acres of land it manages. The Forest Service has designated their wetlands management role as a *National Problem Area for Forest Service Research,* with program implementation to begin in 1993. Research is in progress in some locations on certain issues that the Service considers high priority, such as management of southern bottomland hardwoods for wildlife habitat, and the role of riparian areas as critical habitat and buffers between upland and water-dominated systems. Research program planning in the Forest Service identifies five areas of focus including ecosystem dynamics, restoration and rehabilitation, wetlands resource management, socioeconomic values, and landscape-scale linkages. The Service will be issuing a color brochure entitled *Wetlands in the Forest Service* describing the agency's efforts in wetlands research management and technology transfer.

Other Department of the Interior Agencies

The Department of the Interior issued a booklet entitled *Wetlands Stewardship* (1992) which provides a brief description of policies and programs within the National Park Service, the Bureau of Land Management, the Bureau of Mines, the Bureau of Reclamation, the Geological Survey, the Office of Surface Mining, the Minerals Management Service, and the Bureau of Indian Affairs. These agencies oversee millions of acres of wetlands. Their efforts focus primarily on wetlands inventory, restoration and enhancement, and habitat management. Within the National Park system, the extent of interpretive programs and materials on wetlands varies among parks. Within the Bureau of Land Management, there are plans to develop wetlands educational materials and outreach programs. The Bureau of Mines is involved in research and application of constructed wetlands for the biological treatment of acid mine drainage.

United States Environmental Protection Agency

The Environmental Protection Agency has established a Wetlands Hotline (1-800-832-7828) that provides information on wetlands issues or provides the information link to other sources. The Wetlands Hotline provides, to a limited extent, copies of materials generated by Environmental Protection Agency, and is informed on current wetlands regulatory, educational, and outreach efforts of the Agency. Those materials unavailable through the Hotline should be sought from the Environmental Protection Agency Publications and Information Center (EPIC, 513-569-7980), or the Environmental Protection Agency Public Information Center (202-260-7751).

Within the Environmental Protection Agency's Office of Wetlands, Oceans, and Watersheds, Wetlands Division, there is a Regulatory Branch, and a Wetlands Strategies and State Programs Branch. There are outreach efforts directed towards understanding the regulatory programs of the Agency and outreach efforts directed towards increasing wetlands awareness and conservation. Within this latter outreach effort there have been a number of cooperative efforts including work with the Environmental Law Institute, the National Audubon Society, and the World Wildlife Fund. The Agency has awarded grant monies toward developing wetlands preservation guidance. In addition to materials produced through the Washington, D.C. headquarters, regional Environmental Protection Agency wetlands offices have developed informational materials. These materials, although regionally focused, most often have a much broader applicability since their content includes general wetlands information, and addresses aspects of federal wetland programs. For example, Environmental Protection Agency Region 5, in coordination with Purdue University, has developed a series of interactive software programs on a variety of environmental issues, including their Wetlands Education System.

The Environmental Protection Agency also administers the National Estuary Program which now includes eighteen estuaries. The mission of the National Estuary Program is to identify nationally significant estuaries and to oversee a process for improving and protecting water quality, and enhancing living resources through development of a Comprehensive Conservation and Management Plan. If an estuary meets the program guidelines, Environmental Protection Agency convenes a Management Conference. Conferences include representatives of citizen and user groups, scientific and technical institutions, and government agencies and resource managers at the federal, state, and local levels. Representatives oversee development of Comprehensive Conservation and Management Plan. A national volunteer monitoring network was formed as the result of Environmental Protection Agency sponsored national workshops. The National Estuaries Programs issue regular newsletters and incorporate public education efforts.

The Environmental Protection Agency also has a number of Environmental Research Laboratories. Research at laboratories in Narragansett, Rhode Island; Gulf Breeze, Florida; and Athens, Georgia focus on estuarine and marine water and sediment quality criteria, and estuarine and marine ecology. Research at the laboratory in Corvallis, Oregon focuses on wetlands, particularly the assessment and restoration of degraded environments, cumulative impacts, and watershed and landscape ecology. Each of these laboratories have associated libraries housing technical documents and databases, and they provide reference services to the general public. Their library services and contacts are listed in Access Environmental Protection Agency Library and Information Services (EPA/IMSD-91-106), which can be acquired through EPIC.

United States Department of Defense

U.S. Army Corps of Engineers

The wetlands research, education, and outreach efforts of the Corps of Engineers are conducted by the Waterways Experiment Station. Waterways Experiment Station developed the Wetland Evaluation Technique (WET), a wetland assessment methodology, and provides training in its use. They recently issued a document entitled *Wetland Evaluation Technique, Volume 1: Literature Review and Rationale,* which reviews the literature and evaluates the rationale on which WET is based. The Waterways Experiment Station also conducts training, primarily for federal employees, in federal wetlands delineation methodology and conducts research and training in wetlands restoration and creation. Agencies can also request training for nonfederal employees. The Corps of Engineers generally make their documents available through the National Technical Information Service.

State Agencies

Many of the regional and state offices of the federal agencies listed above, for example, the Sea Grant and Land Grant programs, the Soil Conservation Service, and the Fish and Wildlife Service, are involved in federal, state, and local partnerships. The Conservation Districts are a local subdivision of state government. Although the agencies listed in the section above are federally administered or have national associations, their offices have a state or regional identity and provide wetlands education and outreach efforts to those geographic audiences.

Government emphasis on wetlands issues varies among states. Notably, Michigan, Minnesota, North Carolina, Pennsylvania, Virginia, Washington, and Wisconsin Departments of Natural Resources have published special wetlands issue magazines, protection guides, curricula, video, and filmstrips on wetlands. Some states have touched upon wetlands-related information in publications issued within the Water Resources (e.g., Kentucky's Water Watch program) and Environmental Education divisions. Although these publications focus on the subject state's wetland resources, the documents most often include general information about wetlands and certainly shouldn't be overlooked because of their state focus.

Many state agencies, often coordinated by the Fish and Game or Natural Resources Departments, conduct workshops for Aquatic Project WILD, a curriculum sponsored by the Western Regional Environmental Education Council that includes wetlands activities. Many of these agencies issue newsletters, again, through the Natural Resources or Fish and Game Departments, that provide information on wetland contacts, resources and materials, field trips, and training workshops. The *Carolina Notebook* of the North Carolina Wildlife Resources Commission and *EE News: Environmental Education in Wisconsin* of the Wisconsin Department of Natural Resources Bureau of Information and Education are two examples. Some have extensive volunteer programs which, although not solely focused on wetlands, provide opportunities for field experiences in wetlands. For example, the New Jersey Fish and Game Department sponsor an Environmental Conservation Corps.

The Council of State Governments, Center for the Environment and Natural Resources, in cooperation with the Environmental Protection Agency, has established an electronic WETLANDS database of state wetland protection programs and contacts. The Council also provides loan services from their collection of wetlands informational materials. The Environmental Law Institute has published a Directory of State Environment Agencies, and the National Wildlife Federation annually publishes their Conservation Directory listing state agency offices and contacts.

Nonprofit Environmental Organizations

The primary organizations which generate or incorporate wetlands educational materials and efforts include the National Audubon Society, regional and local National Audubon Society chapters, the National Wildlife Federation and its affiliate organizations, and Ducks Unlimited. Organizations involved with the acquisition and management of lands, including wetlands, for conservation such as the Nature Conservancy and the Trust for Public Land, are involved primarily with conservation and minimally with outreach and education efforts. Contact information for these and other environmental organizations can be found in the National Wildlife Federation's Conservation Directory.

The National Wildlife Federation and National Audubon Society have produced a number of wetlands educational materials, such as curricula and documentary specials. The current major emphasis of these organizations is the promotion of local grassroots involvement and public activist campaigns. The National Wildlife Federation previously produced the Nature Scope series for elementary and junior high school age children, and conducted workshops based upon on the series including *Wading into Wetlands*. Although the National Wildlife Federation has discontinued production of Nature Scope and the associated workshops, past issues, including *Wading Into Wetlands*, are still available for purchase. Also, National Wildlife Federation coordinates with affiliated organizations in most states and publishes a monthly newsletter, *The Leader*. Audubon has a network of regional offices and state chapters in addition to the national offices. There are a few states with independent state Audubon Societies, such as Massachusetts, New Hampshire and New Jersey, that are not affiliated with the National Audubon Society. National Audubon Society regional offices and state chapters, and the National Wildlife Federation and its affiliated organizations, are active in generating educational materials such as slide and video presentations for a broad audience.

There are a number of organizations whose members include hunting and fishing enthusiasts, which have a direct interest in educating their subscribers about wetlands and preserving wetland habitat. Ducks Unlimited has established its Green Wing program for elementary through high school level participants, which supports wetlands and wildlife-related activities and field trips. Ducks Unlimited publishes a children's magazine, Puddler, in cooperation with the Fish and Wildlife Service. Ducks Unlimited interacts in an advisory and financial capacity with local land users in efforts for private land to be managed in the interest of wetland habitat conservation, and they have generated, either in-house or in cooperation with federal agencies, outreach materials geared toward a private landowner's interest in wetland conservation.

Professional Associations and Professional Training

Associations

The Western Regional Environment Education Council has sponsored the Project WILD and Aquatic Project WILD curricula, and is developing Project WET, which focuses on water resources including wetlands, and *Watercourse*, an adult guidebook addressing wetlands topics. Aquatic Project WILD has been implemented in most states. Educators associations produce or endorse a selection of environmental education materials, some of which include materials specifically on wetlands. The National Science Teachers Association distributes two posters,

Wetlands and *Who Lives in Mud*, and a board game *Energy Flow in a Wetland*. The National Marine Educators Association issues *Current: The Journal of Marine Education* and the *National Marine Educators Association News*. The National Geographic Society has education materials catalogs which include a few filmstrips and videos on specific regional wetland ecosystems, for example the Everglades and Okefenokee Swamp, and ecosystem types such as ponds and tidal marshes. There are a number of education media distribution companies, such as Educational Images, Ltd. and Bullfrog Films, which distribute wetland specific films and videos.

There are a number of associations within the academic community and among wetlands professionals involved in the exchange and dissemination of information about wetlands research, regulation, policy, and wetlands management. The Environmental Law Institute provides a number of publications for wetlands professionals and environmental law professionals. Their National Wetlands Newsletter is a primary source of current wetlands information. The Environmental Law Institute conducts training courses and hosts professional gatherings and conferences. The Society of Wetland Scientists publishes a journal and a newsletter which address wetland research, symposia and conferences, events, and publications. Society of Wetland Scientists holds annual national conferences, and the local chapters of Society of Wetland Scientists also conduct regional symposia and meetings, and publish regional newsletters. Also, the Estuarine Research Federation publishes the journal *Estuaries*, issues the Estuarine Research Federation newsletter, and holds biannual conferences. The Association of Wetland Managers and the National Wetlands Technical Council host symposia and conferences and publish proceedings on wetlands management practices and the synthesis of wetlands science. The International Association of Ecology, Intecol, hosts an annual international conference on wetlands science and management.

Training

There are many universities and colleges supporting academic programs with an emphasis on wetland and estuarine science. The *University Curricula Guide of the Marine Technology Society*, the *National Wetlands Newsletter*, and the *Bulletin of the Society of Wetland Scientists* are good information sources for program listings. Many universities and colleges offer noncredit continuing education courses. These courses generally focus on wetland regulations, wetlands identification and delineation, and wetlands restoration and creation, as opposed to the general ecology of wetlands. A number of consulting groups also offer training courses on the delineation, functional assessment, restoration and creation of wetlands. One group, Environmental Concern, provides services for initiating and implementing wetlands education efforts.

Information Networks and Clearinghouses

Other resources do not generate their own information, but serve as a database, network, and clearinghouse. Generally, the information bases are broad but are inclusive of information on wetlands and estuaries. The Alliance for Environmental Education operates the Network for Environmental Education and issues the *Network News*. The Alliance Network includes a system of environmental education and training centers with programs and services that include teacher and professional training, outreach, program development and environmental research. Centers operate at the campuses of colleges, universities, and other institutions. Environmental centers affiliated with the AEE are mandated to serve schools and members of the public at a grassroots level. The Alliance serves as an arena for exchange of materials developed by its participating membership.

Similarly, an information network which does not generate materials, but provides users with access to scientific and education literature, is the Educational Resources Information Center (ERIC). ERIC collects, analyzes, and distributes information from local, state, federal and international agencies, and private sources. They also provide literature searches and synopses. Ohio State University operates the clearinghouse on science, mathematics, and environmental education, and may be accessed through CD-ROM technology located in most universities and many colleges.

Aquariums, Zoos, and Museums

Many states house aquariums, zoos, and natural history museums. Although wetlands information is not always the singular focus of exhibits, there are many aquariums, zoos, and museums which either address wetlands habitat pertinent to their zoological collections or have dedicated an exhibit to wetlands. Examples of the latter include the National Aquarium in Baltimore, the New York Aquarium, the Virginia Marine Science Museum, the Los Angeles County Museum, and the creation of wetlands habitat at the National Zoo in Washington, D.C. Nearly all aquariums support an education program. Some include program elements specifically on wetlands, which are listed in *Education Programs at North American Aquariums*. Contact with facility education directors can steer the inquirer to varied wetlands education opportunities. The Executive Office and Conservation Department of the American Association of Zoos, Parks, and Aquariums (AAZPA) provides information on member organizations, and contact with that office offers information on exhibit contents. The AAZPA also publishes the *Directory of Zoological Parks and Aquariums in the Americas*. The American Association of Museums publishes an *Official Directory* which provides a listing by state of their members. Along with museums, the Directory lists aquariums, botanical and aquatic gardens, bird sanctuaries, environmental education centers, marine museums, wildlife refuges, and zoos with information on contacts, facilities, activities, fields of research, and publications.

WETLANDS INFORMATION

Nearly all the materials listed in appendices A through D provide an introduction to wetlands, including definition, types, functions, causes of loss and rates of loss. Whether a target audience consists of a concerned public, those involved in public or extracurricular education (teachers and leaders, students and youth), those that impact wetlands (farmers, foresters, developers), or those that influence activities in wetlands (local authorities, planners, realtors, architects, engineers), there is generally pertinent information available.

Wetlands Awareness

Information Focus: Audiences:

What are wetlands?   All
The variety of wetlands
The functions and values of wetlands
The status of wetland loss

Nearly all wetlands informational materials address the message of wetlands awareness (Figure 2). Some materials provide this information as introduction to messages on regulation, management or conservation. The primary objective of informational materials listed in Appendix A is to advise the audience of the occurrence of wetlands and their value. They are directed towards fostering a new perspective on wetlands as a resource to be protected and wisely managed.

Application to a Wetlands Education Effort

Teaching Materials

The curricula are the essential educational tools which can be enhanced or aided by the selection of videos, posters, booklets, and activity books. The curricula list a great assortment of activities, projects, field experiences, materials and logistical considerations for presenting the educational message. The Resources listing in Appendix A can aide in further research for leader planning and youth assignments.

Generally, the fundamental objectives of these curricula are to understand what wetlands are, including plant and animal inhabitants, functions, and, with some, the source of impacts. Note that application of some of the curricula, for example Aquatic Project WILD, requires participation in a training workshop in order to receive and use the curriculum. Most of these curricula are inexpensive, costing $10.00 or less. Materials produced by the Sea Grant program can be examined through request for loan from the National Sea Grant Depository. Additional listings to note in Appendix A are: 1) EPA's interactive software program, Wetlands Education System, which can be acquired without cost through submittal of two suitable computer discs, and 2) Decision Making and the Chesapeake Bay (University of Maryland Sea Grant Program). The latter document introduces the tool of simulated decision-making and considerations of the wetland resource within the greater arena of social and political issues: concepts which transcend its regional focus.

Figure 2. Nearly all wetlands informational materials address the functions and values of wetlands, such as provision of fisheries habitat and recreational opportunities. Photo courtesy of the U.S. Environmental Protection Agency.

Posters, coloring posters, and activity sheets from the federal and state agencies can most often be acquired free of charge. Fish and Wildlife Service videos are available for loan through the Audiovisual Department. Videos for purchase from state and federal agencies and environmental organizations are generally less than $30.00. Video and film distribution companies offer rental and often allow review of their product prior to purchase. Appendix A also lists some magazine sources on wetlands. The National Geographic Society has a magazine index service (e.g., "Our Disappearing Wetlands", Mitchell, John G., October 1992, National Geographic Magazine), and offers purchase of back issues. Magazines issued by state agencies may be out of print and harder to acquire, but are well worth the effort. Although the text may be advanced for a young age group, the photographic contents are excellent.

Field Trips

No doubt the most lasting learning experiences are those reinforced by field trips (Figure 3). The most fortunate situation is to be located near a refuge, park,

Figure 3.   Field trips to a local wetland are an effective way to reinforce classroom educational programs.   Photo courtesy of the U.S. Fish and Wildlife Service.

research reserve, environment center, field research station, museum, or aquarium that has an established wetlands interpretive or curriculum program. The National Oceanic and Atmospheric Administration's Reserves and Sanctuaries Division provides a brochure on the locations of, and contacts for, the Estuarine Research Reserves, and the Fish and Wildlife Service Publications Unit provides a brochure listing information on National Wildlife Refuges.   There are also numerous environmental centers associated with the Alliance for Environmental Education. The National Audubon Society and its affiliated chapters own or manage a number of sanctuaries, although some do not allow visitation. National Audubon Society offers two directories to these sanctuaries and their education centers.

National Audubon Society chapters, National Wildlife Federation affiliate organizations, the Littoral Society, and local environmental organizations often lead field trips to wetlands areas.   The most direct access to information on NAS chapters is through contact with the regional offices, and information on these can be received from the NAS Regional and Government Affairs office.   The NWF's Conservation Directory provides contact information (federal, state, private, academic) to which inquiry can be made about environment centers, research stations, and local organizations.

Opportunities with the aforementioned organizations are not essential, as there may indeed be an easily accessible local wetland. As *WOW: The Wonders of*

Wetlands points out, a little perseverance and the effective use of a few key resources will lead to the discovery of an appropriate local wetland. The National Wetlands Inventory map hotline (1-800-USA-MAPS), places callers in contact with the Earth Science Information Center, U.S. Geological Survey, which assists in determining the National Wetlands Inventory and topographic maps needed for an area, and provides information on how to acquire the maps. A number of state Departments of Natural Resources have inventoried and mapped wetlands within their states. The section on wetlands mapping (below) lists a number of information sources used in wetland delineation efforts. There is a chance that the local Cooperative Extension Service, Soil Conservation Service and Conservation District offices have worked with local landowners of wetlands and thus are familiar with wetlands locations.

Extracurricular and Volunteer Opportunities

The older youth and adult age levels open the opportunity for participation in volunteer programs. Ducks Unlimited offers the Green Wing program, which invites youth participation in activities and field trips. The FWS offers the Junior Duck Stamp program which includes wetlands activities. The Izaak Walton League and a number of state Department of Natural Resources or Water Resources administer Adopt-A-Wetland programs. Environmental Protection Agency Region 1 is working on an Adopt-A-Wetland guide. Each of these organizations distributes information packages, and provide an opportunity to sign up with established adoption programs. Information gathered by the volunteering individual or group is often kept on record by resource agencies. The National Park Service, EPA's National Estuary Programs, the FWS Refuge System, many state Game and Fish Departments, Ducks Unlimited, and aquariums and museums have established volunteer programs that include training. Although the extent of direct contact with wetlands depends on wetland availability, they are sources to investigate for potential volunteerism. The Rhode Island Sea Grant Program issues a National Directory of Citizen Volunteer Environmental Monitoring Programs booklet which currently includes some wetlands-oriented programs. With regard to volunteer programs, a provision was made in passage of the National Environmental Education Act for the organization of a Senior Environment Corps. The Senior Environment Corps hopes to provide a computer data base linkage for volunteerism by retired and senior citizens.

The Cooperative Extension Service administers the 4-H program which teaches information on natural resource management, and although its focus is not solely wetlands, it includes wetlands information in its programs. At the high school level, schools in a number of states are participating in the Conservation District's Envirothon, a program administered by the local Conservation Districts (see the Federal Agencies section above), which provides education in a variety of natural resources, including wetlands.

Wetland Regulations and Legislation

Information Focus: Audiences:

Wetlands regulations   Wetland professionals
Identification of regulators   Those that directly impact
Current policies   wetlands
Information contacts   Those that influence
                                                activities in wetlands
                                            Concerned and activist
                                                public

The informational materials listed in Appendix B are designed to assist audiences in understanding the regulations implementing Section 404 of the Clean Water Act, and the associated state laws and programs with authority over activities in wetlands. The workbooks and guidebooks provide contact information. The Environmental Law Institute's National Wetlands Newsletter is a primary information source on current developments in, and influences on, wetlands regulations and legislation. The primary objective of this category of materials is to avoid wetland impact and loss, to avoid delays and frustration experienced by the regulated community, and to provide guidance to those who wish to participate in the public regulatory process.

Application to a Wetlands Education Effort

Wetlands are a resource subject to multiple levels of regulation from local government bodies and state and federal agencies. The Environmental Protection Agency is attempting to strengthen the role of states in wetlands regulation and protection, primarily through the Office of Wetlands State Grants programs and other grants related to preservation planning. A number of states have developed strategies for wetland protection that include strengthened regulatory elements, tax incentives, area-wide, interagency conservation agreements, and education and outreach. Also, there are a number of state and federal programs with incentive or disincentive programs for conservation and restoration of wetlands. An informed landowner or planner can save considerable time and expense through familiarity with the laws and regulations affecting wetlands alterations.

Wetlands Mapping

Wetlands regulations are applicable to proposals for alteration of wetlands as defined by local, state, and federal agencies. Wetland determinations often vary among regulating agencies, even though many agencies have incorporated the federal definition and approach for wetlands delineation. The Corps of Engineers information pamphlets state that regional offices will determine wetlands jurisdiction. However, the response time is often unrealistic for those who wish to know of extent of jurisdictional wetlands immediately. Consulting professionals provide a much faster service of delineating wetlands. Also, consulting professionals, as well as many college and university continuing education programs, offer courses in wetlands delineation methods and wetlands regulations. The Corps of Engineers Waterways Experiment Station conducts such courses, but generally they are restricted to federal employees and state wetlands managers.

Many states have inventoried and mapped their wetlands. Inquiry should be made with the appropriate state Departments of Natural Resources. *Statewide Wetland Strategies* reviews information sources often used in wetland delineation efforts and provides contact information for acquisition of maps. Regional Corps of Engineers and Environmental Protection Agency offices have mapped wetlands within local areas. Information regarding local delineations can be obtained from the National Wetlands Inventory Coordinator in the regional office of the Fish and Wildlife Service. Other mapping efforts have been undertaken in support of Soil Conservation Service County Soil Survey maps, Soil Conservation Service State Soil Geographic Data Base maps and National Soil Geographic Data Base maps, Soil Conservation Service Swampbuster maps, Federal Emergency Management Agency Flood Hazard maps, U.S. Geological Survey Land-Use Land Analysis maps, and the National Oceanic and Atmospheric Administration Coastal Ocean Program/Crosswatch Change Analysis Program data base.

Wetlands Regulations

Although state and federal agencies define and implement wetlands regulations, the consulting professions and academic continuing education courses are the means for using and interpreting those regulations. As mentioned above, consulting professionals offer courses in wetlands regulations. Many state agencies and regional federal offices have issued informational materials on wetlands regulations. These materials provide contact information for further inquiry with the appropriate state and federal wetlands personnel. Appendix B lists a number of informational materials addressing wetlands regulations. Several documents are of particular interest. *Federal Wetlands Protection Program in New England: A Guide to Section 404 for Citizens and States* is applicable beyond its targeted audience of the New England states. Other documents which focus on the Section 404 program include *Wetlands: The Corps of Engineers' Administration of the Section 404 Program*, and the National Wildlife Federation's *Citizens' Guide to Wetlands Protection*. Washington State Department of Ecology has produced a Wetlands Regulation Guidebook jointly covering federal and Washington State regulations, and, although regional, can serve as an introduction to the state and federal regulatory processes. The Council of State Governments recently updated their WETLANDS database which provides information on state wetlands protection programs and contacts.

The regulatory program of Section 404 of the Clean Water Act is probably familiar to many. There are other federal regulatory programs, the Coastal Zone Management Act for example, and a host of other legislation and federal support and development programs affecting activities in wetlands. There are many wetlands restoration efforts spurred by the Food Securities Act, Land and Water Conservation Fund Act, Sport Fish and Wildlife Restoration Acts, and numerous others. These regulations, legislation, and programs have not received as great a public awareness as Section 404 of the Clean Water Act. Information on these programs is a significant educational message for future planners and resource managers. This maze of information is best summarized in Statewide Wetlands Strategies: *A Guide to Protecting and Managing the Resource, Wetlands Overview: Federal and State Policies Legislation and Programs*, and the *Wetlands Protection Workbook*.

Enforcement

A plea of ignorance about state and federal wetlands regulations in the event of violation is inadequate grounds for dismissal of the violation. If wetlands are filled without a Department of the Army permit, Sections 308 and 309 of the Clean Water Act provide the authority for civil and criminal litigation and directive by the Environmental Protection Agency and Corps of Engineers for removal of the fill, compensation and fines. Similarly, state regulatory programs carry penalty authorities for discharge of fill, and the Conservation and Wetlands Reserve programs carry penalty authorities for violation of easement commitments. An awareness of the occurrence of wetlands and a knowledge of the regulations affecting them can save property owners time and expense. Such awareness can in fact save a township money through coordination with other regulatory requirements, such as water quality planning, flood control planning, and recreational planning.

Wetlands Conservation and Planning

Information Focus:

Overview of wetland functions and
    values and trends of loss
Guidance on implementing a
    preservation ordinance
Guidance on wetlands acquisition
Guidance for private lands
    preservation mechanisms

Audiences:

Land trusts
State and local
    legislators
Planners
Owners of wetlands
Concerned public

A relatively recent type of informational material is that which provides guidance on how to plan and administer wetlands conservation efforts. With the limited ability of regulatory programs to stop wetlands loss has come a focus on mechanisms to preserve and restore wetland resources at a grassroots level. The materials listed in Appendix C provide guidance in developing and implementing state and local preservation ordinances. Samples of successful ordinances are included, and fundamental considerations in planning ordinances are reviewed. There are a number of federal laws, as well as support and development programs, which affect activities in wetlands and can provide mechanisms and much sought after finances for wetlands preservation and restoration. Materials in Appendix C are directed towards clarifying the many legislative frameworks for wetlands preservation and restoration and the mechanisms for private donorship of wetlands.

Application in a Wetlands Education Effort

As mentioned, individuals and groups concerned about conserving remaining wetlands have determined that effective protection will stem from an understanding of the resource as valuable, and the fostering of a personal desire to protect and manage wetlands. Once an audience is informed about the value and assets of wetlands there remains the complicated task of implementing conservation plans. This effort requires a great deal of ingenuity in acquiring limited and much sought after funding. Or, in the case of private wetlands ownership, it requires decisions on the nature of the conservation contribution.

The informational materials in Appendix C introduce their audience to wetlands, their functions, and, often, the activities affecting wetland loss. They address direct federal, state, and local wetland regulatory influence, and mechanisms for wetlands protection. These include such topics as incentive and disincentive programs (e.g., state and county tax assessment, Swampbuster), inclusive interpretation of laws and regulations which are not directed at wetlands but include wetlands in their spectrum (e.g., floodplain management, erosion and sediment control, recreational improvement, sanitary codes), and protection through acquisition, easements, or donation. A number of materials, including *Statewide Wetlands Strategies: A Guide to Protecting and Managing the Resource, Protecting Nontidal Wetlands, National Wetland Protection Symposium: Strengthening the Role of the States, Our National Wetlands Heritage: A Protection Guidebook* and *Wetlands Conservation Through Local Community Planning* provide examples of, and frameworks for, implementing a wetlands protection ordinance.

To our knowledge, there are no existing workshops or courses for the interpretation and application of conservation and planning materials. Perhaps the best existing exchange of information lies within the proceedings of symposia and the databases of state wetlands programs. These information sources recount the experiences and recommendations of managers and planners in their wetlands protection efforts. *Statewide Wetlands Strategies: A Guide to Protecting and Managing the Resource*, throughout its text, introduces examples of protection efforts. Inquiry can be directed to referenced agencies and organizations on their experiences in wetlands protection.

The topic of wetlands protection planning and an understanding of the federal and state mechanisms, across a spectrum of government programs, awaits a synthesis and application by educators. *Wetlands Overview: Federal and State Policies, Legislation, and Programs, Statewide Wetlands Strategies: A Guide to Protecting and Managing the Resource*, and the *Wetlands Protection Workbook* are very useful information sources, but admittedly are a lot for educators to digest, synthesize, and prepare for presentation. That type of effort, in coordination with coverage of actual examples of wetlands protection ordinances, agreements, and strategies (provided in the many brief examples in Statewide Wetlands Strategies) would be a useful contribution to the body of wetlands information.

Wetlands Science and Management

Information Focus:	Audiences:
Current wetland science	Wetland professionals
Current wetland science in regulatory and policy decision making	Those involved in wetlands research
	Policy making and regulating community
	Those involved in wetlands mitigation and restoration

Informational materials in Appendix D represent some of the efforts in understanding the ecology and function of wetlands and the effort of academic and professional associations in bridging the gap between wetlands science and wetlands management. The legislative and regulatory programs for wetlands protection were established before wetlands research provided a base for effective decision making. There remains a great need for scientific foundation in decision making and also a great need for relating the current findings about wetlands to a broader nonscientific audience.

The restoration and creation of wetlands is a major area of current research and application. NOAA, the USACE, EPA, and the Forest Service are involved in on-going, long-term research. There are many state and federal agencies involved in the applications of wetlands restoration and creation efforts, for example, the USACE dredge spoil management projects, the SCS and FWS involvement with the Conservation and Wetlands Reserve programs, the Bureau of Mines biological treatment of acid mine drainage, and the cooperative wetland habitat restoration efforts of the FWS with the Bureau of Reclamation, Bureau of Land Management, Bureau of Indian Affairs, and Ducks Unlimited. Many consulting firms have had long-term involvement in wetlands restoration and creation efforts. One such firm, Environmental Concern of St. Michaels, Maryland, has initiated an educational program based on such efforts.

EDUCATIONAL GAPS

With authorization of the Clean Water Act and implementation of the Section 404 (b)(1) Guidelines, came a new focus on wetlands. They were to be accepted, through Congressional decree and authority, as a resource to be protected. There were now regulating authorities and a regulated community. Wetlands had to be defined, classified, and delineated for regulatory purposes. Regulators grappled with definition of the resource and implementation of the regulations. Methods were devised to evaluate wetland functions in order to assess wetland value, and to evaluate the impacts from their alteration. In the political arena, the issues of wetlands definition, delineation, and value have sparked a level of controversy that introduced much of the general public to wetlands, and the public has come to know wetlands very much from this regulatory perspective.

There is now greater effort to inform the public, including teachers and youth, private landowners, businesses, and local officials whose decisions affect land use and management, on broader issues. Curricula, activity books, videos, and posters are providing information on what wetlands are, the types of wetlands, and experiences which help to foster an understanding of how wetlands perform their ecological functions.   Informational materials are going beyond introducing wetlands to their audience, to invoking a call to personal action and responsibility for wetlands in land use planning and management. The listing of materials in the attached appendices attests to these efforts, and bears witness to the beginnings of the new direction of wetlands education and outreach efforts.

Wetland Topics in Need of Focus

From review of the variety of informational materials listed in the attached appendices, there seems little material available that links the layman (the landowner, businessman, concerned citizen) with a solid understanding of what wetlands do and how they do it, that is, basic wetlands science. Wetlands are a complex natural resource, and a variety of factors influence their structure, occurrence and how effectively they perform their functions. Common public perceptions of wetlands still invoke images of swamps like the Everglades and Okefenokee, or ponded prairie sloughs occupied by hundreds of ducks. Yet most of the nations wetlands include the much less obvious wooded wetlands, palustrine emergents, and those sloughs which appear very dry in the summer season. One educator noted the susceptibility of educational materials' reference to glamorous wetlands. In an effort to convey wetlands information, glamorous wetlands, either ecologically or aesthetically, exemplify the message, and the audience loses out on introduction to the more geographically-true "neighborhood" wetlands.

The best efforts at basic wetlands science are curricula and activity guides for youth. Children set up jars with sponges and see filtration at work, or pour water over soil in baking pans to watch erosion. In order for citizens to respect wetlands in their local area enough to want to conserve them, people must be able to relate to them and, to some degree, understand their functions and role in the landscape. Outreach materials list and briefly describe wetland functions. However, most people cannot relate those functions to their local wetlands. Audiences need to be convinced of the effectiveness of wetlands in performing their functions, and how that affects their personal or community livelihood. Educators are starting to collect and present information on the economic contribution of wetlands, both in generating and saving revenues. The link of wetlands to the million dollar industries of hunting, birdwatching, tourism, commercial fisheries and sport fishing, and the shellfish industry has been drawn. The potential savings from application of wetlands in wastewater treatment, pollution attenuation, and disaster prevention have been touched upon. Wetlands attributes need to be presented in terms of the costs incurred from not protecting the wetlands resource. This approach provides wetlands a social context.

A basic understanding of wetland functions can lead to a respect for their natural value and can spark the comprehension of their potential for economic savings and social contribution. Bridging the gap between science and outreach is not a new issue. The partnership of science and outreach occurs, for example, through field research stations that offer academic-led programs in public participation. It is an issue that needs resolution through a changed mindset among academics and agencies, and a dedication of personnel and fiscal resources. Each grant towards wetlands research should have a rider dedicating environmental education funds to disseminate the research information to a much broader audience. The Sea Grant and Cooperative Extension Service models, embracing the linkage of research and outreach, need application within all those agencies conducting wetlands research and outreach.

There are few materials that address decision-making processes and the consideration of wetlands in those decisions. The University of Maryland Sea Grant Program's *Decision Making and the Chesapeake Bay* should not be overlooked because of its regional focus. It is an issues-oriented study set in a simulated decision-making format. Wetland value and an understanding of the ecology of the wetlands resource is considered amongst a host of social and economic issues. The curriculum could serve as a model for many regional wetland programs and provide the framework for participation by a wide spectrum of local participants.

Considerations for Implementing Education and Outreach Efforts

Education and outreach require dedication of a variety of efforts. Once materials are generated there remains the aspect of distribution. Often, materials of great use are printed in limited supply, undoubtedly due to budgetary constraints, and with the passing of time become inaccessible. The Sea Grant program and FWS are exemplary in their centralization and availability of materials. Other agencies are working towards such systems. The lack of centralization and accessibility has caused a great deal of redundancy in effort. Education and outreach efforts could be much more productive if educators could contact central clearinghouses and use existing information, and then tailor it to local and specific educational needs. A federal interagency task force, assigned by the Domestic Policy Council, drafted the suggestion of implementing a coordinating body to oversee networking in the distribution and implementation of existing materials. Needless to say, with such centralization and subsequent coordination among federal agencies, a much more efficient outreach effort could be underway.

Greater availability and simplicity in acquiring wetlands informational materials could aide in an educational strategy to impart greater understanding and retention of the educational message. A variety of information could be provided through different sources, in different ways, over time. Educational efforts could save time and money through use of appropriate existing information or, with relatively little effort, tailoring of existing information. In so doing, those efforts could embark on the strategy of recurring wetlands education.

A potential oversight of education and outreach efforts is that of generating materials independently of an educational strategy. Often, outreach materials are subject to the mass distribution mindset: produce a multitude of informational material and simply distribute it. Yet the development and production of educational materials should be considered in the context of their application. In so doing their effectiveness can be assessed prior to the expenditure of energies in development and production. The creative temptation exists to produce materials, yet time and monies for wetlands education and outreach are limited. Effort should be directed initially toward developing a wetlands education plan that will effectively relate the educational message and make materials easily accessible.

Many informational materials address the concerned public, yet that audience remains an under-utilized educational resource. When accurately informed about wetlands and wetlands regulations, and with appropriate oversight, the public can and does spread the word. People who participate in activist campaigns and volunteer programs are committed to contributing to wetlands conservation. Often, there are limited mechanisms to employ their enthusiasm and few tangible rewards. Motivation and direction to these potential grassroots educators could be provided through distribution of free materials and funded training workshops. It is essentially the environmental organizations that have filled this gap. There are the beginnings of cooperative efforts among federal agencies and environmental organizations, such as the joint Environmental Protection Agency/National Audubon Society Audubon's America program. Although involvement of the volunteer public requires a commitment of staff and materials, annual U. S. Fish and Wildlife Service reports suggest an enormous benefit and return from that commitment. A focus of education efforts should be to evaluate where and how volunteers, both young and old, can be incorporated into wetlands education efforts. Contact with long-established volunteer programs can provide beneficial insight for troubleshooting and problem solving necessary for such efforts.

GUIDELINES FOR IMPLEMENTING A WETLANDS EDUCATION EFFORT

Wetlands educators are attempting to inform the uninformed, foster a new perspective on the wetlands resource, and compel the advance from simple awareness to stewardship. This will require persistent and compelling education efforts. Implementing a wetlands education program introduces the necessity of developing a plan. Planning to accomplish any task involves addressing a set of fundamental steps. These steps include consideration of the educational message and the target audience, clarification of the goals and objectives, planning for implementation, and program evaluation.

The Message and the Audience

The starting point involves deciding on the message and identifying the audience (Figure 4). The listing of materials in the appendices provides a perspective on what the wetlands messages have been. Wetlands education programs will necessarily have specific messages, which might include single topics or multiple topics. Coincident with determination of the message will be decisions on the appropriate level of detail of information addressing the messages. Wetlands awareness is a consistent message in essentially all information contexts. Nevertheless, a synthesis of messages, for example wetlands perform functions of social significance, they have been deemed a valuable resource and thus protected through regulation and legislation, and there is now an effort to develop conservation plans and provide guidance in that effort of planning, is important in the advance from awareness to stewardship. Many of the guidebooks have used this approach.

A decision on the target audience will direct decisions on the level of information, its tone, and its format for presentation. Informational materials may need to be specifically developed. Fortunately, many existing wetlands informational materials can effectively inform both youth and adult audiences. Videos, magazines, and posters relay their messages to youth, with teacher interpretation, through adult age groups. As the Washington State Department of Ecology, in *Designing Community Environmental Education Programs: A Guide for Local Government* points out, understanding the target audience and knowing what motivates them is important. The booklet suggests inviting representatives of the target audience to help in forming the education program. They can provide perspective and suggestion in the selection or development of materials and the format for presentation. Conducting a local survey can identify audience perceptions and behaviors. A survey can determine how informed the audience is, the information they already have, and what the roadblocks may be to perceiving information.

Goals and Objectives

Given the rate of decline of wetlands, most could agree that one essential goal is to foster a personal wetlands stewardship ethic. This involves informing the uninformed and changing past perspectives of wetlands as a hindrance rather than an environmental asset. In implementing a wetlands education program, the goals will be more specific. Given the target audience and the message, what response does the education program want to elicit? For example, is the goal to have realtors inform potential property buyers that development in wetlands is regulated? Old Woman Creek National Estuarine Research Reserve has implemented wetlands

Figure 4. A key to implementing a wetland educational program is deciding on the message and identifying the audience. The messages here are wetlands awareness and functions and values, and the target audience is teenagers and adults. Photo courtesy of the Washington State Department of Ecology.

informational materials in an elective course in the local realtor certification program. Clear definition of the education or outreach goal helps to maintain focus throughout the program. The goals may be simple or multiple. If they are multiple, they should be prioritized.

Objectives lay out the specifics. Projects and tasks are defined, time frames are projected. For the goal of realtors informing potential property buyers about regulatory restraints the objectives may include selecting a representative focus group of the target audience (for audience insight), developing or acquiring informational materials on wetlands regulations, assessing how to have these materials incorporated into the realtor certification program, and evaluating mechanisms for extensive materials distribution and application.

Action Plans

With the message, the audience, the goals, and the objectives clarified, one must determine how to get it all done. Initial considerations include the budget, level of work effort, and available personnel. There should be review of what has been done. For example, have other such educational efforts been undertaken? Do pertinent materials exist? Further considerations include determining the role that the agency or organization is going to play (funder, key player, facilitator, partner). And, how are the objectives going to be most effectively implemented, e.g., tailoring existing materials, becoming involved in a cooperative effort with another agency or organization, enlisting volunteers? The final action plans should list each objective and its relevant tasks, clarify the expense (finalize the budget) and the level of work effort, and set completion dates.

Evaluation

Evaluation of an education effort will help to assess a level of success in reaching the goal and, equally significantly, will provide an idea of the education needs still remaining. The evaluation may be passive. There may be enormous request for the materials used, inquiry from other geographical areas about implementing the program, etc. Alternatively, survey measures may be needed to determine reaction from the target audience. There are a variety of aspects of an educational program to evaluate. Did the actual costs meet or exceed the proposed costs? If costs were more, was the level of success in reaching the goal worth the costs? Is there consistent inquiry on particular aspects or topics of the education program that signal inadequate communication of the message or, conversely, particular pertinence of the message and its presentation? Certainly, it is useful to retain all news articles, telephone notes, and letters addressing the program.

CONCLUSION

There are many wetlands educators, a term defined within this chapter as the agencies, organizations, and associations that produce wetlands informational materials or are otherwise involved in wetlands education efforts. They have produced many wetlands informational materials. Thus, there is an existing body of available materials (many of which are free or of minimal cost) and experiences which the potential wetlands educator can call upon. Familiarity with these, and a consideration of the apparent gaps in current wetlands educational efforts, can help in the necessary steps of wetlands education program planning and implementation. The need remains to centralize existing wetlands informational materials and make them widely and easily accessible. In addition, there is a need for further interagency and organization coordination and cooperative efforts in developing and implementing wetlands educational efforts. There must be a dedication of personnel and fiscal resources to accomplish effective education and outreach programs; those that have done it well have excellent materials and outreach programs. Information in this chapter is intended to streamline the budgets and work efforts in that dedication.

WETLANDS AWARENESS

Curricula/Activity Materials

Curricula

Aquatic Project WILD[1]

Beneath the Shell Teacher's Guide to Nonpoint Source Pollution[2]

A Curriculum Guide for Wetland Education Grades K-8[3]

Decision Making and the Chesapeake Bay[4]

Discover Wetlands[5]

Environmental Education Curriculum Manual[6]

The Estuary Guide Level 1 (K-3), Level 2 (4-8), Level 3 (9-12)[7]

Project Estuary[8]

Marine and Estuarine Ecology Volume 1[9]

Marine Science Education Workbooks: Tides and Marshes[4]

Marine Science Education Workbooks: Food Webs in an Estuary[4]

Marshes, Estuaries and Wetlands[10]

Salt Marsh Manual - An Educator's Guide[11]

Sea Sampler: Aquatic Activities for the Field and Classroom (Elementary and Secondary)[12]

South Slough Estuarine Sanctuary[13]
- Estuary: an Ecosystem and a Resource - Teacher's Manual
- In Search of the Treasures of the South Slough - Level 1 Grades 4/5
  Teacher's Manual
- The Secret of the Medallion - Level 2 Grade 6 Teacher's Manual
- The Lore of South Slough - Level 3

A Teacher's Guide to the Study of Wetlands[14]
- Willa in Wetlands Play

U.S. Fish and Wildlife Service and the Alaska Department of Fish and Game[15]
- Wetlands and Wildlife Alaska Wildlife Curriculum:
  Primary Teachers's Guide K-3
  Upper Elementary Teacher's Guide 4-6
  Junior and Senior High Teacher's Guide 7-12
  Field Trip Manual
  Teacher Information Manual Part 1 & 2

Wading into Wetlands. Ranger Rick Nature Scope[16]

Wetlands are Wonderlands[17]
- Member/Youth Guide
- Leader/Teacher Guide

Wetlands Project Manual and Project Record[18]

Wildlife Habitat Conservation Teacher's Pac Series[19]
- Estuaries and Tidal Marshes
- Freshwater Marshes
- Wetlands Conservation and Use

WOW! The Wonders of Wetlands-An Educator's Guide[20]

Activity Materials

Energy Flow in a Wetland (board game)[21]

My Wetland Coloring Book[22]

U.S. Fish and Wildlife Service and the Alaska Department of Fish and Game[15]
- Wetland Cards - illustrations and text
- The Caribou Game (board game)
- Brandt for the Future (board game)

Welcome to the Wetlands (coloring poster)[23]

Wetlands (coloring poster)[5]

Wetlands Coloring Book[24]

Wetlands, Wildlife, and You![25]

Wetland Wonderland[26]

Posters

A Child's Wish for Wetlands[27]

Energy Flow in an Alaska Wetland[15]

Estuaries....where rivers meet the sea[28]

Federation of Ontario Naturalists[29]
- Lake Ecosystem
- Life in an Ontario Wetland
- Ontario's Amphibians and Reptiles
- Soil Ecosystem

National Audubon Society[30]
- Rivers of Life
- Wetlands: Wonders Worth Saving, Life in a Salt Marsh
- Wetlands: Wonders Worth Saving, Life in a Freshwater Marsh

National Science Teachers Association[21]
- Who Lives In Mud?
- Wetlands

U.S. Fish and Wildlife Service[24]
- America's Endangered Wetlands
- Wetlands are Important
- Wetlands, Wildlife and You

U.S. Wetlands: What Do We Do With Them?[19]

Wetlands[5]

Wetlands [31]

A Wetland Year[32]

Videos
Bullfrog Films[33]
- The Wasting of a Wetland
- Natural Wastewater Treatment
- Estuary

Cooperative Extension Service[34]
- Many, with an array of topics including wetland values,
  restoration, protection. For example:
  "Do Your Part: A Wetlands Discovery Adventure!" (Colorado)
  "The Value of Hoosier Wetlands" (Indiana)
  "Wetlands" (Maine)
  "Prairie Potholes are for People" (North Dakota)

Coronet/MTI Film & Video[35]
- Birds of Shore and Marsh
- The Marsh
- We Explore the Marsh
- One Day at Teton Marsh
- Prairie Slough

Ducks Under Siege[36]

Educational Images, Ltd.[37]
- Salt Marshes - A Special Resource
- Freshwater and Salt Marsh
- Bog Ecology
- Florida Bay and the Everglades

The Living Tidal Marsh (Limited circulation to NJ PSE&G service area)[38]

Louisiana Wetlands: America's Coastal Crisis[22]

The Marsh Community Our Vanishing Marshland[39]

National Geographic Society[40]
- A Swamp Ecosystem
- A Tidal Flat and Its Ecosystem

U.S. Fish and Wildlife Service[41]
- America's Wetlands
- Wetlands in Crisis
- Do Your Part: A Wetlands Discovery Adventure!
- Wetlands, Wildlife and You: The Duck Stamp Story
- Wetlands for the Future
- In Celebration of America's Wildlife

Washington State Department of Ecology[5]
- Fabulous Wetlands
- Yellowlegs, Eelgrass and Tideflats
- Wetlands Nightmare: Solutions for Local Government
- Washington's Wetlands

The Wealth in Wetlands[42]

Wetlands Video Disc[43]

Why Wetlands[29]

Slides

Educational Images, Ltd.[37]
- Freshwater and Salt Marsh
- Salt Marsh Biome
- Ecology of a Bog
- Bog Ecology (different than above)
- Ecology of a Swamp
- Ecology of the Everglades
- The Everglades (different than above)

Salt Marshes: Nurseries of the Sea[44]

Wetlands[45]

Magazines

Diversity in the Wetland. DU Magazine Special photo issue January/February 1992[46]

The Last Wetlands. Audubon Magazine July 1990, Vol. 92, No. 4[45]

Our Disappearing Wetlands. National Geographic October 1992, Vol. 182, No. 4[40]

Puddler Magazine[46]

Virginia Marine Resource Bulletin. Untitled issue on wetlands. Spring 1989, Vol. 21, No. 1[47]

Wetlands, Wonderlands. Wisconsin Natural Resources Magazine Special Supplement[32]

Wetlands. Michigan Natural Resources Magazine photo issue.
January/February 1991.[48]

Other Documents

At Home with Wetlands: A Landowner's Guide. 1991. Washington State
Department of Ecology, Publication #90-31[5]

Coastal America: A Partnership for Action[49]

Designing Community Environmental Education Programs: A Guide for Local
Government. 1992. Washington State Department of Ecology #92-99[5]

Ducks Unlimited Green Wing Program pamphlet[46]

Field Trips...Logistics is Key to it All[50]

Freshwater Wetlands: A Citizen's Primer[51]

Get Your Feet Wet: Get Into Wetlands. Environmental Education in Wisconsin,
Environmental Education News, Spring 1989[52]

National Marine Educators Association (NMEA)[53]
- Current: The Journal of Marine Education
- NMEA News

National Wetlands Newsletter[54]

Recognizing Wetlands. 1987. (USACE pamphlet available from EPA Wetlands
Hotline)[55]

Saving the Wetlands booklet[45]

Status Report on Our Nation's Wetlands. October 1987. National Wildlife
Federation[16]

U.S. Environmental Protection Agency[55]
- EPA Wetlands Hotline 1-800-832-7828
- Documents Available from the U.S. EPA's Wetlands Protection Hotline. April
  1992
- Wetlands Education System. Interactive Computer Software Program.[56] (For
  IBM PC or compatible)
- Wetlands Protection Workbook. Draft, September 1992
- Agriculture and Wetlands: A Compilation of Factsheets. August 1992
- American Wetlands Month pamphlet. May 1991

- Wetlands in the Rocky Mountains and Northern Great Plains pamphlet
- Beyond the Estuary: The Importance of Upstream Wetlands in Estuarine Processes. July 1990
- Wetlands Protection Fact Sheet series. 1989
- America's Wetlands Our Vital Link Between Land and Water. February 1988
- Environmental Backgrounder: Wetlands. 1988
- Protecting Our Nations's Wetlands and Your Pocketbook Fact Sheet. June 1988
- EPA Journal "Protecting Our Wetlands" Jan/Feb 1986, Vol. 12, No. 1
- EPA Journal "Saving Our Wetlands" October 1983, Vol. 9, No. 2

U.S. Fish and Wildlife Service[57]
- 1-800-USA-MAPS
- Wetlands Status and Trends in the Conterminous United States, Mid 1970's to Mid 1980's. Dahl, T.E. and C.E. Johnson. 1991 (GPO 024/010/06692/7)
- National Wetlands Inventory Fact Sheets and Information. 1992
- Wetlands Losses in the United States 1780's To 1980's. Dahl, T.E. 1990
- Federal Coastal Wetland Mapping Programs. Biological Report 90(18) December 1990 (NTIS PB92128339)
- Wetlands: Meeting the President's Challenge. 1990 Wetlands Action Plan
- National List of Plant Species that Occur in Wetlands. 1988 National Summary (GPO 024/010/00682/0; NTIS PB89-127914/AS)
- Wetlands of the United States: Current Status and Recent Trends. Tiner, Ralph, W., Jr. March 1984 (NTIS PB90198201)
- Classification of Wetlands and Deepwater Habitats of the United States. Cowardin, L.M., et. al. 1979 (GPO 024/010/00665/0)

Wetlands Stewardship. U.S. Department of the Interior. April 1992[58]

Wetlands Campaign[45]

Books

A Field Guide to Coastal Wetland Plants of the Northeastern United States. 1987. Tiner, R.W., Jr. University of Massachusetts Press, Amherst

Field Guide to Nontidal Wetland Identification. 1988. Tiner, R.W., Jr. Maryland Department of Natural Resources, Annapolis, MD

Freshwater Wetlands: A Guide to Common Indicator Plants of the Northeast. 1981. Magee, Dennis, W. University of Massachusetts Press, Amherst

How Wet is a Wetland? The Impacts of the Proposed Revisions to the Federal Wetlands Delineation Manual. 1992. Environmental Defense Fund and World Wildlife Fund, Island Press, Washington, D.C.[59]

Life and Death of the Salt Marsh. 1969. Teal, John and Mildred. Ballantine Books

The Life of the Marsh. 1966. Niering, William A. McGraw-Hill, Inc.

Pond and Brook: A Guide to Nature in Freshwater Environments. 1990. Caduto, M.J. University Press of New England, Hanover, NH

Pond Life: A Guide to Common Plants and Animals of North American Ponds and Lakes. 1967. Reid, George K. Golden Guide. Golden Press, New York, NY

Walking the Wetlands: A Hikers Guide to Common Plants and Animals of Marshes, Bogs, and Swamps. 1989. Lyons, Janet and Sandra Jordan. John Wiley & Sons, Inc.

Wetlands. 1989. Niering, William A. The Audubon Society Nature Guides. Alfred A. Knopf, New York, NY

Resource Guides

Audubon Sanctuary Directory. NAS Sanctuary Department[60]

Bibliography of Educational Materials Available for the Nation's Sea Grant College Programs. January 1991[9]

Directory of Zoological Parks and Aquariums in the Americas[61]

ECONET Database. Alliance for Environmental Education[62]

Environmental Education Resource Guide[63]

ERIC Database. Environmental Resources Information Center[64]

Estuary - An Ecosystem and a Resource: A Reading Guide for Grades 9-12. 1984[13]

Extension Wetlands Educational Materials. USDA Extension Service for the Cooperative Extension Service[34]

International Directory of Marine Science Libraries and Information Centers[65]

Marine Education Computer Print-out Bibliography. Virginia Sea Grant College Program[47]

National Association of Conservation Districts Directory[42]

National Audubon Society Wildlife Sanctuaries. NAS Sanctuary Department[60]

National Directory of Citizen Volunteer Environmental Monitoring Programs. April 1990[66]

National Sea Grant Depository[67]

National Wildlife Federation Conservation Directory[16]

Official Directory. The American Association of Museums[68]

Sea Grants Abstracts. Woods Hole Database[69]

State University System of Florida Sea Grant College Program[70]
- Florida Marine Education Resources Bibliography. March 1983
- Florida 4-H Department Marine Booklists

U.S. Environmental Protection Agency[55]
- Wetlands Reading List. Draft 2/19/92
- Guide to Wetlands Educational Programs and Materials in Pennsylvania. November 1991
- Environmental Education Materials for Teachers and Young People (Grades K- 12). July 1991
- Wetlands Protection Bibliographic Series. October 1988
- Books for Young People on Environmental Issues. January 1987

U.S. Fish and Wildlife Service Publications List[24]

University Curricula Guide. Marine Technology Association[71]

Wading Into Wetlands[16]
- Bibliography
- Ranger Rick Wetland Index

The Water Quality Catalog: A Source Book of Public Information Materials[72]

WOW: The Wonders of Wetlands. Resources Listing[20]

APPENDIX B

WETLANDS REGULATIONS AND LEGISLATION

Building Near Wetlands: The Dry Facts. September 1991. Wisconsin Department of Natural Resources Bulletin[32]

Council of State Governments
- Wetlands: A National Database of State Wetland Protection. User's Manual Version 1.2 Programs and Contacts[73]
- Wetlands Protection and the States. April 1990. National Conference of State Legislatures[74]
- Suggested State Legislation[75]
- Resource Guide to State Environmental Management. 1990[73]

Developer's Guide to Federal Wetlands Regulations. 1990. Liebesman, L.R. National Association of Homebuilders[76]

Environmental Law Institute[54]
- National Wetlands Newsletter
- Annual Cumulative Index of the National Wetlands Newsletter
- Clean Water Deskbook. 1988.
- Directory of State Environment Agencies. 1985. Hubler, Kathryn and Timothy Henderson, editors
- Implementation of the "Swampbuster" Provisions of the Food Security Act of 1985. 1990

Soil and Water Conservation Society (SWCS)[77]
- Implementing the Conservation Title of the Food Security Act: A Field Oriented Assessment by SWCS. 1990
- Implementing the Conservation Title of the Food Security Act of 1985

U.S. Environmental Protection Agency[55]
- EPA Wetlands Hotline 1-800-832-7828
- Draft Wetlands Protection Workbook. September 1992
- Agriculture and Wetlands information sheet series. October 1991

- Water Quality Standards for Wetlands: National Guidance. July 1990
- Wetlands and 401 Certification: Opportunities and Guidelines for States and Eligible Indian Tribes. April 1989
- Highlights of Section 404: Federal Regulatory Program to Protect Waters of the United States. October 1989

EPA Region 1
- The Federal Wetlands Protection Program in New England: A Guide to Section 404 for Citizens and States. August 1991[78]
- Improving Wetland Protection in New England: A Blueprint for State and Federal Action. July 1991[78]

Wetlands Overview: Federal and State Policies, Legislation, and Programs. Nov. 1991 General Accounting Office (GAO/RCED-92-79FF)[79]

Wetland Regulation Guidebook. July 1988. Washington State Department of Ecology #88-5[5]

Wetlands: Their Use and Regulation. March 1984. Office of Technology Assessment (NTIS PB84-175918)[24c]

Wetlands: The Corps of Engineers' Administration of the Section 404 Program. July 1988. General Accounting Office (GAO/RCED-88-110)[79]

Wetlands and Real Estate Development. 1988. Steinberg, R.E., ed. Government Institutes[80]

APPENDIX C

WETLANDS CONSERVATION AND PLANNING

The Conservation Easement Handbook: Managing Land Conservation and Historic Preservation Easement Programs. 1988. Diehl, J. and T.S. Barrett[81]

Conservation Easement Stewardship Guide: Designing, Monitoring, and Enforcing Easements. 1991. Lind, B. Island Press, Washington, D.C.[59]

Environmental Law Institute[54]
- Our National Wetlands Heritage: A Protection Guidebook. 1983. Kusler, J.A.
- Non-Tidal Wetlands Protection: A Handbook for Maryland Local Governments. 1983
- Wetlands Protection: The Role of Economics. 1990. Scodari, Paul F.
- Directory of State Environmental Agencies. 1985. Hubler, K. and T. Henderson, eds.

Michigan Wetlands Yours to Protect: A Citizen's Guide to Local Involvement in Wetland Protection[82]

National Audubon Society[45]
- Saving the Wetlands: A Guide for Audubon Chapters, Government Agencies, and Educators. April 1990
- Saving Wetlands: A Citizen's Guide for Action in Florida. January 1991[83]

National Wetlands Priority Conservation Plan. April 1989. U.S. Fish and Wildlife Service[24]

National Wetland Protection Symposium: Strengthening the Role of the States. 1985. Kusler, J.A. and R. Hamann, eds.[84]

National Wildlife Federation[16]
- A Citizens Guide to Protecting Wetlands. 1989. Goldman-Carter, J.
- Conservation Directory

Protecting Non-Tidal Wetlands. 1988. Burke, D.G., E.J. Meyers, R.W. Tiner, Jr., H. Groman[85]

Statewide Wetlands Strategies: A Guide to Protecting and Managing the Resource. 1992. [59]

U.S. Environmental Protection Agency[55]
- Wetlands Protection Workbook. Draft, September 1992
- The Private Landowner's Wetlands Assistance Guide: Voluntary Options for Wetlands Stewardship in Maryland. October 1992
- Protecting Coastal and Wetland Resources: A Guide for Local Governments. October 1992
- Guidance on Developing Local Wetlands Projects: A Case Study of Three Counties and Guidelines for Others. November 1991
- Wetlands Conservation Through Local Community Programs. December 1991. Emmer, Rod E., Univ. of New Orleans
- Catalog of State Wetlands Protection Grants Fiscal Year 1990. April 1991
- Financing State Wetlands Programs. November 1990. Apogee Research, Inc.
- Saving Bays and Estuaries: A Handbook of Tactics. 1988. Office of Marine and Estuarine Programs
- Financing Marine and Estuarine Programs: A Guide to Resources. 1988. Office of Marine and Estuarine Programs

Washington State Department of Ecology[5]
- Designing Wetlands Preservation Programs for Local Governments: A Guide to Non-regulatory Protection #92-18. 1992
- Designing Wetlands Preservation Programs for Local Governments: A Summary #92-19. 1992
- Wetlands Preservation: An Information and Action Guide #90-5. 1990

Wetlands: A National Database of State Wetland Protection. User's Manual Version 1.2 Programs and Contacts. Council of State Governments[73]

Wetlands and Water Quality: A Citizen's Handbook for Protecting Wetlands. 1990. Paulson, G.A. Lake Michigan Federation[85]

Wetlands Protection: A Local Government Handbook. September 1991

Wetlands Protection: A Handbook for Local Officials. 1990. Pennsylvania Department of Natural Resources[87]

Wetland Protection Guidebook. 1988. Michigan Department of Natural Resources[88]

APPENDIX D

WETLANDS SCIENCE AND MANAGEMENT

Association of Wetland Managers[84]
- Wetlands Creation and Restoration: The Status of the Science. 1990
- International Symposium: Wetlands and River Corridor Management. July 1989
- National Wetland Symposium: Urban Wetlands. June 1988
- National Wetland Symposium: Wetland Hydrology. September 1987
- National Wetland Symposium: Mitigation of Impacts and Losses. October 1986
- National Wetlands Assessment Symposium. June 1985
- National Wetland Protection Symposium: Strengthening the Role of the States. September 1984

Environmental Law Institute[54]
- National Wetlands Newsletter
- Annual Cumulative Index of the National Wetlands Newsletter
- Environmental Forum
- Environmental Law Reporter

Estuaries Journal[89]

Protecting America's Wetlands: An Action Agenda. 1988. The Conservation Foundation[90]

Society for Wetlands Scientists (SWS)[91]
- Wetlands Journal of SWS
- SWS Bulletin

Statewide Wetlands Strategies: A Guide to Protecting and Managing the Resource. 1992. World Wildlife Fund. Island Press, Washington, D.C.[59]

Wetlands Action Plan: EPA's Short-term Agenda in Response to Recommendations of the National Wetlands Policy Forum. January 1989[55]

Resources

An Overview of Major Wetland Functions. 1984. Sather, J.H. and R.D. Smith. (NTIS PB85119907)[24c]

ECONET Database. Alliance for Environmental Education[62]

EPA's Wetlands Protection Bibliographic Series. October 1988[55]

ERIC Database. Environmental Resources Information Center[64]

Increasing Our Wetland Resources. 1988. Zelasny, J. and J.S. Feierabend, eds. National Wildlife Federation, Washington, D.C.[16]

National Sea Grant Depository[67]

Resource Guide to State Environmental Management. 1990. The Council of State Governments[73]

Sea Grants Abstracts. Woods Hole Database[69]

U.S. Fish and Wildlife Service Publications List[24]

Wetland Evaluation Technique (WET) Volume 1: Literature Review and Rationale. Wetlands Research Program Technical Report WRP-DE-2. October 1991. Adamus, P.R. et. al.[24c]

Wetland Replacement Evaluation Procedure. 1992. Bartoldus, C., E. Garbish, M. Kraus. Environmental Concern[20]

Books

An Approach to Decision Making in Wetland Restoration and Creation. 1992. Hairston, A. J., ed., Island Press, Washington, D.C. (Available also through NTIS)

Constructed Wetlands for Wastewater Treatment: Municipal, Industrial, and Agricultural. 1990. Hammer, D.A., ed. CRC Press, Fort Lauderdale, FL

Creating Freshwater Wetlands. 1992. Hammer, Donald, A. Lewis Publishers, Chelsea, MI

Estuarine Ecology. 1989. Day, John, W., Jr. et al. John Wiley & Sons, Inc., New York

Estuarine Ecosystems: A Systems Approach Volumes I & II. 1986. Knox, George, A. CRC Press, Boca Raton, FL

Restoration of Aquatic Ecosystems. 1992. National Research Council Committee on Restoration of Aquatic Ecosystems: Science, Technology, and Public Policy. National Academy Press, Washington, D.C.

Wetlands. 1986. Mitsch, William, J. and James G. Gosselink. Von Nostrand Reinhold Company, New York

Wetlands Creation and Restoration: The Status of the Science. 1990. Kusler, J.A. and M. Kentula. Island Press, Washington, D.C.

Wetland Economics and Assessment: An Annotated Bibliography. 1989. Leitch, J.A. and B.L. Ekstrom. Garland Publishing, Inc., New York

Wetlands Protection: The Role of Economics. 1990. Scodari, P.F. Environmental Law Institute.

APPENDIX E

RESOURCE CONTACTS

1. Project WILD
   P.O. Box 18060
   Boulder, CO 80308-8060
   (303)444-2390

2. New Jersey Department of Environmental Protection and Energy
   Office of Communication and Environmental Education
   CN 402
   Trenton, NJ 08625-0402

3. Centers for Nature Education, Inc.
   4007 Bishop Hill Road
   Marcellus, NY 13108
   (315)673-1350

4. University of Maryland
   Sea Grant Program
   1224 H.J. Patterson Hall
   College Park, MD 20742
   (301)405-6371

5. Department of Ecology
   Wetland Section
   P.O. Box 47600
   Olympia, WA 98504
   (206)438-7538 or 459-6774

6. Elkhorn Slough National Estuarine Research Reserve
   1700 Elkhorn Road
   Watsonville, CA 95076
   (408)728-2822 or 728-8560

7.   Padilla Bay National Estuarine Research Reserve
     1700 Elkhorn Road
     Watsonville, CA 95076
     (408)728-2822 or 728-8560

8.   North Carolina Division of Coastal Management
     P.O. Box 27687
     Raleigh, NC 27611-7687
     (919)733-2293

9.   Mississippi-Alabama Sea Grant Consortium
     Caylor Building
     Gulf Coast Research Lab
     Ocean Springs, Mississippi 39564
     (601)875-9341

10. Pacific Science Center/Washington Sea Grant
     Marine Education Project
     200 Second Ave. N.
     Seattle, WA 91809
     (206)443-2870

11. South Carolina Sea Grant Consortium
     287 Meeting Street
     Charleston, SC 29401
     (803)727-2078

12. South Slough National Estuarine Research Reserve
     P.O. Box 5417
     Charleston, OR 97420
     (503)888-5559

13. Oregon State University Sea Grant College Program
     AdS A500G
     Corvallis, OR 97331
     (503)737-2714

14. National Children's Theater for the Environment
     910 Prince Street
     Alexandria, VA 22314
     (703)519-6155

15. United States Department of the Interior
     Fish and Wildlife Service
     1011 E. Tudor Road
     Anchorage, Alaska 99503-6199

16. National Wildlife Federation
     1400 Sixteenth St., NW
     Washington, D.C. 20036-2266

17. Illinois-Indiana Sea Grant Program
     University of Illinois at Urbana-Champaign
     65 Mumford Hall
     1301 W. Gregory Drive
     Urbana, IL 61801
     (217)333-9448

18. Thomas Greenawalt
     115 Rolfs Hall
     University of Florida
     Gainesville, FL 32611
     (4-H Materials)

19. National Institute for Urban Wildlife
     10921 Trotting Ridge Way
     Columbia, MD 21044
     (301)596-3311

20. Environmental Concern, Inc.
     P.O. Box P
     St. Michaels, MD 21663
     (410)745-9620

21. National Science Teachers Association
     Special Publications
     1742 connecticut Avenue, NWS
     Washington, D.C. 20009
     (800)722-6782 or (202)328-5800

22. United States Environmental Protection Agency
     Wetlands Outreach 6E-FT
     1445 Ross Avenue
     Dallas, TX 75202-2733
     (214)655-7444

23. EPA Region 5
     Public Affairs Office
     77 West Jackson Street
     Chicago, IL 60604-3590
     (312)353-6198

24.   Hierarchy of pursuit for FWS Publications:

      a) Publications Unit
United States Fish and Wildlife Service
Room 130 Arlington Square
1849 C Street, NW
Washington, D.C. 20240
(703)358-1711

      b) Superintendent of Documents
Government Printing Office
Washington, D.C. 20402
(202)783-3238

      c) National Technical Information Service
Department of Commerce
5285 Port Royal Road
Springfield, VA 22161
(703)487-4780

25.   Dr. James Pease
Extension Wildlife Specialist
Department of Animal Ecology
Iowa State University Extension
124 Science II
Ames, Iowa 50011
(515)294-6148 or 7429

26.   Ducks Unlimited Canada
1190 Waverly Street
Winnipeg, Manitoba
Canada R3T 2E2
(204)467-3000

27.   U.S. Environmental protection Agency
Region 4 Water Management Division
345 Cortland Street, N.E.
Atlanta, GA 30365
(404)347-2126

28.   NOS/NOAA
Sanctuaries and Reserve Division
Office of Ocean and Coastal Resource Management
1825 Connecticut Ave., NW
Washington, D.C. 20235
(202)606-4380 or (202)606-4126

29.   Federation of Ontario Naturalists
      355 Lesmill Road
      Don Mills
      Ontario, Canada M3B 2W8
      (416)444-8419

30.   National Audubon Society
      Route 1, Box 171
      Sharon, CT 06069
      (203)364-0520

31.   U.S. Environmental Protection Agency
      Region 8 Water Management Division
      999 18th Street
      Suite 500
      Denver Place
      Denver, CO 80202-2405
      (303)293-1552

32.   Wisconsin Department of Natural Resources
      Bureau of Information and Education
      P.O. Box 7921
      Madison, WI 53707-7921
      (608)266-6790

33.   Bullfrog Films, Inc.
      Oley, PA 19547
      (215)779-8226

34.   USDA Extension Service
      Natural Resources and Rural Development
      14th & Independence Ave., SW
      Room 3871 South Building
      Washington, D.c. 20250-0900
      (202)720-5468: Mr. James Miller

35.   Coronet/MTI Film & Video
      108 Wilmot Road
      Deerfield, IL 60015
      (800)621-2131

36.   National Audubon Society Productions
        666 Pennsylvania Ave., S.E.
        Washington, D.C. 20003
        (202)547-9009
        For Purchase: PBS Video 1-800-424-7963

37.   Educational Images, Ltd.
        P.O. Box 4356
        West Side Station
        Elmira, NY 14905
        (607)732-1090

38.   Public Service Electric and Gas Company
        P.O. Box 570
        T10C
        Newark, NJ 07101
        (201)430-8699: Ms. Gail Davis

39.   Encyclopedia Britannica Educational Corp.
        310 South Michigan Ave.
        Chicago, IL 60604
        (800)554-9862 Ext. 6514 or (800)621-3900

40.   National Geographic Society Educational Services
        P.O. Box 98019
        Washington, D.C. 20090
        1-800-368-2728
        Film and Video preview service: (717)822-8899
        Magazine Index: (202)857-7059

41.   U.S. Fish and Wildlife Service
        Audiovisual Office
        Room 3444
        1849 C Street
        Washington, D.C. 20240
        (202)208-5611

42.   National Association of Conservation Districts
        P.O. Box 855
        League City, TX 77574-0855
        (800)825-5547

National Association of Conservation Districts
Washington Headquarters
509 Capitol Court, N.E.
Washington, D.C. 20002
(202)547-6223

43.   Optical Data Corporation
        30 Technology Drive
        Box 4919
        Warren, NJ
        (908)668-0022

44.   University of New Hampshire Sea Grant Extension Program
        Kingman Farm
        Durham, New Hampshire 03824-3512
        (603)749-1565

45.   National Audubon Society
        Education Division
        700 Broadway
        New York, NY 10003
        (212)979-3000

46.   Ducks Unlimited
        One Waterfowl Way
        Long Grove, Il 60047
        (901)758-3825

47.   Virginia Sea Grant College Program
        Marine Advisory Services
        Virginia Institute of Marine Science
        Gloucester Point, VA 23062
        (804)642-7170

48.   Michigan Department of Natural Resources
        Box 30034
        Lansing, MI 48909
        (800)248-5848

49.   Coastal America
        722 Jackson Place, NW
        Washington, D.C. 20503
        (202)395-3706

50.   Texas A&M University
        Sea Grant College Program
        P.O. Box 1675
        Galveston, TX 77553-1675
        (409)762-9800

51.   The Catskill Center
        route 208
        Arkville, NY 12406
        (914)586-2611

52.   Wisconsin Department of natural Resources
        Bureau of Information and Education
        P.O. Box 7921
        Madison, WI 53707
        (608)267-5239

53.   National Marine Educators Association
        P.O. Box 51215
        Pacific Grove, CA 93950
        (408)648-4841: Nora Deens

54.   Environmental Law Institute
        1616 P Street, NW
        Washington, D.C. 20036
        (202)328-5150

55.   Hierarchy of pursuit for U.S. Environmental Protection Agency
        Documents:

        a)   Wetlands Hotline (800)832-7828

        b)   U.S. Environmental Protection Agency
                Publications and Information Center
                11029 Kenwood Road
                Cincinnati, OH 45242
                (513)569-7980

        c)   United States Environmental Protection Agency
                Public Information Center
                PM-211B
                401 M Street, SW
                Washington, D.C. 20460
                (202)260-7751

d)   Superintendant of Documents
Government Printing Office
Washington, D.C. 20402
(202)783-3238

e)   National Technical Information Service
Department of Commerce
5285 Port Royal Road
Springfield, VA 22161
(703)487-4780

56.   U.S. Environmental Protection Agency
Region 5
Alfred Krause WCP-15J
77 West Jackson
Chicago, IL 60604
(312)886-0246

57.   For Document Availability see 24 above

For the National Wetlands Inventory Study:
United States Fish and Wildlife Service
National Wetlands Inventory
9720 Executive Center Drive
Suite 101 Monroe Building
St. Petersburg, FL 33702-2440
(813)893-3624

58.   U.S. Department of the Interior
Wetlands Policy Group
18th & C Streets, N.W.
Washington, D.C. 20240
(202)208-4678

59.   Acquisition of the Statewide Wetlands Strategies
A Guide to protecting and Managing the Resource:
Island Press
1718 Connecticut Ave., N.W.
Suite 300
Washington, D.C. 20009
(800)828-1302

For the Authors:
Environmental Defense Fund
257 Park Avenue South
New York, NY 10010
(212)505-2100

World Wildlife Fund
1250 24th Street, N.W.
Washington, D.C. 20037
(202)778-9688

60.   National Audubon Society
       Sanctuary Department
       93 West Cornwall Road
       Sharon, CT 06069
       (203)364-0048

       National Audubon Society
       Education Division Headquarters Office
       325 Cornwall Bridge Road
       Sharon, CT 06069
       (203)364-0520
       (Nature Education Centers

       National Audubon Society
       Regional & Government Affairs Office
       666 Pennsylvania Avenue, S.E.
       Washington, D.C. 20003
       (202)547-9009

See number 47 above for NAS Education Division New York, NY

61.   American Association of Zoological Parks and Aquariums
       Ogleby Park
       Wheeling, West Virginia 26003
       (304)242-2160

       American Association of Zoological Parks and Aquariums
       Executive Office and Conservation Department
       7970-D Old Georgetown Road
       Bethesda, MD 20814
       (301)907-7777

62.   Alliance for Environmental Education
       51 Main Street
       P.O. Box 368
       The Plains, VA 22171
       (703)253-5812

63.   Friends of the Earth
       218 D Street, SE
       Washington, D.C. 20003
       (202)544-2600

64.   ERIC Clearinghouse for Science,
       Mathematics, and Environmental Education
       Room 310
       1200 Chambers Road
       Columbus, OH 43212-1792
       (614)292-6717

65.   Woods Hole Oceanographic Institution
       Sea Grant Program
       Woods Hole, MA 02543
       (508)548-1400 Ext. 2398

66.   Rhode Island Sea Grant
       University of Rhode Island, Bay Campus
       Narragansett, RI 02882-1197
       (401)792-6800 Publications: (401)792-6842

67.   National Sea Grant Depository
       Pell Library Building
       University of Rhode Island
       Narragansett Bay Campus
       Naragansett, RI 02282-1197
       (401)792-6114

68.   American Association of Museums
       1225 Eye Street, N.W.
       Suite 200
       Washington, D.C. 20005
       (202)289-1818

69.   Sea Grant Abstracts
       P.O. Box 125
       Woods Hole, MA 02543
       (508)548-2743

70.   State University System of Florida
        Sea Grant College Program
        Building 803
        Gainesville, FL 32611
        (904)392-5870

71.   Marine Technology Association
        1828 L Street, N.W.
        Washington, D.C. 20036
        (202)775-5966

72.   Water Environment Federation
        Public Education Program
        601 Wythe Street
        Alexandria, VA 22314-1994
        (703)684-2438 or (800)666-0206

73.   Council of State Governments
        The Centers for Health and Environment
        P.O. Box 11910
        Iron Works Road
        Lexington, KY 40578-1910
        (606)231-1923 or 1866

74.   National Conference of State Legislatures
        444 North Capitol Street, N.W.
        Suite 500
        Washington, D.C. 20001
        (202)624-5400

75.   Council of State Governments
        State Policy and Innovations Group
        P.O. Box 11910
        Iron Works Road
        Lexington, KY 40578-1910
        (606)231-1840

76.   National Association of Homebuilders, Washington, D.C.
        1201 15th Street, N.W.
        Washington, D.C. 20005-2800
        (202)822-0485

77.   Soil and Water Conservation Society
        7515 NE Ankeny Road
        Ankeny, Iowa 50021
        (515)289-2331

78.   New England Interstate Environmental Training Center
      2 Fort Road
      South Portland, Maine 04106
      (207)767-2539

79.   United States General Accounting Office
      Document Distribution Office
      P.O. Box 6015
      Gaithersburg, MD 20877
      (202)512-6000

80.   Government Institutes
      4 Research Place
      Suite 200
      Rockville, MD 20850
      (301)291-2300

81.   The Land Trust Alliance
      900 17th Street, N.W.
      Suite 410 Washington, D.C. 20006
      (202)785-1410

82.   Tip of the Mitt Watershed Council
      Box 300
      Conway, MI 49722
      (616)347-1181

83.   National Audubon Society
      102 E. 4th Avenue
      Tallahassee, FL 32303
      (904)222-2473

84.   Association of Wetland Managers
      P.O. Box 2463
      Berne, NY 12023-9746
      (518)872-1804

85.   American Planning Association
      1313 E. 60th Street
      Chicago, IL 60637
      (312)955-9100

American Planning Association
1776 Massachusetts Ave., N.W.
Washington, D.C. 20036
(202)872-0611

86.   Lake Michigan Federation
      59 E. Van Buren
      Suite 2215
      Chicago, IL 60605
      (312)939-0838

87.   Pennsylvania Department of Environmental Resources
      Office of Public Liason
      P.O. Box 2063
      Harrisburg, PA 17105-2063

88.   Michigan Department of Natural Resources
      Land and Water Management Division
      Box 30028
      Lansing, MI 48909

89.   Estuarine Research Federation
      P.O. Box 544
      Crownsville, MD 21032-0544

90.   The Conservation Foundation
      1250 Twenty-fourth Street, N.W.
      Washington, D.C. 20037

91.   Society for Wetland Scientists
      P.O. Box 296
      Wilmington, NC 28402

92.   Earth Science Information Center
      U.S. Geological Survey
      National Headquarters
      507 National Center
      Reston, VA 22092
      (703)648-6045

93.   International Association of Ecology
       c/o Mr. William Mitsch
       School of Natural Resources
       2021 Coffey Road
       Ohio State University
       Columbus, OH 43210
       (614)292-9773

94.   United States Forest Service
       Forest Environment Research
       14th & Independent, S.W.
       P.O. Box 96090
       Washington, D.C. 20090-6090
       (202)205-1524

95.   Izaak Walton League of America
       1401 Wilson Blvd. Level B
       Arlington, VA 22209
       (703)528-1818

96.   Senior Environment Corps
       6410 Rockledge Drive
       Suite 203
       Bethesda, MD 20817
       (301)571-9792